Mathematics and Python Programming

J.C. Bautista

ISBN: 978-1-326-01796-5
Original title:
Matemáticas y Programación en Python, Copyright © 2014 by José Carlos Bautista
English translation by José Carlos Bautista, Copyright © 2014

Trademarked names, logos, and images may appear in this book. Rather than use a trademark symbol with every occurrence of a trademarked name, logo, or image we use the names, logos, and images only in an editorial fashion and to the benefit of the trademark owner, with no intention of infringement of the trademark.

The use in this publication of trade names, trademarks, service marks, and similar terms, even if they are not identified as such, is not to be taken as an expression of opinion as to whether or not they are subject to proprietary rights.

The information in this book is distributed on an "as is" basis, without warranty. Although every precaution has been taken in the preparation of this work, neither the author(s) nor the publisher shall have any liability to any person or entity with respect to any loss or damage caused or alleged to be caused directly or indirectly by the information contained in this work.

This book intends to show Python applications to the study of Mathematics. Programs have been developed with the aim of showing the possibilities of this language: they have not been thoroughly tested for all possible cases, and have not been evaluated by usability testing. The use of these programs to perform calculations can in some cases produce unexpected or wrong results, since the programs have been developed exclusively as examples for this book. The use of the software is strictly the responsibility of the reader.

To Carmen and Diego.

Contents

0 Fundamentals of Python — **3**
- 0.1 What is Python? — 3
- 0.2 Our first Python program — 5
- 0.3 Operations with numbers — 5
- 0.4 NumPy and SciPy. The if statement — 6
- 0.5 The statements for and while — 10
- 0.6 Operations with polynomials — 14
- 0.7 More on the while statement — 15
- 0.8 Matplotlib — 18
- 0.9 Text Files. Statistics — 23

1 Sets — **32**
- 1.1 Equivalence relations — 32
- 1.2 Countable Sets: **N**, **Z**, **Q** — 33
- 1.3 Three examples of Number Theory — 35
  - 1.3.1 Prime Numbers — 35
  - 1.3.2 Euclidean algorithm — 37
  - 1.3.3 Pythagorean triple — 38
- 1.4 Equivalent sets. Ordered sets. — 40
- 1.5 The set of real numbers — 41

2 Fields — **44**
- 2.1 The field of real numbers — 44
- 2.2 Square matrices — 46
- 2.3 The field of complex numbers — 48

3 Metric spaces — **58**
- 3.1 Metric spaces — 58
- 3.2 Open sets, closed sets, bounded sets — 59
- 3.3 Sequences — 61

4 Limit — **68**
- 4.1 Convergent sequences. Limit — 68
- 4.2 Properties of the sequences — 70
- 4.3 Monotonic sequences — 71
- 4.4 Cauchy sequence — 73
- 4.5 Nested intervals — 76
- 4.6 Series — 79

5 Function. Continuity — **83**
- 5.1 Function — 83
- 5.2 Limit of a function — 84
- 5.3 Continuity — 85
- 5.4 Continuous Functions — 91
- 5.5 Graphics programs in this chapter — 95

6 Conics — **99**
- 6.1 Degenerate conics — 99
- 6.2 Ellipses and Hyperbolas — 101
- 6.3 Parabolas — 110

7 Exponential function — 113
- 7.1 Power series — 113
- 7.2 Exponential function $exp(z)$ — 114
- 7.3 Graphics programs in this chapter — 123

8 Differentiation — 126
- 8.1 Derivative of a function — 126
- 8.2 Geometric meaning — 129
- 8.3 Obtaining derivatives — 132
- 8.4 Differential Calculus Theorems — 134
- 8.5 Taylor series — 141
- 8.6 Applications of the derivative — 151
  - 8.6.1 Representation of polynomial functions — 151
  - 8.6.2 Physics: parabolic trajectory — 155
  - 8.6.3 Physics: Planck's law — 158
  - 8.6.4 Economics: production function — 161
  - 8.6.5 Physics: Newton's law of cooling — 161

9 Integral — 162
- 9.1 Riemann integral — 162
- 9.2 Indefinite integral — 166
- 9.3 Definite Integral — 168
- 9.4 Fundamental theorem of integral calculus — 178
- 9.5 Lengths, areas and volumes — 181
  - 9.5.1 Area of a curvilinear sector — 181
  - 9.5.2 Arc length — 183
  - 9.5.3 Area of a surface of revolution — 185
  - 9.5.4 Volume of a solid of revolution — 186
- 9.6 Applications of the integral to Physics — 194
  - 9.6.1 W done by a constant force — 196
  - 9.6.2 W done to compress a spring — 196
  - 9.6.3 W during an isothermal expansion — 197
  - 9.6.4 W during an adiabatic expansion — 197
  - 9.6.5 Carnot Cycle — 198
- 9.7 The logarithm function — 202
- 9.8 Numerical integration — 206

10 Vectors — 214
- 10.1 Vector space — 214
- 10.2 Equations of a straight line in the plane — 223
- 10.3 Vector Functions — 231
- 10.4 Limit, continuity and derivative of a vector function — 233
- 10.5 Scalar Fields — 238
- 10.6 Gradient — 241
- 10.7 Integration of a vector function — 243
  - 10.7.1 Line integral of a vector field — 244
  - 10.7.2 Circulation of a conservative vector field — 245
  - 10.7.3 Curl of a vector field — 245
  - 10.7.4 Multiple integrals — 248
  - 10.7.5 Flux of a vector field. Divergence — 249
- 10.8 Field lines. Examples of vector fields — 252

Chapters Scheme

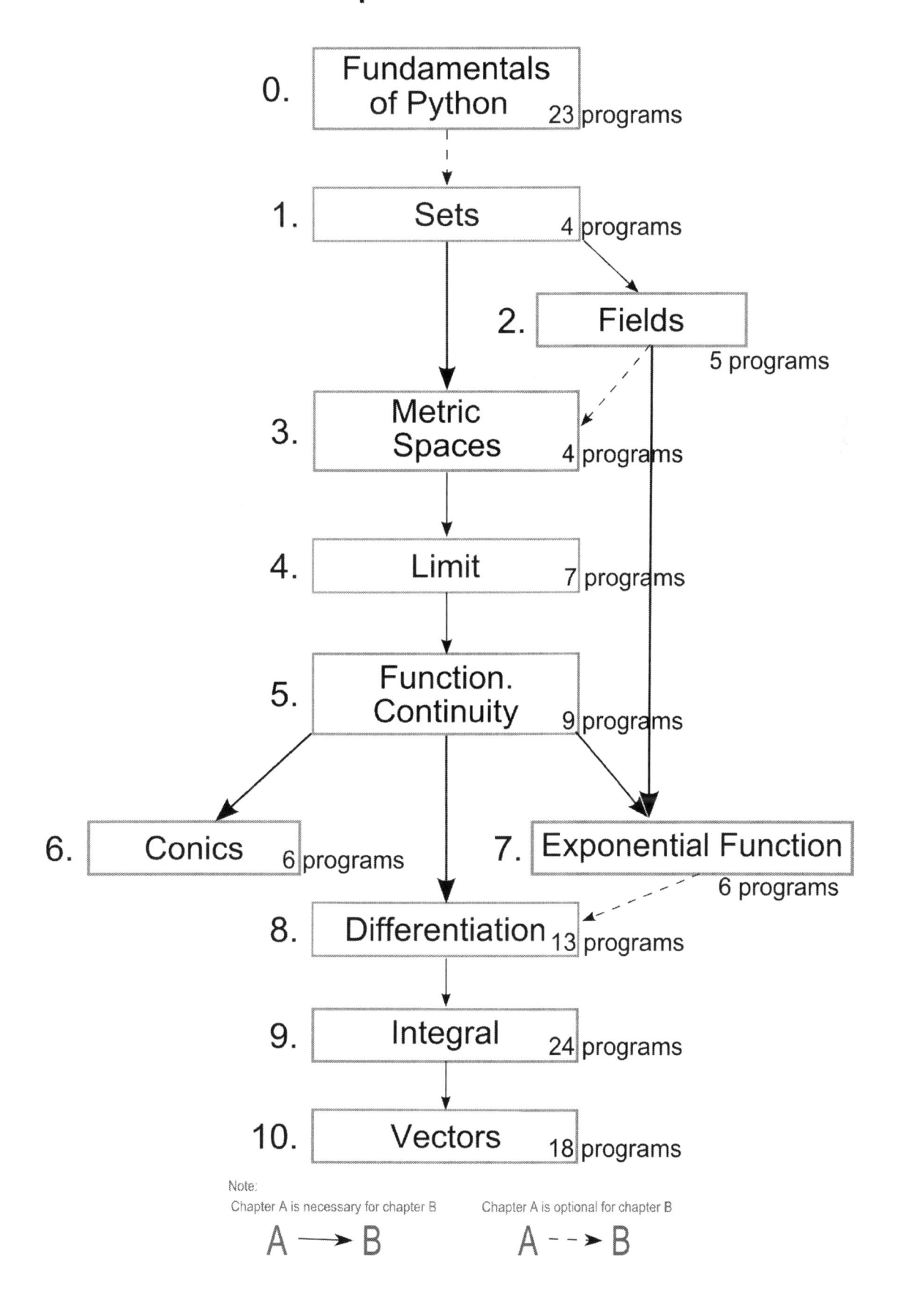

- **0.** Fundamentals of Python — 23 programs
- **1.** Sets — 4 programs
- **2.** Fields — 5 programs
- **3.** Metric Spaces — 4 programs
- **4.** Limit — 7 programs
- **5.** Function. Continuity — 9 programs
- **6.** Conics — 6 programs
- **7.** Exponential Function — 6 programs
- **8.** Differentiation — 13 programs
- **9.** Integral — 24 programs
- **10.** Vectors — 18 programs

Note:
Chapter A is necessary for chapter B: A ⟶ B
Chapter A is optional for chapter B: A --▶ B

Table 1: List of Python programs

nº	py	Description
1	pa.py	Calculates the permutations of the five vowels
2	pb.py	First program: write 'hellow'
3	pc.py	Operations with integers and real numbers
4	pd.py	Specifies the number of decimal places shown
5	pe.py	Operations with variables
6	pf.py	Definition of a function. Mathematical operations. Python statement `if`
7	pg.py	Statement `if - else`
8	ph.py	Statement `if - elif - else`
9	pi.py	Statement `for`
10	pj.py	Statement `for`, moving from 5 to 5
11	pk.py	Statement `for` using turtle graphics
12	pl.py	Statement `for`: Archimedes spiral with turtle graphics
13	pm.py	Plots the function $y = a \cdot sen3\alpha$ with turtle graphics
14	pn.py	Numpy. Operations with polynomials
15	po.py	Example of an endless loop `while`
16	pq.py	Statement `while`: Decomposition in prime factors.
17	pr.py	Statement `while` y lists: Decomposition in prime factors.
18	ps.py	Adjustment of experimental data with Numpy and Matplotlib
19	pt.py	2D graphical representation the equation of Van der Waals
20	pu.py	Plots the former graph as a 3D surface
21	pv.py	Counts letters of a text file
22	pw.py	Counts and represents frequencies of letters in texts files
23	px.py	Prepares statistics and represents histogram
24	p1a.py	Calculates 30 rational numbers between two randomly chosen rational numbers
25	p1b.py	Sieve of Eratosthenes
26	p1c.py	Euclidean algorithm to calculate the greatest common divisor
27	p1d.py	Computes and represents Pythagorean triangles
28	p2a.py	Matrix multiplication
29	p2b.py	Calculation of the inverse matrix through row operations with Sympy
30	p2c.py	Operations with complex numbers
31	p2d.py	Represents graphically complex numbers
32	p2e.py	Resolution of the quadratic equation and graphical representation of the solutions
33	p3a.py	Calculates terms and the sum of an arithmetic progression
34	p3b.py	Calculates terms and the sum of a geometric progression
35	p3c.py	Calculates the 100 first terms of the Fibonacci sequence
36	p3d.py	Calculates the terms of the Fibonacci sequence as binary numbers
37	p4a.py	Calculates terms of a^n, with $a < 1$, until approaching at a distance $< \epsilon$ from the limit
38	p4b.py	Calculates the square root of a number by using terms of a convergent sequence
39	p4c.py	Calculates 5000 terms of the sequence which defines the number e
40	p4d.py	Calculates how many terms of the sequence of the number e are necessary to make the difference between two consecutive terms less than a value ϵ
41	p4e.py	Calculates the sequence of nested intervals whose limit is the number e
42	p4f.py	Represents graphically the sequence of nested intervals whose limit is the number e
43	p4g.py	Shows graphically that the sequence $a^n/n!$ approaches zero
44	p5a.py	Shows the continuity of the function $y = x^2$ from the definition of continuity
45	p5b.py	Represents graphically a discontinuous function
46	p5c.py	Second theorem of Bolzano-Cauchy
47	p5d.py	Represents graphically the function $f(x) = \frac{x^2-9}{x-1}$
48	p5e.py	Represents graphically the functions $senx$ y $cosx$
49	p5f.py	Represents graphically a piecewise-defined continuous function
50	p5g.py	Represents graphically the function $y = x$, $y = senx$, $y = senx/x$ for values of x approaching zero.
51	p5h.py	Represents graphically the function $f(x) = \frac{x \cdot senx}{5}$
52	p5i.py	Represents graphically the function $f(x) = x^2$
53	p6a.py	Determines the type of conic if its discriminant is zero
54	p6b.py	Represents graphically the elliptical orbit of the moon around the Earth

Continues on the next page

Table 1 – *Continues from the previous page*

nº	py	Description
55	p6c.py	Represents graphically the elliptical orbits of the atomic model of Sommerfeld for n=4
56	p6d.py	Represents and determines the type of ellipse if the conic has $A_{33} > 0$
57	p6e.py	Computes and represents graphically the hyperbola if the conic has $A_{33} < 0$
58	p6f.py	Computes and represents graphically a parabola
59	p7a.py	Power series of the exponential function
60	p7b.py	Fits the data of the population of Spain to an exponential function
61	p7c.py	Represents graphically the roots of $z = 1.05 e^{j\pi}$
62	p7d.py	Represents graphically the functions $y = e^x$; $y = e^{-x}$; $y = e^{-x^2}-$
63	p7e.py	Represents graphically the powers of $z = e^{\frac{\pi}{6}j}$
64	p7f.py	Represents graphically the powers $z = 1.02 \cdot e^{j\frac{\pi}{10}}$
65	p8a.py	Calculates the value of the derivative of a polynomial function at a given point on the basis of the definition of the derivative, and compares it with its exact value calculated with Numpy
66	p8b.py	Represents graphically the theorems of Lagrange and Rolle for polynomial functions
67	p8c.py	Represents graphically the theorem of Cauchy for the astroid
68	p8d.py	Represents the Taylor series for the function $y = 3sen2x$, in a neighbourhood of $x = a$
69	p8e.py	Maclaurin series of a function, using Sympy
70	p8f.py	Represents polynomial functions, marking their maximum, minimum, inflection point and the points where the function intercepts x-axis and y-axis
71	p8g.py	Represents a parabolic shot for different angles, and calculates the security parabola
72	p8h.py	Simplifies Planck equation for the black-body radiation
73	p8i.py	Represents graphically $f(x) = x^3 \exp(-x/T)$ as a simple version of Planck equation
74	p8j.py	Represents the Leibniz series for $\pi/4$
75	p8k.py	Represents the meaning of the derivative and differential
76	p8l.py	Represents hyperbolic sine, cosine and tangent
77	p8m.py	Represents equilateral hyperbola, showing Sht and Cht
78	p9a.py	Calculates and represents graphically the Riemann sum of a function
79	p9b.py	Calculates anti-derivatives of functions using Sympy
80	p9c.py	Represents graphically $f(x) = \sqrt{(1-x^2)}$ and calculates its area in the interval $x \in [0, 1]$
81	p9d.py	Represents graphically a polynomial function, and calculates the area bounded by its curve and the X axis
82	p9e.py	Represents graphically two functions and the area bounded by them
83	p9f.py	Represents graphically $f(x) = 1$ and $g(x) = senx$
84	p9g.py	Represents the mean value theorem for integrals, for the function $f(x) = x^2$
85	p9h.py	Represents the fundamental theorem of integral calculus
86	p9i.py	Calculates the area bounded by a curve expressed y polar coordinates
87	p9j.py	Calculates the length of the cycloid. Graphical representation
88	p9k.py	3D representation of the surface of revolution of the cycloid
89	p9l.py	Represents the area bounded by $f(x) = x^2$ and $g(x) = x^5$, and calculates the volume of revolution
90	p9m.py	3D representation of the revolution volume of the previous program
91	p9n.py	Represents the area bounded by the clepsydra curve and Y axis, and calculates this area and the volume of revolution
92	p9o.py	3D representation of the cone, cylinder and clepsydra, and calculates its volume, size and the time it takes to be emptied
93	p9p.py	Calculates and represents graphically work, as an aplication of integral to Physics
94	p9q.py	Calculates and represents graphically the Carnot cycle
95	p9r.py	Given a table of data, calculates and represents the numerical integral, taking the left end of each subrange
96	p9s.py	Given a table of data, calculates and represents the numerical integral, taking the right end of each subrange
97	p9t.py	Given a table of data, calculates and represents the numerical integral, taking the midpoint of each subrange
98	p9u.py	Given a table of data, calculates and represents the numerical integral by the method of trapezes
99	p9v.py	Given a table of data, calculates and represents the numerical integral by the method of the parabolas

Continues on the next page

CONTENTS

Table 1 – *Continues from the previous page*

nº	py	Description
100	p9w.py	Represents graphically lne as the area under the curve $y = f(x)$ from $x = 1$ to $x = e$
101	p9x.py	Represents the curve of the natural logarithm
102	p10a.py	Calculates the dot product of two vectors
103	p10b.py	Calculates the length of a vector
104	p10c.py	Normalizes a vector
105	p10d.py	Calculates the angle formed by two vectors
106	p10e.py	Calculates and represents in 3D the cross product of two vectors
107	p10f.py	Calculates the equations of a straight line, and represents it on a plane
108	p10g.py	3D representation of a vectorial function
109	p10h.py	Represents graphically the vector function $\mathbf{r}$, and its derivative $d\mathbf{r}/dt$
110	p10i.py	Represents graphically a parabolic shot: $\mathbf{r}, \mathbf{v}, \mathbf{a}$
111	p10j.py	Represents graphically the scalar field $z = 0.03(x^2 - y^2)$
112	p10k.py	3D representation of the previous scalar field
113	p10l.py	Represents the contour lines for the electric potential V created by a point electric charge
114	p10m.py	Calculates the gradient
115	p10n.py	Represents graphically the scalar field $z = 0.03(x^2 - y^2)$ and the gradient vector at some points
116	p10o.py	Calculates the curl of a vector using Sympy
117	p10q.py	Represents graphically the speeds vector field $\mathbf{v} = -3y\mathbf{i}$
118	p10r.py	Calculates the divergence of a vector
119	p10s.py	3D representation of the net flux of a vector field through a closed surface
120	p10t.py	Represents graphically a vector field

Preface

I'll start quoting philosopher Karl Popper: *"I give the most value to express myself easily and understandably. But this does not mean, unfortunately, that my explanations can be understood very easily"*. Thus, my main criterion in writing the book has been attempting to be as clear as possible. In the first place I'll explain briefly my thesis: teaching of mathematics in the secondary level of education and first college science courses does not employ enough computing resources but it continues to devote too much time to the resolution of mechanical and manual exercises by the students.

During my more than fifteen years of experience as a teacher, I have observed that in spite of the increasing presence of computers at home and in classrooms, the teaching of mathematics is actually not using many computing resources that are available to both students and teachers free of charge but it still maintains the old teaching schema: definition-theorem-proof, and then the student must do many exercises which just consist in doing repetitive operations on paper. Instead, I propose to increase the use of computers so that students can achieve a faster and better understanding of mathematical concepts and spend less time developing manual skills that require several minutes or even hours of calculations, when a short computer program (written in most cases by the student) can solve the problem in a few tenths of a second. In fact, when these students would perform calculations on their work within a few years, they very probably will use a computer.

I have tried to put my two cents to support this way of thinking about mathematics teaching and I created www.pysamples.com, and I wrote my book *Matemáticas y Programación en Python*, whose English translation you are reading. This book has three objectives:

1. To present mathematical analysis in a useful way so that it can serve as a bridge between calculus that is studied at the end of secondary education and the mathematics syllabus studied in the first courses of any science college.

2. To spend the available school time in a more rational way: more time to get that the student will be able to understand the concepts and less time to learn how to do mechanical operations.

3. To learn a modern, free, multipurpose, easy and useful programming language: Python.

I have made every effort to translate into English the second Spanish edition of this book, anyway, not being English my native language, any suggestion concerning the translation by the English-speaking readers will be welcome. I have revised the layout of this English edition, the code of the programs and I have corrected some typos. Some 3D renderings have been modified so that they appear more clear with the print on paper in black and white. The layout of the book has been improved to reduce the number of pages without reducing its content, which will make its price more affordable, specially for students.

I hope that this book is useful for both students and teachers to make easier the teaching and learning of mathematics and the access to free modern computing tools, so that they can create their own programs.

José Carlos Bautista Marugán, Madrid, August 2014.

0 | Fundamentals of Python

0.1 What is Python?

In the drafting of the entire text, program and graphics development, image editing, etc. we have use only free software, so that to develop the programs it is not necessary to purchase any software package: all the required software is free and legally available.

As Guido van Rossum, creator of the Python language, said in his blog [1]:«*Python is currently one of the most popular dynamic programming languages... Today, Python is used for everything from throw-away scripts to large scalable web servers that provide uninterrupted service 24x7. It is used for GUI and database programming, client- and server-side web programming, and application testing. It is used by scientists writing applications for the world's fastest supercomputers and by children first learning to program.*»

Compared with other languages and programming environments, Python has some advantages:

- Python is a modern open source language. You can download and install Python free and legally in both Windows and Linux and Apple.

- A simple syntax, without braces {} or semicolons.

- Its versatility allows you to use it to perform calculations, word processing, graphics, etc.

- Octave and Scilab environments are specifically designed to perform mathematical calculations and represent functions; Python is a general-purpose language, but with specific modules to perform the same tasks as Octave and Scilab, and with a simpler syntax.

- The learning curve for Python is significantly faster than other languages. The creation of a first program to display a word on the screen can require several steps in other programming languages, this task is immediate using Python.

- There are several programming environments in Python, so that the user can choose whichever suits him/her best. In this book, we have used Spyder, of which we shall speak later.

For a more extensive discussion about the advantages of Python, check out the post by Prof. Lorena Barba, of George Washington University: *"Why I push for Python"* http://lorenabarba.com/blog/why-i-push-for-python/.

As an example of the speed and simplicity of the syntax of Python, we show the following program, which calculates the permutations of the five vowels. We will not explain at this time the details of the code, we just want to show that we can calculate the 120 permutations in less than one thousandth of a second, writing only eleven lines of code:

```
# -*- coding: utf-8 -*-
"""
Mathematics and Python Programming      www.pysamples.com
```

---

[1]http://python-history.blogspot.com.es/2009/01/introduction-and-overview.html

pa.py
```
"""

import itertools

elements = ['a', 'e', 'i', 'o', 'u']
permutations = list(itertools.permutations(elements))
result = ''
for permutation in permutations:
    word = ''
    for i in range(0, 5):
        word += permutation[i]
    result += word + ', '
result = result[0:len(result) - 2]
print result
```

```
──────────────────────────── output of the program ────────────────────────────
aeiou, aeiuo, aeoiu, aeoui, aeuio, aeuoi, aieou, aieuo, aioeu, aioue, aiueo, aiuoe,
aoeiu, aoeui, aoieu, aoiue, aouei, aouie, aueio, aueoi, auieo, auioe, auoei, auoie,
eaiou, eaiuo, eaoiu, eaoui, eauio, eauoi, eiaou, eiauo, eioau, eioua, eiuao, eiuoa,
eoaiu, eoaui, eoiau, eoiua, eouai, eouia, euaio, euaoi, euiao, euioa, euoai, euoia,
iaeou, iaeuo, iaoeu, iaoue, iaueo, iauoe, ieaou, ieauo, ieoau, ieoua, ieuao, ieuoa,
ioaeu, ioaue, ioeau, ioeua, iouae, iouea, iuaeo, iuaoe, iueao, iueoa, iuoae, iuoea,
oaeiu, oaeui, oaieu, oaiue, oauei, oauie, oeaiu, oeaui, oeiau, oeiua, oeuai, oeuia,
oiaeu, oiaue, oieau, oieua, oiuae, oiuea, ouaei, ouaie, oueai, oueia, ouiae, ouiea,
uaeio, uaeoi, uaieo, uaioe, uaoei, uaoie, ueaio, ueaoi, ueiao, ueioa, ueoai, ueoia,
uiaeo, uiaoe, uieao, uieoa, uioae, uioea, uoaei, uoaie, uoeai, uoeia, uoiae, uoiea
0.32901763916  ms
────────────────────────────────────────────────────────────────────────────────
```

This book has been written using free software. All the applications that are used can be obtain free and legally on the internet. For many of them exists a version for Windows, Mac and Linux. The version of Python that has been used in this book is 2.7 , which has been installed along with the libraries Numpy and Scipy. At the time of completing this book, the most recent version is 2.7.6 , released in November 2013 (http://www.python.org/download/releases/2.7.6/). You can view a chronological list of the versions of Python in this direction: http://docs.python.org/2/license.html.

I preferred version 2.7 because today is the supported version by Google Application Engine and it is currently used in courses such as MIT's *Introduction to Computer Science and Programming Using Python*, (https://www.edx.org/course/mitx/mitx-6-00-1x-introduction-computer-1122. The programs in this book are written in the version 2.7 of Python, and if you decide to use version 3 of the language, you'll need retyping or use a converter available in the address http://docs.python.org/2/library/2to3.html, which in my opinion it is not necessary if you stay in the version 2.7 . The differences between both versions are mainly syntactic rather than functional. In any case, if you don't know which version decide, you can see these recommendations: http://wiki.python.org/moin/Python2orPython3.

I preferred writing the code of the programs as clear as possible. That is, the programs are elaborated trying to show as clearly as possible their operation illustrating a mathematical concept, and they do not seek to carry out the task with the fewest lines of code, nor to use procedures possibly more efficient and in a shorter time, but which will be probably more difficult to understand.

In some programs, their running time has been timed by using the `timeit` module whose form of employment can be found in http://docs.python.org/2/library/timeit.html. The time is given simply as an illustration of the speed of Python to perform tasks, or to compare two different ways to schedule a same task, and it can be different when you run the program on different computer.

Let's talk now about the programming environment in Python: there are several free environments that can be downloaded and installed easily. I will just name a few: **IDLE**, which is installed by default with Python and which also you can find at http://www.python.org/download/; **Spyder**, which I have been using to write the programs of the book. Both of them are excellent, but in my opinion **Spyder** is far superior when you are looking for errors in the code, helping you to write words of the language, etc. The name of Spyder is an acronym that leaves already clear its guidance to scientific programming: "Spyder is the Scientific PYthon Development EnviRonment". You can download free of charge at the address https://code.google.com/p/spyderlib/. Along with the former IDEs is also worth noting **IPython**, which can be downloaded for free at http://ipython.org/install.html. For Windows users, the easiest option probably is winpython: http://winpython.sourceforge.net/.

Almost all of the graphics contained in the book have been developed through the corresponding Python programs that are also listed in the book. The few exceptions to this rule are some geometric proofs, which have been developed using **Geogebra** (http://www.geogebra.org/cms/es/), and the diagrams, which have been developed with **Inkscape** (http://www.inkscape.org/es/). **Gimp** has also been used for editing images (http://www.gimp.org/). These three applications are also available free of charge and legally on the Internet.

All Python programs, and the book draft in Latex, have been made using a laptop Samsung R540, whose operating system is Linux Mint http://www.linuxmint.com/. The programs have been written in the version 2.7 of Python, and their syntax follows the PEP 8 recommendations which can be found at http://www.python.org/dev/peps/pep-0008/.

The pages that follow constitute a brief introduction to Python programming, and are especially oriented to mathematical calculations. You can extend this information with any of the free books and tutorials available at the following web addresses:

- Byte of Python: http://files.swaroopch.com/python/byteofpython_120.pdf
- Google's Python Class: https://developers.google.com/edu/python/
- Dive into Python: http://www.diveintopython.net/toc/index.html
- Learn Python The Hard Way: http://learnpythonthehardway.org/book/
- Python for you and me: http://kushal.fedorapeople.org/book/
- Python Documentation: http://docs.python.org/2/download.html

0.2 Our first Python program

Unlike other languages, the typical first program written in Python is very simple:

```
────────────────────────── first python program ──────────────────────────
>>> print 'hellow'
hellow
```

It is possible to use Python as a calculator, but this would be like using a sledgehammer to crack a nut. To take full advantage of Python we need to write a Python program. Python programs are saved as files with the extension .py.

0.3 Operations with numbers

We will see below several short programs that we'll discuss briefly. In this first program we will define two variables that we call e1 and e2, corresponding to two integers. The following lines make mathematical operations with them. Note that if the two integers are divided directly, the result is an integer. For the division to have decimal places, we have the first integer multiplied by 1.0 in the last line of the program. The first line (coding: utf-8) is required to be able to print characters such as accented letters, etc.

```
# -*- coding: utf-8 -*-
'''
Mathematics and Python Programming     www.pysamples.com
pc.py
'''

e1 = 23
e2 = -2
r1 = 0.01
r2 = 2.5e-3
print 'operations with integers:'
print 'e1 + e2 = ', e1 + e2
print 'e1 - e2 = ', e1 - e2
print '-e2 = ', -e2
print 'e1 * e2 = ', e1 * e2
print 'fifth power of e2 = ', e2 ** 5
print 'division of integers: e1/e2 = ', e1 / e2
print ('exact division of integers: e1/e2 = ' + str(1.0 * e1 / e2))
print 'operations with real numbers:'
print 'e1 * r1 = ', e1 * r1
print 'e1 / r1 = ', e1 / r1
print 'r1 / r2 = ', r1 / r2
print 'fourth power of r2 = ', r2 ** 4
print 'division: 7.5/2 = ', 7.5 / 2
print 'truncated division 7.5//2 = ', 7.5 // 2
print ('modulus: 7.5 % 2 = ' + str(7.5 // 2))
```

───────────────────────── output of the program ─────────────────────────
```
operations with integers:
e1 + e2 =  21
e1 - e2 =  25
-e2 =  2
e1 * e2 =  -46
fifth power of e2 =  -32
division of integers: e1/e2 =  -12
exact division of integers: e1/e2 = -11.5
operations with real numbers:
e1 * r1 =  0.23
e1 / r1 =  2300.0
r1 / r2 =  4.0
fourth power of r2 =  3.90625e-11
division: 7.5/2 =  3.75
truncated division 7.5//2 =  3.0
modulus: 7.5 % 2 = 3.0
```

In this program we just need to comment the last two lines: the truncated division, that eliminates the decimal of the division; and the last `print` statement, which is too long, so we frame its argument between parentheses and unite the elements we want to print using signs . The `str()` instruction converts the result of the operation module into a text string so that we can add it to the string to be printed with the print statement.

It is also possible to perform operations with complex numbers. We show an example in section 2.3.

0.4 NumPy and SciPy. The `if` statement

If we need to perform operations more complicated that the sum, multiplication, etc. , Python has "packages" which already include coded instructions for performing these operations, so we don't have to indicate step by step, for example, operations such as matrix multiplication: but we can do that with a single statement.

0.4. NUMPY AND SCIPY. THE IF STATEMENT

Throughout the book, we have used mainly NumPy, which is the main package for scientific calculations using Python. You can find more information about NumPy at the following address: http://www.numpy.org/. We just need to add the import Numpy statement at the beginning of the Numpy program to include Numpy in our program.

Let's take a look at how numbers are displayed with a given number of decimal places, and how to truncate a number.

```
# -*- coding: utf-8 -*-
'''
pd.py
'''

import numpy as np

pi = np.pi
print pi
print str(pi)
print "%10.2f" % pi
print "%10.8f" % pi
print np.trunc(pi)
print "%20.18f" % pi
```

```
_____ output of the program _____
3.14159265359
3.14159265359
      3.14
3.14159265
3.0
3.141592653589793116
```

Operations with variables. We must often perform addition, subtraction, etc. with variables, as well as increase or decrease its value in a given quantity. The following program shows how to do this. Note the changing the value of the variable z.

```
# -*- coding: utf-8 -*-
'''
Mathematics and Python Programming    www.pysamples.com
pe.py
'''

x = 1
z = 100 * x
print 'z = 100x = ', z
z += 2
print 'z increases by 2 units: z+=2 : ', z
z -= 52
print 'z decreases in 52 units: z-=2 : ', z
z *= 3
print 'z triples: z*=3 :', z
z /= 10
print 'z divided by 10: z/=10 : ', z
z %= 4
print 'remainder of dividing z by 4: z%=4 : ', z
```

```
_____ output of the program _____
z = 100x =  100
z increases by 2 units: z+=2 :  102
z decreases in 52 unidades: z-=2 :  50
z triples: z*=3 : 150
z divided by 10: z/=10 :  15
remainder of dividing z by 4: z%=4 :  3
```

Python packages are available to perform calculations. We have already mentioned NumPy. The following program uses NumPy to perform operations such as logarithm, sine, cosine, etc. The program shows how to define a function which we call `degreestogms`. This function gets a number of degrees in decimal format, and transforms it in degrees, minutes and seconds. The function returns the text string `strgms` which we can include in any part of the program. We only need to include its name and degree value to use this feature, for example: `degreestogms(10.55)` gives as a result the text string 10.0º 33.0' 0.00".

```python
# -*- coding: utf-8 -*-
'''
Mathematics and Python Programming    www.pysamples.com
pf.py
'''

import numpy as np


def degreestogms(degrees):
    g = np.floor(degrees)
    m = np.floor((degrees - g) * 60)
    s = round((((degrees - g) * 60) - m) * 60)
    if s == 60:
        s = 0
        m += 1
    if m == 60:
        m = 0
        g += 1
    strgms = str(g) + '    ' + str(m) + "' " + "%5.2f" % s + "''"
    return strgms

x = 6
y = 0.5
radian = 1
print radian, ' radian = ', np.rad2deg(radian), ' degrees'
print radian, ' radian = ' + degreestogms(np.rad2deg(radian))
degrees = 180
print degrees, ' degrees = ', np.deg2rad(degrees), ' radians'
print 'value of number e: ', np.e
print 'value of pi: ', np.pi
print 'e^x = ', np.power(np.e, x)
print 'x^y = ', np.power(x, y)
print 'log10(1024) = ' + "%8.5f" % np.log10(1024)
print 'ln(1024) = ' + "%8.5f" % np.log(1024)
print 'log2(1024) = ' + "%4.1f" % np.log2(1024)
print '6! = ', np.math.factorial(6)
alfaradianes = np.pi / 6
print ('alfa = ' + "%8.5f" % alfaradianes +
       ' radians = pi/' + str(x))
alfagrados = np.rad2deg(alfaradianes)
print ('sin(' + degreestogms(alfagrados) + ') = ' + "%8.5f" % np.sin(alfaradianes))
print ('cos(' + degreestogms(alfagrados) + ') = ' + "%8.5f" % np.cos(alfaradianes))
print ('tan(' + degreestogms(alfagrados) + ') = ' + "%8.5f" % np.tan(alfaradianes))
print ('asin(1) = ' + "%8.5f" % np.arcsin(1) +
       ' radians = ' + degreestogms(np.rad2deg(np.arcsin(1))))
print ('acos(1) = ' + "%8.5f" % np.arccos(1) +
       ' radians = ' + degreestogms(np.rad2deg(np.arccos(1))))
print ('atan(1) = ' + "%8.5f" % np.arctan(1) +
```

0.4. NUMPY AND SCIPY. THE IF STATEMENT

```
                  ' radians = ' + degreestogms(np.rad2deg(np.arctan(1))))
print degreestogms(10.55)
```

In this program we have included a statement of which we had not discussed so far: the if statement. The syntax for the if statement is as follows:

```
if condition:
    instructions to execute if the condition is met
```

In this example:

```
if s == 60:
    s = 0
    m += 1
```

if the seconds (s) are 60, puts the seconds to zero (s = 0), and increases the number of minutes in a unit (m += 1). The statements to execute if if statement is met, are indented. Likewise, the statements belonging to the function degreestogms are indented after the statement:

```
def degreestogms(grados):
```

Python uses indentation instead of braces to delimit the functions, unlike other languages such as Java and C. The result when you run the program is the following:

```
―――――――――――――――――――――――――――――――――――― output of the program ――――――――
1  radian =  57.2957795131  degrees
1  radian = 57.0° 17.0' 45.00''
180  degrees =  3.14159265359  radians
value of number e:  2.71828182846
value of pi:  3.14159265359
e^x =  403.428793493
x^y =  2.44948974278
log10(1024) =  3.01030
ln(1024) =  6.93147
log2(1024) = 10.0
6! =  720
alfa =  0.52360 radians = pi/6
sen(30.0° 0'  0.00'') =  0.50000
cos(30.0° 0'  0.00'') =  0.86603
tan(30.0° 0'  0.00'') =  0.57735
asen(1) =  1.57080 radians = 90.0° 0.0'  0.00''
acos(1) =  0.00000 radians = 0.0° 0.0'  0.00''
atan(1) =  0.78540 radians = 45.0° 0.0'  0.00''
10.0° 33.0'  0.00''
―――――――――――――――――――――――――――――――――――――――――――――――――――――――――――――――
```

Let's see another example of the if statement. The following program prompts you to enter an integer, then determines if the number is odd or even. The if statement checks whether that number is divisible by two:

```
 if (number % 2) == 0:
```

and otherwise, there is only one possibility (else): the number is odd.

```
# -*- coding: utf-8 -*-
'''
Mathematics and Python Programming    www.pysamples.com
pg.py
'''

number = int(raw_input('Write an integer: '))

if (number % 2) == 0:
    print 'Number ', number, ' is even'
else:
    print 'Number ', number, ' is odd'
```

```
------- output of the program -------
Write an integer:: 4
Number  4  is even
```

```
------- output of the program -------
Write an integer: 45
Number  45  is odd
```

What happens if there exists more than two options? We can include several nested statements, or code every possibility with an instruction `elif`, and include the default option if all the above fail: the `else` statement:

```python
# -*- coding: utf-8 -*-
'''
Mathematics and Python Programming     www.pysamples.com
ph.py
'''

letra = raw_input('Write a letter: ')
print letra

if (letra in 'aeiou'):
    print 'You wrote vowel: ', letra
elif letra in 'bcdfghjklmn pqrstvwxyz BCDFGHJKLMN PQRSTVWXYZ':
    print 'You wrote consonant: ', letra
elif letra in '1234567890':
    print 'You wrote a number: ', letra
else:
    print letra, ' is not a letter nor a number'
```

```
------- several outputs of the program -------
Write a letter: e
e
You wrote vowel:  e

Write a letter: H
H
You wrote consonant:  H

Write a letter: 8
8
You wrote a number:  8

Write a letter: $
$
$  is not a letter nor a number
```

0.5 The statements for and while

The instructions `for`, and `while` are used to repeat the execution of a piece of code a number of times. Let's start with the simplest, but also more limited statement: the `for` statement:

```python
# -*- coding: utf-8 -*-
'''
Mathematics and Python Programming     www.pysamples.com
pi.py
'''

alist = ''
for i in range(1, 20):
    alist = alist + str(i) + ','
print alist
```

0.5. THE STATEMENTS FOR AND WHILE

──────────────── output of the program ────────────────
```
1,2,3,4,5,6,7,8,9,10,11,12,13,14,15,16,17,18,19,
```

Note that although the indicated range is (1.20), the variable i take only 19 values. If we want the loop to take n values, we must indicate a range (0,n) or (1, n+1). We can also be very specific about counting by twos, fives, etc.

```python
# -*- coding: utf-8 -*-
'''
Mathematics and Python Programming    www.pysamples.com
pj.py
'''

alist = ''
for i in range(0, 101, 5):
    alist = alist + str(i) + ','
print alist
```

──────────────── output of the program ────────────────
```
0,5,10,15,20,25,30,35,40,45,50,55,60,65,70,75,80,85,90,95,100,
```

Another example of use of the for statement, now importing the module turtle, which allows you to create graphics by a moving tortoise, in a similar manner as it is done using the Logo language. You can obtain more information about these graphics at http://docs.python.org/2.7/library/turtle.html.

The drawing will be plotted with movements of a cursor which is called turtle, and it lets you visually understand how the for statement works, although in this example the program does not resolve any mathematical issue. With each stroke of the turtle, the color changes and the value of variable i, which is used as a counter, is printed.

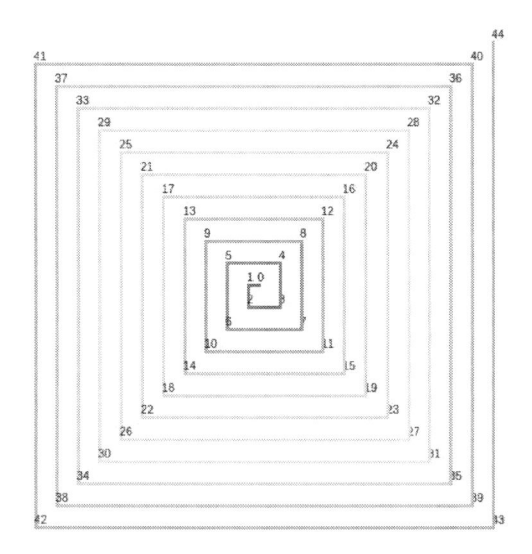

```python
# -*- coding: utf-8 -*-
'''
Mathematics and Python Programming    www.pysamples.com
pk.py
'''

import turtle as tt
```

```python
tt.mode('logo')
tt.reset()
tt.home()
screen = tt.getscreen()
screen.colormode(255)
r = 255
g = 50
b = 30
tt.pencolor(r, g, b)
tt.pensize(3)
tt.write('0')
for i in range(0, 45):
    tt.forward(10 * i)
    tt.left(90)
    tt.pencolor('black')
    tt.write(str(i))
    if i < 15:
        g += 12
    elif i < 30:
        b += 12
    else:
        r -= 12
    tt.pencolor(r, g, b)
tt.hideturtle()
```

We can also use it to represent a function, for example, the spiral of Archimedes $r = a\varphi$, and to conclude, we show a second program that represents the function $r = a \cdot \sin 3\alpha$ with turtle graphics.

```python
# -*- coding: utf-8 -*-
'''
Mathematics and Python Programming    www.pysamples.com
p1.py
'''
import numpy as np
import turtle as tt

tt.mode('logo')
tt.reset()
tt.home()
screen = tt.getscreen()
screen.colormode(255)
r = 221
g = 25
b = 127
tt.pencolor(r, g, b)
tt.speed('fastest')
degrees = 0
a = 5
tt.pensize(5)
for degrees in range(0, 360 * 5):
    radians = np.deg2rad(degrees)
    r = a * radians
    x = int(r * np.cos(radians))
    y = int(r * np.sin(radians))
    tt.setpos(x, y)
    tt.forward(r - 5)
    tt.pendown()
    tt.forward(5)
```

0.5. THE STATEMENTS FOR AND WHILE

```
        tt.penup()
        tt.left(1)
tt.hideturtle()
```

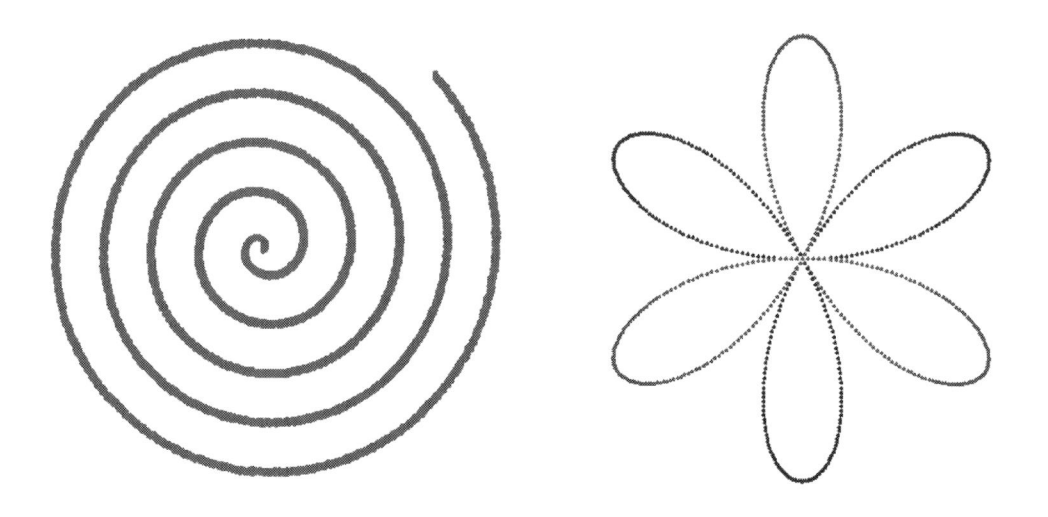

```
# -*- coding: utf-8 -*-
'''
Mathematics and Python Programming    www.pysamples.com
pm.py
'''

import numpy as np
import turtle as tt

tt.mode('logo')
tt.reset()
tt.home()
tt.speed('fastest')
a = 200
degrees = 0
while degrees <= 180:
    radians = np.deg2rad(degrees)
    r = a * np.sin(3 * radians)
    x = int(r * np.cos(radians))
    y = int(r * np.sin(radians))
    y2 = - y
    tt.goto(x, y)
    tt.pendown()
    tt.pencolor('blue')
    tt.dot(4)
    tt.penup()
    tt.goto(x, y2)
    tt.pendown()
    tt.pencolor('red')
    tt.dot(4)
    tt.penup()
    degrees += 0.5
tt.hideturtle()
```

0.6 Operations with polynomials

Numpy has instructions to perform operations with polynomials. Let's see an example:

```python
# -*- coding: utf-8 -*-
'''
Mathematics and Python Programming    www.pysamples.com
pn.py
'''

import numpy as np

p1 = np.poly1d([1, 2, 3])
print 'p1 = '
print p1
print 'Value of the polynomial p1 for x=5:', np.polyval(p1, 5)
p2 = np.poly1d([1, -2, 1, 3, 2])
print 'p2 = '
print p2
print 'addition: '
print np.polyadd(p1, p2)
print 'subtraction p2-p1:'
print np.polysub(p2, p1)
print 'product: '
p = np.polymul(p1, p2)
print p
print 'polynomial division: p2/p1: '
quotient = np.polydiv(p2, p1)
print quotient
print 'quotient = '
print quotient[0]
print 'remainder = '
print quotient[1]
print 'p2 and its derivatives: '
for i in range(0, 6):
    print np.polyder(p2, m=i)
print 'integral of the first derivative of p2:'
p2int = np.polyint([4, -6, 2, 3])
print np.poly1d(p2int)
print 'binomials and product: '
polinomio = np.poly1d([0, 1])
for j in range(-2, 2):
    print np.poly1d([1, j])
    polinomio = np.polymul(polinomio, np.poly1d([1, j]))
print polinomio
print 'roots of the polynomial'
print np.sort(np.roots(polinomio))
```

──────────────────── output of the program ────────────────────
```
p1 =
   2
1 x + 2 x + 3
Value of the polynomial p1 for x=5: 38
p2 =
   4     3     2
1 x - 2 x + 1 x + 3 x + 2
addition:
   4     3     2
1 x - 2 x + 2 x + 5 x + 5
subtraction p2-p1:
   4     3
```

```
1 x - 2 x + 1 x - 1
product:
     6     3     2
1 x - 1 x + 11 x + 13 x + 6
polynomial division: p2/p1:
(poly1d([ 1., -4.,  6.]), poly1d([  3., -16.]))
quotient =
   2
1 x - 4 x + 6
remainder =

3 x - 16
p2 and its derivatives:
   4     3     2
1 x - 2 x + 1 x + 3 x + 2
   3     2
4 x - 6 x + 2 x + 3
    2
12 x - 12 x + 2

24 x - 12

24

0
integral of the first derivative of p2:
   4     3     2
1 x - 2 x + 1 x + 3 x
binomials and product:

1 x - 2

1 x - 1

1 x

1 x + 1
   4     3     2
1 x - 2 x - 1 x + 2 x
roots of the polynomial
[-1.  0.  1.  2.]
```

0.7 More on the while statement

Now let's look at the while statement in an example that breaks down a given number into its prime factors. We show the elapsed time after the results, just to get an idea of the speed of Python. Of course, this time is shown only for informational purposes and it will depend on the characteristics of the computer you use.

In the first place, let's take a look at a incorrect while loop, because it never meets the condition which makes the loop to end: the number to divide does not decrease, and the variable PrimeFactor does not stop growing indefinitely:

```python
# -*- coding: utf-8 -*-
'''
Mathematics and Python Programming      www.pysamples.com
p0.py
endless while statement
'''

number = int(raw_input('Write a natural number: '))
result = str(number) + ' = '
PrimeFactor = 2
divisions = 0
print 'number: ', number
while PrimeFactor <= number:
    remainder = number % PrimeFactor
    print number, PrimeFactor, remainder
```

```
        divisions += 1
        if remainder == 0:
            result = result + str(PrimeFactor) + '.'
            #number = number / PrimeFactor
            divisions += 1
        else:
            if PrimeFactor > 2:
                PrimeFactor += 2
            else:
                PrimeFactor += 1
result = result.rstrip('.')
print result
print divisions, ' divisions made'
```

When you run it, we divide every time the same number, and the loop never ends:

```
———————————————————————— output of the program ————————————————————————
120 2 0
120 2 0
120 2 0
120 2 0
120 2 0
...
```

The error is in the commented line

```
#number = number / PrimeFactor
```

This instruction is what makes smaller the number with each division, so that there will come a time when the loop finishes. It has been commented on purpose to show that in a while loop we must always make sure that we will arrive at the end of the loop.

Now let's look at the program uncommenting that line, and commenting the line

```
print number, PrimeFactor, remainder
```

The program works, but as we shall see, it is not very efficient. We show the execution time in milliseconds.

```
———————————————————— several outputs of the program ————————————————————
number:  499
499 = 499
251  divisions made
0.17786026001  ms

number:  2013
2013 = 3.11.61
36  divisions made
0.0619888305664  ms

number:  123456780
123456780 = 2.2.3.3.5.47.14593
7310  divisions made
4.15802001953  ms
```

Now note that we don't indicate in the while statement how many divisions must be done, since we don't know it because it will depend on the number that you entered. What we are doing is to indicate that the whole block of instructions under the while statement must be run while the prime factor by which we are dividing be less than the given number. With each division, it checks if the number is divisible by the same factor. If so, it keeps on dividing by that factor again while the factor is less than the number. If the remainder is not zero, the number is not divisible by that factor, and then we increase the factor in two units if it was odd, or in a unit if we were dividing by 2. Unlike the wrong program, at each step, the number decreases or the factor by which we divided get increased, so that the loop ends at some point.

The procedure has the disadvantage of that it try to divide by all the multiples of 3, 6, 9, etc, and in all these cases it will result that the number is not divisible, as has been previously divided by the lesser prime factor 2, 3, ... , and we waste some time. To get an idea of the speed of the

0.7. MORE ON THE WHILE STATEMENT

program, it has been executed to break down a list of 1000 random integer numbers, and it has been clocked a time of 191,375 ms, making a total of 284705 divisions.

An improvement of the program consists of using a list of prime numbers and use it to take the prime factors. This is what we are going to do now, in the next version of our program of decomposition of a number into its prime factors. It includes a list of prime numbers that will be the factors which we will use to begin to divide by, rather than to increase the factor by adding two units. In this way we will reduce the number of divisions required, and the processing time of our computer. In addition, if the given number is in the prime list, now the program won't make any division:

```python
# -*- coding: utf-8 -*-
'''
Mathematics and Python Programming    www.pysamples.com
pr.py
'''

number = int(raw_input('Write a natural number: '))
result = str(number) + ' = '
divisions = 0
PrimeList = [2, 3, 5, 7, 11, 13, 17, 19, 23, 29, 31, 37, 41, 43, 47, 53, 59, 61, 67, 71,
73, 79, 83, 89, 97, 101, 103, 107, 109, 113, 127, 131, 137, 139, 149, 151, 157, 163, 167,
173, 179, 181, 191, 193, 197, 199, 211, 223, 227, 229, 233, 239, 241, 251, 257, 263, 269,
271, 277, 281, 283, 293, 307, 311, 313, 317, 331, 337, 347, 349, 353, 359, 367, 373, 379,
383, 389, 397, 401, 409, 419, 421, 431, 433, 439, 443, 449, 457, 461, 463, 467, 479, 487,
491, 499, 503, 509, 521, 523, 541, 547, 557, 563, 569, 571, 577, 587, 593, 599, 601, 607,
613, 617, 619, 631, 641, 643, 647, 653, 659, 661, 673, 677, 683, 691, 701, 709, 719, 727,
733, 739, 743, 751, 757, 761, 769, 773, 787, 797, 809, 811, 821, 823, 827, 829, 839, 853,
857, 859, 863, 877, 881, 883, 887, 907, 911, 919, 929, 937, 941, 947, 953, 967, 971, 977,
983, 991, 997, 1009]

i = 0
PrimeFactor = PrimeList[i]

if number in PrimeList:
    result = str(number) + ' is prime'
    divisions = 0
else:
    while PrimeFactor <= number:
        resto = number % PrimeFactor
        #print PrimeFactor
        divisions += 1
        if resto == 0:
            result = result + str(PrimeFactor) + '.'
            number = number / PrimeFactor
            divisions += 1
        else:
            if PrimeFactor < 1009:
                i += 1
                PrimeFactor = PrimeList[i]
            else:
                PrimeFactor += 2
        #print PrimeFactor, ', ', number
result = result.rstrip('.')
print result
print divisions, ' divisions made.'
```

```
―――――――――――――――――――――― several program outputs ――――――――――――――――――――――
Write a natural number: 499
499 is prime
0   divisions made.
0.0569820404053  ms

Write a natural number: 2013
2013 = 3.11.61
23   divisions made.
0.0579357147217  ms

Write a natural number: 123456780
123456780 = 2.2.3.3.5.47.14593
6974   divisions made.
4.02808189392  ms
```
――

This version of the program is more efficient: 0 divisions instead of 251 divisions in the first program to break down the number 499; 23 compared to 36 divisions to break down the number 2013; and 6974 divisions compared to 7310 for the last number. If the number is greater than the prime numbers in the list, the program is less efficient and its running times are close to those of the previous version. This improved version has been also timed to decompose 1000 random numbers using both programs to compare their performance, and we obtained the following results:

- Without any list of prime numbers: 191,375 ms, making a total of 284705 divisions.

- With a list of prime numbers: 144,076 ms, making a total of 173176 divisions.

It achieves a reduction of time of around 25%, and a reduction of 39% in the number of divisions.

0.8 Matplotlib

Although it is fun use turtle to draw graphics, Python has a library specially designed to represent functions and graphics: Matplotlib, that we will see below with an example that will serve us to represent data with Matplotlib and to compare two interpolation methods.

It is very common in scientific work to need to consult tables of values, and to interpolate a value measured experimentally. Suppose we want to determine the sugar content of a sample of grapes. If we measure the density of the juice of these grapes, we may know its sugar content querying a table like the following, which we'll use in a file called densities.csv, which can be read with any text editor. The complete list of values appears in the Python program below:

――――――――――――――――――――――――――――――――― file densities.csv ―――――――――――――――――――――――――――――――――
```
"density","sugars (g/l)"
1015,10
1020,23
1025,36
...
1165,410
1170,423
```
――

It is a list of pairs of values of density and sugar content, separated by a comma. In the example we used a short table. Normally the file will be much longer. In the sample of juice we measure its density and experimentally we get the value, for example, 1068. This value is not listed in the table, so how can we know how much sugar (mainly glucose and fructose) it contains?

The following Python program does it by two methods: linear interpolation, and by the adjustment of the statistical data of the table to a straight line by the method of least squares. In addition, it represents the points of the table, the adjusted line and the interpolated point (in red):

```
# -*- coding: utf-8 -*-
'''
```

0.8. MATPLOTLIB

```
ps.py
'''

import csv
import matplotlib.pyplot as plt
from scipy import stats
import numpy as np

d = []
gl = []
with open('densities.csv', 'rb') as csvfile:
    spamreader = csv.reader(csvfile, delimiter=',', quotechar='"')
    i = 0
    for r in spamreader:
        if i > 0:
            d.append(int(r[0]))
            gl.append(int(r[1]))
        i += 1
print d
print gl

# linear interpolation
delta = 5    # distance between density data
density = 1068   # experimental value of density
i = 0
for datum in d:
    if datum < density:
        dant = datum
        j = i
    i += 1
print 'previous density: ', dant
glant = gl[j]
print 'previous sugars: ', glant
f = np.round(glant + ((density - dant) * (gl[j + 1] - gl[j]) / 5.0), 1)
print 'measured density: ', density
print 'linearly interpolated sugars: ', "%.1f" % f

#least squares adjust
print 'least squares adjust:'
slope, intercept, r_value, p_value, std_err = stats.linregress(d, gl)
print 'r^2: ', "%8.6f" % (r_value ** 2)
print ('straight line: y = ' + "%6.2f" % slope + 'x + ' + "%6.2f" % intercept)
print 'measured density: ', density
f2 = np.round(slope * density + intercept, 1)
print ('sugars adjusted by least squares method: ' + "%.1f" % f2)

#graphic
f0 = slope * d[0] + intercept
fn = slope * d[-1] + intercept
plt.plot([d[0], d[-1]], [f0, fn], 'k-', lw=1.5)  # straight line adjust
plt.plot(d, gl, 'wo')
plt.xlim(1010, 1180)
plt.plot(density, f2, 'ro')
plt.plot([density, density], [0, f2], 'r--', lw=1.0)
plt.plot([1010, density], [f2, f2], 'r--', lw=1.0)
plt.xlabel('density (g/l)')
```

```
plt.ylabel('sugars (g/l)')
plt.show()
```

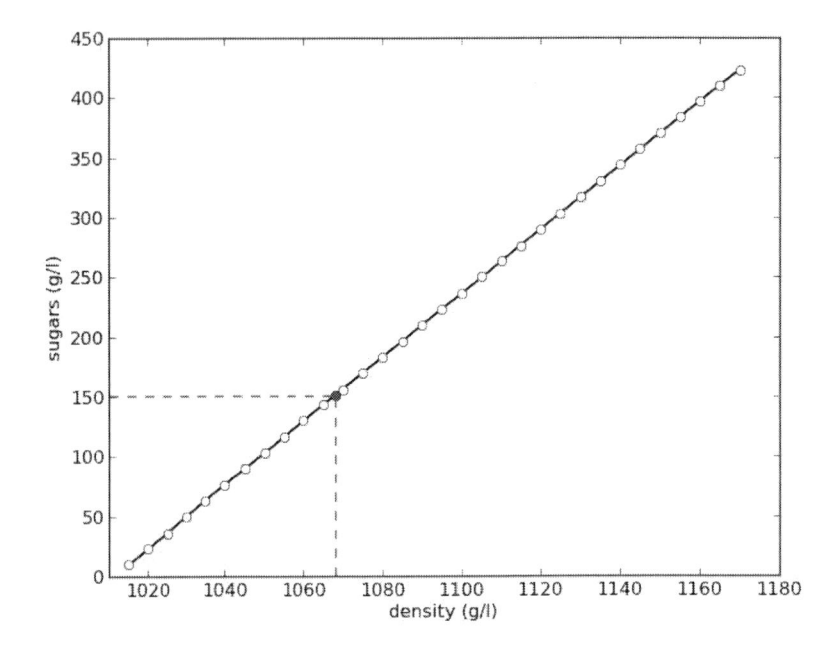

```
_____ output of the program _____
[1015, 1020, 1025, 1030, 1035, 1040, 1045, 1050, 1055, 1060, 1065, 1070, 1075, 1080, 1085, 1090,
1095, 1100, 1105, 1110, 1115, 1120, 1125, 1130, 1135, 1140, 1145, 1150, 1155, 1160, 1165, 1170]
[10, 23, 36, 50, 63, 76, 90, 103, 116, 130, 143, 156, 170, 183, 196, 210, 223,
236, 250, 263, 276, 290, 303, 317, 330, 344, 357, 370, 383, 397, 410, 423]
previous density: 1065
previous sugars:  143
measured density: 1068
linearly interpolated sugars:  150.8
least squares adjust:
r^2:  0.999993
straight line: y =    2.67x + -2700.00
measured density: 1068
sugars adjusted by least squares method: 151.1
```

If querying a more complete table, the value that is listed for the density of 1068 is just 151, the value that we have obtained.

Matplotlib also allows you to create more complex graphics. As an example, we are going to represent the equation of Van der Waals for several temperatures:

$$(P + \frac{a}{V^2})(V - b) = RT$$

for gas CO_2. Each line corresponds to a temperature value: they are isotherms.

0.8. MATPLOTLIB

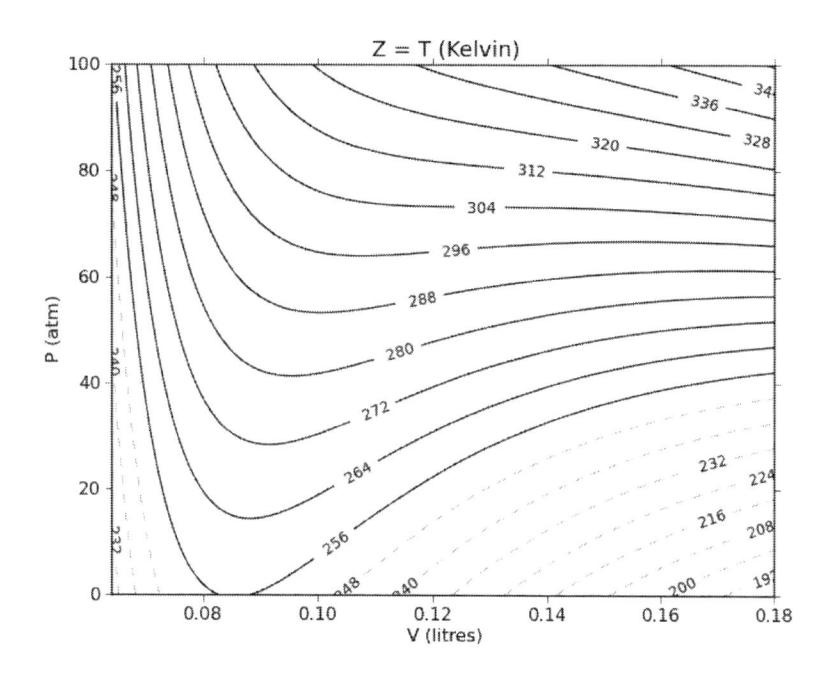

Z = T (Kelvin)

[Contour plot: P (atm) vs V (litres), with temperature contours labeled 256, 264, 272, 280, 288, 296, 304, 312, 320, 328, 336, 344 and dashed contours 192, 200, 208, 216, 224, 232, 240, 248]

```
# -*- coding: utf-8 -*-
'''
Mathematics and Python Programming    www.pysamples.com
ps.py
'''

import csv
import matplotlib.pyplot as plt
from scipy import stats
import numpy as np

d = []
gl = []
with open('densities.csv', 'rb') as csvfile:
    spamreader = csv.reader(csvfile, delimiter=',', quotechar='"')
    i = 0
    for r in spamreader:
        if i > 0:
            d.append(int(r[0]))
            gl.append(int(r[1]))
        i += 1
print d
print gl

# linear interpolation
delta = 5      # distance between density data
density = 1068 # experimental value of density
i = 0
for datum in d:
    if datum < density:
        dant = datum
        j = i
    i += 1
print 'previous density: ', dant
glant = gl[j]
```

```python
print 'previous sugars: ', glant
f = np.round(glant + ((density - dant) * (gl[j + 1] - gl[j]) / 5.0), 1)
print 'measured density: ', density
print 'linearly interpolated sugars: ', "%.1f" % f

#least squares adjust
print 'least squares adjust:'
slope, intercept, r_value, p_value, std_err = stats.linregress(d, gl)
print 'r^2: ', "%8.6f" % (r_value ** 2)
print ('straight line: y = ' + "%6.2f" % slope + 'x + ' + "%6.2f" % intercept)
print 'measured density: ', density
f2 = np.round(slope * density + intercept, 1)
print ('sugars adjusted by least squares method: ' + "%.1f" % f2)

#graphic
f0 = slope * d[0] + intercept
fn = slope * d[-1] + intercept
plt.plot([d[0], d[-1]], [f0, fn], 'k-', lw=1.5)  # straight line adjust
plt.plot(d, gl, 'wo')
plt.xlim(1010, 1180)
plt.plot(density, f2, 'ro')
plt.plot([density, density], [0, f2], 'r--', lw=1.0)
plt.plot([1010, density], [f2, f2], 'r--', lw=1.0)
plt.xlabel('density (g/l)')
plt.ylabel('sugars (g/l)')
plt.show()
```

The 2D representation corresponds to the following 3D representation created by the program py.py. The Z-axis corresponds to the pressure.

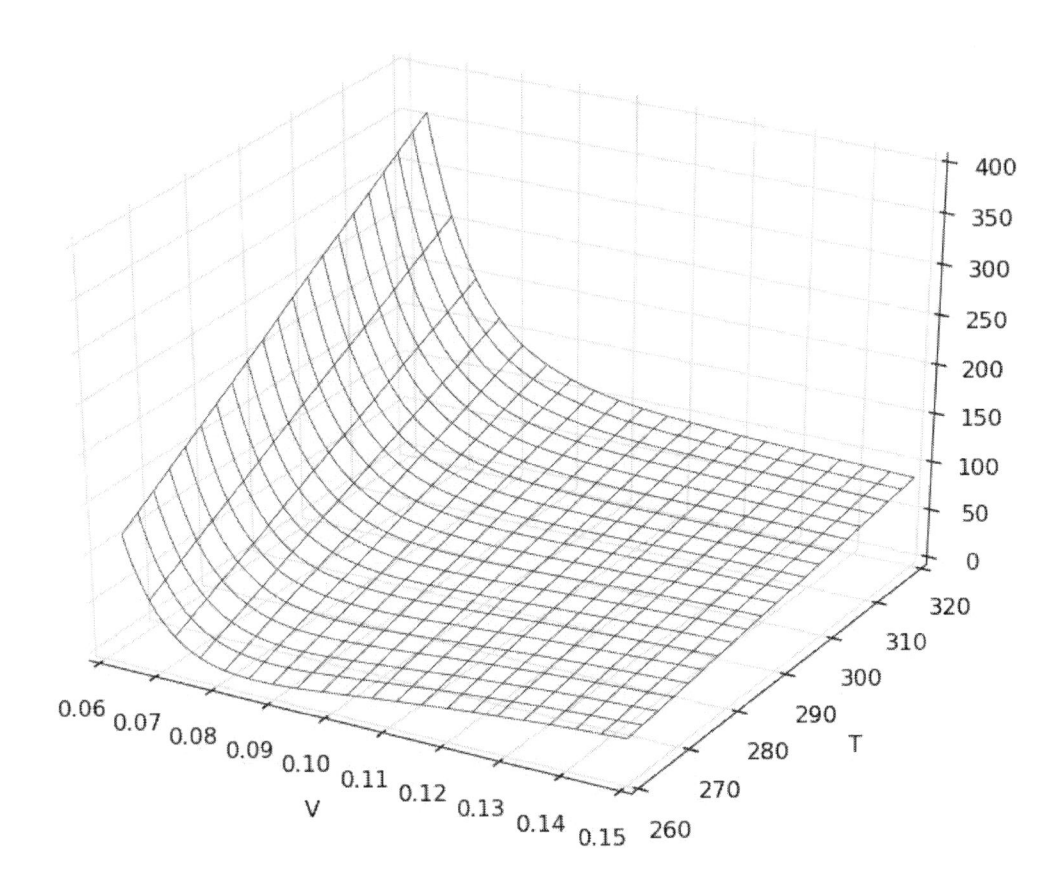

```python
# -*- coding: utf-8 -*-
"""
Mathematics and Python Programming    www.pysamples.com
pu.py
"""

from mpl_toolkits.mplot3d import Axes3D
import matplotlib.pyplot as plt
import numpy as np

fig = plt.figure()
ax = fig.add_subplot(111, projection='3d')
ax.w_xaxis.set_pane_color((1.0, 1.0, 1.0, 1.0))
ax.w_yaxis.set_pane_color((1.0, 1.0, 1.0, 1.0))
ax.w_zaxis.set_pane_color((1.0, 1.0, 1.0, 1.0))
a = 3.592   # atm l2 mol-2
b = 0.0427  # l mol-1
R = 0.08206 # atl l K-1 mol-1
pointsnumber = 500
v = np.linspace(0.064, 0.15, pointsnumber)  # x = V
t = np.linspace(260, 320, pointsnumber)
X, Y = np.meshgrid(v, t)
Z = (R * Y / (X - b)) - (a / X ** 2)
ax.plot_wireframe(X, Y, Z, rstride=25, cstride=25, lw=0.5, color='b')
plt.xlabel('V')
plt.ylabel('T')
plt.show()
```

0.9 Text Files. Statistics

Now we will check the ability of Python to open text files and read their contents. We are going to use the text of The Life and Adventures of Robinson Crusoe, by Daniel Defoe, which can be obtained free at http://www.gutenberg.org/. The following program opens the text file The Life and Adventures of Robinson Crusoe.txt and counts how many letters a are in the text, how many b letters, etc., and stores the results in a dictionary. Subsequently we take the dictionary data, order them and transform them into a list. Finally, the program calculates the percentage of occurrence of each letter in the text and produces a graph. At the end of the results the elapsed time it takes to program to make the count of letters of the book in milliseconds has been added.

```python
# -*- coding: utf-8 -*-
"""
Mathematics and Python Programming    www.pysamples.com
pv.py
"""

vocals = 'aeiouAEIOU'
consonants = 'bBcCdDfFgGhHjJkKlLmMnN pPqQrRsStTvVwWxXyYzZ'

dictvoc = {}
dictcons = {}

input = file('The Life and Adventures of Robinson Crusoe.txt')
text = input.read()
utext = unicode(text, 'utf-8')
lentext = len(utext)
print lentext
```

```
dictvoc['a'] = (utext.count('a') + utext.count(' ') +
                utext.count('A') + utext.count(' '))
dictvoc['e'] = (utext.count('e') + utext.count(' ') +
                utext.count('E') + utext.count(' '))
dictvoc['i'] = (utext.count('i') + utext.count(' ') +
                utext.count('I') + utext.count(' '))
dictvoc['o'] = (utext.count('o') + utext.count(' ') +
                utext.count('O') + utext.count(' '))
dictvoc['u'] = (utext.count('u') + utext.count(' ') +
                utext.count('U') + utext.count(' ') +
                utext.count(' ') + utext.count(' '))

nvocals = dictvoc.values()
number_vocals = 0
for i in nvocals:
    number_vocals += i
print 'vocals total: ', number_vocals
listvocals = dictvoc.items()
listvocals.sort()
print listvocals


def countcons(text, letra):
    recuento = (text.count(letra) +
                text.count(letra.upper()))
    return recuento
dictcons['b'] = countcons(utext, 'b')
dictcons['c'] = countcons(utext, 'c')
dictcons['cedilla'] = countcons(utext, ' ')
dictcons['d'] = countcons(utext, 'd')
dictcons['f'] = countcons(utext, 'f')
dictcons['g'] = countcons(utext, 'g')
dictcons['h'] = countcons(utext, 'h')
dictcons['j'] = countcons(utext, 'j')
dictcons['k'] = countcons(utext, 'k')
dictcons['l'] = countcons(utext, 'l')
dictcons['m'] = countcons(utext, 'm')
dictcons['n'] = countcons(utext, 'n')
enemays = ' '.encode('utf-8')
nene = utext.count(' ') + utext.count(enemays)
dictcons['ntilde'] = nene
dictcons['p'] = countcons(utext, 'p')
dictcons['q'] = countcons(utext, 'q')
dictcons['r'] = countcons(utext, 'r')
dictcons['s'] = countcons(utext, 's')
dictcons['t'] = countcons(utext, 't')
dictcons['v'] = countcons(utext, 'v')
dictcons['w'] = countcons(utext, 'w')
dictcons['x'] = countcons(utext, 'x')
dictcons['y'] = countcons(utext, 'y')
dictcons['z'] = countcons(utext, 'z')

nconsonants = dictcons.values()
number_consonants = 0
for i in nconsonants:
```

0.9. TEXT FILES. STATISTICS

```
        number_consonants += i
print 'consonants total: ', number_consonants
listcons = dictcons.items()
listcons.sort()
print listcons
#
array100vocals = [0.0, 0.0, 0.0, 0.0, 0.0]
for i in range(0, 5):
    array100vocals[i] = float("%7.3f" %
                        (100.0 * listvocals[i][1] /
                        (number_vocals + number_consonants)))
percentagevocals = 0
for i in array100vocals:
    percentagevocals += i
print array100vocals, ' = ', percentagevocals, '%'
#
array100cons = [0.0, 0.0, 0.0, 0.0, 0.0, 0.0, 0.0,
                0.0, 0.0, 0.0, 0.0, 0.0, 0.0, 0.0,
                0.0, 0.0, 0.0, 0.0, 0.0, 0.0, 0.0,
                0.0, 0.0]
for i in range(0, 23):
    array100cons[i] = float("%7.3f" %
                      (100.0 * listcons[i][1] /
                      (number_vocals + number_consonants)))
percentagecons = 0
for i in array100cons:
    percentagecons += i
print array100cons, ' = ', percentagecons, '%'
```

```
─────────────────────────────── results ───────────────────────────────
1225554
vocals total:  353824
[('a', 76513), ('e', 114605), ('i', 62614), ('o', 74068), ('u', 26024)]
consonants total:  578460
[('b', 14413), ('c', 20263), ('cedilla', 0), ('d', 43319), ('f', 21955), ('g', 18499),
('h', 62726), ('j', 762), ('k', 6340), ('l', 33363), ('m', 25850), ('n', 62436),
('ntilde', 0), ('p', 14327), ('q', 613), ('r', 50874), ('s', 53858), ('t', 91614),
('v', 10222), ('w', 26208), ('x', 1071), ('y', 19391), ('z', 356)]
[8.207, 12.293, 6.716, 7.945, 2.791]  =   37.952 %
[1.546, 2.173, 0.0, 4.647, 2.355, 1.984, 6.728, 0.082, 0.68, 3.579, 2.773, 6.697, 0.0,
1.537, 0.066, 5.457, 5.777, 9.827, 1.096, 2.811, 0.115, 2.08, 0.038]  =   62.048 %
107.511043549  ms
```

In little more than one tenth of a second, the program has made the counting and classification of more than one million letters contained in the work of Defoe.

The program has been run to analyse five texts: The Life and Adventures of Robinson Crusoe; 10 episodes of BBC's English We Speak; The Tragedy of Hamlet, by William Shakespeare; A Treatise on Electricity and Magnetism, by J.C. Maxwell; and Adventures of Huckleberry Finn, by Mark Twain.

For each book we get the count of letters, which the following program represents as points. As you can see, it is possible to represent the five sets of points on the same graph, each set with a different bookmark; and the program represents one more set, corresponding to the average of the other five sets. The average is represented as a line. Also note that the labels for the X-axis are the corresponding letters instead of a simple numbering.

```
# -*- coding: utf-8 -*-
"""
Mathematics and Python Programming     www.pysamples.com
pw.py
"""

import matplotlib.pyplot as plt

fig = plt.figure()
ax = fig.add_subplot(1, 1, 1)
xticks = ['$a$', '$e$', '$i$', '$o$', '$u$',
          '$b$', '$c$', '$d$', '$f$', '$g$',
          '$h$', '$j$', '$k$', '$l$', '$m$',
          '$n$', '$p$', '$q$', '$r$', '$s$',
          '$t$', '$v$', '$w$', '$x$', '$y$', '$z$']
x = [0, 1, 2, 3, 4, 5, 6, 7, 8, 9, 10, 11, 12, 13, 14, 15,
     16, 17, 18, 19, 20, 21, 22, 23, 24, 25]
defoe = [8.207, 12.293, 6.716, 7.945, 2.791, 1.546, 2.173, 4.647, 2.355,
         1.984, 6.728, 0.082, 0.68, 3.579, 2.773, 6.697, 1.537, 0.066,
         5.457, 5.777, 9.827, 1.096, 2.811, 0.115, 2.08, 0.038]
BBC_EWS = [7.314, 11.614, 7.228, 8.817, 3.072, 2.114, 2.718, 2.889, 2.055,
           2.679, 5.351, 0.184, 1.346, 4.202, 3.132, 6.565, 1.503, 0.02,
           5.633, 6.703, 8.692, 0.709, 2.436, 0.276, 2.731, 0.02]
hamlet = [7.381, 13.36, 6.492, 8.536, 3.759, 1.394, 1.933, 3.747, 2.034,
          1.83, 6.712, 0.001, 0.998, 4.458, 3.245, 6.222, 1.5, 0.159,
          5.844, 6.187, 8.875, 0.407, 2.363, 0.112, 2.411, 0.04]
maxwell = [7.232, 13.571, 7.944, 7.242, 2.595, 1.37, 4.537, 3.373, 3.349,
           1.287, 5.281, 0.136, 0.228, 3.713, 2.013, 6.813, 2.03, 0.37,
           5.755, 6.051, 10.61, 0.971, 1.496, 0.375, 1.513, 0.145]
finn = [8.409, 11.176, 6.439, 8.349, 3.181, 1.692, 1.844, 5.468, 1.778,
        2.457, 6.117, 0.27, 1.314, 4.012, 2.366, 7.505, 1.325, 0.043,
        4.537, 5.792, 9.663, 0.672, 3.076, 0.117, 2.356, 0.044]
print len(defoe), len(BBC_EWS), len(hamlet), len(maxwell), len(finn)
media = []
for i in range(0, 26):
    promedio = float("%7.3f" % ((defoe[i] + BBC_EWS[i] + hamlet[i] +
                      maxwell[i] + finn[i]) / 5))
    media.append(promedio)
```

0.9. TEXT FILES. STATISTICS

```
print media

plt.plot(x, media, color='#F00707', lw=1.5)

p1, = plt.plot(x, defoe, 'yo')
p2, = plt.plot(x, BBC_EWS, 'k+')
p3, = plt.plot(x, hamlet, 'rD')
p4, = plt.plot(x, maxwell, 'b*')
p5, = plt.plot(x, finn, 'gs')
plt.ylabel('% of each letter')
plt.xlabel('letters')
plt.legend(('average', 'robinson', 'BBC', 'hamlet', 'maxwell', 'finn'),
           loc='upper center')
plt.grid(axis='x')
plt.xticks(x, xticks)
ax.set_xlim(-1, 26)
plt.show()
```

We are going to take advantage of the program that we have just seen to get a list of data with which to draw up statistics. We apply the program to 20 texts in English from different eras and authors and we want to find what percentage of vocal letters in the total number of letters is given in the texts in English. The code in the following program will be explained by pieces of code. In the first place, we import the packages that we are going to need and show the data of vowels and consonants of the 20 analysed texts[1]:

```
# -*- coding: utf-8 -*-
"""
Mathematics and Python Programming    www.pysamples.com
px.py
"""

import numpy as np
import matplotlib.pyplot as plt

# data from English texts mentioned in the book:

vocals = [353824, 5795, 48352, 256568, 160562, 1229, 8825, 415116,
          175791, 38536, 2881, 2657, 92734, 183069, 276660, 245519,
          32740, 188099, 446497, 234732]
consonants = [578460, 9437, 73971, 408384, 267005, 1951, 13922, 685427,
              286981, 64981, 4703, 4480, 147707, 293164, 459023, 402004,
              56058, 309459, 751081, 379425]

nvocals = np.sum(vocals)
nconsonants = np.sum(consonants)
books = len(vocals)
print 'data from ' + str(books) + ' books'
print ('Sample of ' + str(nvocals + nconsonants) +
       ' letters, of which ' + str(nvocals) + ' are vocals.')
per1000 = np.zeros(books, float)
```

---

[1] The Life and Adventures of Robinson Crusoe, by D. Defoe; 10 episodes of BBC English We Speak; The Tragedy of Hamlet, by W. Shakespeare; A Treatise on Electricity and Magnetism, by J.C. Maxwell; Adventures of Huckleberry Finn, by M. Twain; J.F. Kennedy's Berlin speech; Winston Churchill's Iron Curtain Speech; The Canterbury Tales and Other Poems, by Geoffrey Chaucer; 1984, by G. Orwell; The Old man and the Sea, by E. Hemingway; USA Independence Declaration; 'I have a dream' speech, by Martin Luther King; USMC Manual; English Grammar in Familiar Lectures, by Samuel Kirkham; The Crisis, by W. Churchill; America's Great Depression, by M. N. Rothbard; Shakespeare sonnets; Fairy and Folk Tales of the Irish Peasantry, by W.B. Yeats; Ulysses, by James Joyce; Maxims And Opinions Of Field-Marshal His Grace The Duke Of Wellington, Selected From His Writings And Speeches During A Public Life Of More Than Half A Century, by A. Wellesley, Duke of Wellington.

```python
for i in range(0, books):  # vocals por cada mil letras
    per1000[i] = float("%8.2f" % (1000.0 * vocals[i] /
                      (vocals[i] + consonants[i])))
print 'data: '
print per1000
creciente = np.sort(per1000)
print 'Data arranged from lowest to highest: '
print creciente
#max, min, recorrido, numero de datos
maximo = float("%8.2f" % np.max(per1000))
print 'maximum: ', maximo
minimo = float("%8.2f" % np.min(per1000))
print 'minimum: ', minimo
recorrido = float("%8.2f" % (maximo - minimo))
print 'range: ', recorrido
n = len(per1000)
print 'n: ', n
suma = np.sum(per1000)
print 'sum: ', suma
```

────────────────── output of the program ──────────────────
```
data from 20 books
Sample of 8367809 letters, of which 3170186 are vocals.
data:
[ 379.52  380.45  395.28  385.84  375.52  386.48  387.96  377.19  379.87
  372.27  379.88  372.29  385.68  384.41  376.06  379.17  368.7   378.04
  372.83  382.2 ]
Data arranged from lowest to highest:
[ 368.7   372.27  372.29  372.83  375.52  376.06  377.19  378.04  379.17
  379.52  379.87  379.88  380.45  382.2   384.41  385.68  385.84  386.48
  387.96  395.28]
maximum:  395.28
minimum:  368.7
range:  26.58
n:  20
sum:  7599.64
```

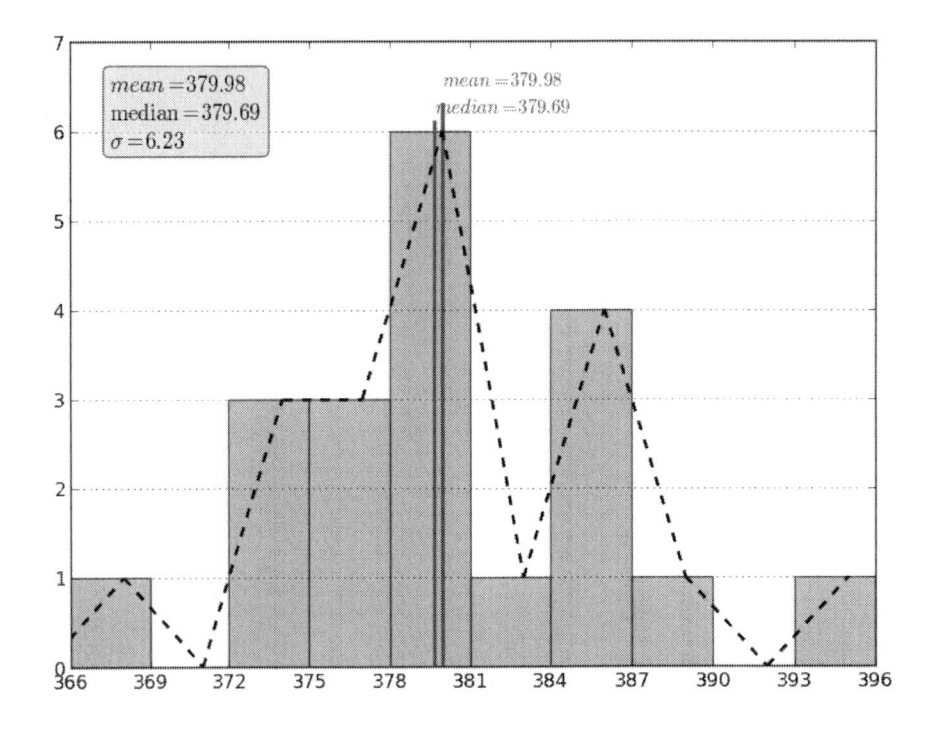

We distributed the data in intervals. It has been chosen to take 10 intervals:

0.9. TEXT FILES. STATISTICS

```
nintervals = 11   # 10 + 1
amplitud = np.ceil((recorrido / (nintervals - 1)))   # 2
print 'width of an interval: ', amplitud

liminf = np.trunc(np.min(per1000)) - (amplitud / 2.0)
intervals = np.zeros(nintervals, int)
intervals[0] = liminf   # -1
classmarks = np.zeros(nintervals, float)
classmarks[0] = liminf + (amplitud / 2.0)
for i in range(1, nintervals):
    intervals[i] = intervals[i - 1] + amplitud
for i in range(1, nintervals - 1):
    classmarks[i] = float("%8.2f" %
                          (classmarks[i - 1] + amplitud))
print str(nintervals - 1) + ' intervals: ' + str(intervals)
print 'class marks: ' + str(classmarks)
fintervals = np.zeros(nintervals - 1, int)
for i in range(1, nintervals):
    for j in range(0, n):
        if ((creciente[j] < intervals[i]) and (creciente[j] >= intervals[i - 1])):
            fintervals[i - 1] += 1
print 'f of the intervals: ', fintervals
```

```
──────────────────────────────── output of the program ────────────────────────────────
width of an interval:  3.0
10 intervals: [366 369 372 375 378 381 384 387 390 393 396]
class marks: [ 368. 371. 374. 377. 380. 383. 386. 389. 392. 395.   0.]
f of the intervals:  [1 0 3 3 6 1 4 1 0 1]
```

Now, we are going to calculate the mean, median, etc.:

```
arithmeticmean = float("%8.2f" % np.mean(per1000))
print 'arithmetic mean: ', arithmeticmean
median = float("%8.2f" % np.median(per1000))
print 'median: ', median
product = 1.0
for i in range(0, n):
    product = product * creciente[i]
geometric_mean = float("%8.2f" % product ** (1.0 / n))
print 'geometric mean: ', geometric_mean
print 'arithmetic mean must be greater than geometric mean:'
print arithmeticmean, ' > ', geometric_mean
```

```
──────────────────────────────── output of the program ────────────────────────────────
arithmetic mean:  379.98
median:  379.69
geometric mean:  379.93
arithmetic mean must be greater than geometric mean:
379.98  >  379.93
```

The following block of code calculates the measures of variability, as well as of asymmetry and kurtosis:

```
diferencias = np.zeros(25, float)
for i in range(0, n):
    diferencias[i] = float("%8.2f" %
                    (np.absolute(per1000[i] - arithmeticmean)))
print '|xi - mean|: ', diferencias
standard_deviation = float("%8.2f" % (np.sum(diferencias) / n))
print ('standard deviation of ungrouped data: ' +
        str(standard_deviation))
```

```python
diferenciasAG = np.zeros(nintervals - 1, float)
for i in range(0, nintervals - 1):
    diferenciasAG[i] = float("%8.2f" % (fintervals[i] *
                        np.absolute(classmarks[i] - arithmeticmean)))
print 'fi |mark_i - mean|: ', diferenciasAG
standard_deviationAG = float("%8.2f" % (np.sum(diferenciasAG) / nintervals))
print 'standard deviation of grouped data: ', standard_deviationAG
standard_deviation = float("%8.2f" % np.std(per1000))
print 'standard deviation s: ', standard_deviation
variance = float("%8.2f" % np.var(per1000))
print 'variance s2: ', variance
print 'If the data follow the normal distribution, '
print 'the mean deviation should be approximately (4 * s / 5):'
print ('mean deviation: ' + str(standard_deviation) +
        '; 4s/5: ' + str(0.8 * standard_deviation))
if arithmeticmean != 0:
    variacionPearson = (100 * standard_deviation / arithmeticmean)
    print ('Pearson coefficient: C.V. =' +
            "%4.2f" % variacionPearson)
asimetriaPearson = float("%8.2f" %
        (3 * (arithmeticmean - median) / standard_deviation))
if asimetriaPearson == 0:
    phrase1 = 'arithmetic mean = median'
elif asimetriaPearson < 0:
    phrase1 = 'arithmetic mean < median'
else:
    phrase1 = 'arithmetic mean > median'
print 'Pearson asymmetry: ', asimetriaPearson, ': ', phrase1
m4 = 0
for i in range(0, n):
    m4 += diferencias[i] ** 4
m4 = float("%8.2f" % (m4 / n))
print 'm4 = ', m4
print ('a: C.V. =' + "%4.2f" % variacionPearson)
kurtosis = float("%8.2f" % (m4 / (variance ** 2)))
if kurtosis == 0:
    phrase2 = 'normal curve'
elif kurtosis < 0:
    phrase2 = 'platykurtic'
else:
    phrase2 = 'leptokurtic'
print 'kurtosis = ', kurtosis, ' : the curve is ', phrase2
```

```
_____ output of the program _____
|xi - mean|:  [  0.46    0.47   15.3    5.86    4.46   6.5    7.98   2.79   0.11   7.71
    0.1    7.69    5.7    4.43    3.92   0.81  11.28   1.94   7.15   2.22   0.
    0.     0.     0.     0.  ]
standard deviation of ungrouped data: 4.84
fi |mark_i - mean|:  [ 11.98   0.    17.94   8.94   0.12   3.02  24.08   9.02   0.   15.02]
standard deviation of grouped data:  8.19
standard deviation s:  6.23
variance s2:  38.81
If the data follow the normal distribution,
the mean deviation should be approximately (4 * s / 5):
mean deviation: 6.23; 4s/5: 4.984
Pearson coefficient: C.V. =1.64
Pearson asymmetry:  0.14 : arithmetic mean > median
m4 =  4491.18
a: C.V. =1.64
kurtosis =  2.98  : the curve is  leptokurtic
```

Finally, the program draws up a histogram with the data grouped in intervals, the polygon of frequencies, the arithmetic mean and the median:

```python
fig, ax = plt.subplots(1)
textstr = ('$mean=%.2f$\n$\mathrm{median}=%.2f$\n$\sigma=%.2f$' %
           (arithmeticmean, median, standard_deviation))
props = dict(boxstyle='round', facecolor='#FCE945', alpha=0.6)
ax.text(0.05, 0.95, textstr, transform=ax.transAxes,
        fontsize=14, verticalalignment='top', bbox=props)
binsx = np.linspace(intervals[0], intervals[nintervals - 1],
                    nintervals)
ax.hist(per1000, binsx, alpha=0.75, color='#F2AE04')

plt.plot([arithmeticmean, arithmeticmean],
         [0, np.max(fintervals) + 0.3], 'r', lw=2.5)
plt.text(arithmeticmean, np.max(fintervals) + 0.5,
         '$mean=%.2f$' % arithmeticmean,
         horizontalalignment='left', color='red')
plt.plot([median, median],
         [0, np.max(fintervals) + 0.1], 'g', lw=2.5)
plt.text(median, np.max(fintervals) + 0.2,
         '$median=%.2f$' % median,
         horizontalalignment='left', color='green')

plt.plot([np.min(intervals) - 1, classmarks[0]],
         [0, fintervals[0]], 'k—', lw=2)
for i in range(0, (len(fintervals) - 1)):
    plt.plot([classmarks[i], classmarks[i + 1]],
             [fintervals[i], fintervals[i + 1]], 'k—', lw=2)

plt.ylim(0, np.max(fintervals) + 1)
plt.xlim(np.floor(np.min(intervals)), np.ceil(np.max(intervals)))
plt.grid(axis='y')
xticks = intervals
plt.xticks(xticks)
plt.show()
```

$

1 | Sets

If one should try to use an abacus to calculate the number of this great multitude, though he spent as many kalpas as Ganges sands he could never know the full sum.
The Lotus Sutra

This is one of the difficulties that arise when we try, with our finite minds, to discuss the infinite, assigning to it some properties that we give to the finite and limited.
Galileo Galilei, Discourses and Mathematical Demonstrations Relating to Two New Sciences, 1638.

It is not surprising that our language should be incapable of describing the processes occurring within the atoms, for, as has been remarked, it was invented to describe the experiences of daily life, and these consist only of processes involving exceedingly large numbers of atoms. Furthermore, it is very difficult to modify our language so that it will be able to describe these atomic processes, for words can only describe things of which we can form mental pictures, and this ability, too, is a result of daily experience. Fortunately, mathematics is not subject to this limitation, and it has been possible to invent a mathematical scheme (...) that seems entirely adequate.
Werner Heisenberg, The Physical Principles of the Quantum Theory, 1930.

1.1 Equivalence relations

We will adopt the definition of sets given by the founder of the theory of sets, the German mathematician Georg Cantor (1845-1918): *a set is a collection of different objects, defined by our intuition or our thinking, which can be conceived as a whole. Those objects are called elements of the set.*

Therefore, for a given set and any object a, this object a will be an element of the set, or it will not. The set without elements is the empty set, $\emptyset$. Let's take any two sets X and Y. Any subset $R \subset X \times Y$ is called a binary relationship between X and Y. If $Y = X$, it is said that R is a binary relation on X. If this relation R meets three properties that we will discuss below, then it is called a relation of equivalence, and it will provide the set X of a structure that we often see in the following chapters. These three properties are:

1. Reflexivity: $aRa, \forall a \in X$.

2. Symmetry: if aRb, then bRa.

3. Transitivity: if aRb and bRc, then aRc.

We can use the symbol $\sim$ to specify that R is a relation of equivalence. Any given element $x \in X$, the set of all elements that are related to x, and therefore are equivalent to x, is called the equivalence class of element x, and it is denoted as $\bar{x}$. Any element belonging to the same equivalence class of x is called a representative of class $\bar{x}$. In this way, a relation of equivalence in a set X establishes a partition of the set X in disjoint subsets called equivalence classes. The union of all of them is the same set X.

Any two equivalence classes are disjunct or match, as if there were an element x belonging to two different equivalence classes $\bar{a}$ and $\bar{b}$, we would have $x \sim a$ and $x \sim b$, but $\bar{a}$ and $\bar{b}$ would not be related, because they supposedly belong to different classes of equivalence, which is a contradiction because the relationship of equivalence is transitive and if $x \sim a$, and $x \sim b$, necessarily it implies that $a \sim b$, and therefore $\bar{a}$ matches $\bar{b}$. And vice versa: if you have a partition of a set X in disjoint subsets, then each of these subsets will be a class of equivalence, and there exist a relationship of equivalence in the set X, since then we would say that two elements are related if they belong to the same set. Therefore, every element is related to itself (reflexivity); if an element y belongs to the same subset that x, this element x also belongs to the same subset that y (symmetry); if y belongs to the same set that x, and z belongs to the same subset that y, then z belongs to the same subset that x (transitivity), so we have proved that the partition of a set establishes a relationship of equivalence.

1.2 Countable Sets: N, Z, Q

Let's take a finite set. If we take some elements of this set in the following way: firs we take an element x_1, then an element x_2 of the complement x_1; then another element x_3 of the complement of $\{x_1, x_2\}$, there will come a time in which we depleted all of the elements in the set. On the other hand, if the set is infinite this will not happen, and we can take x_1, x_2, x_3, ... different elements of the set.

The simpler finite set is the set of natural numbers, **N**. If we take any infinite set X and it is possible to establish a one-to-one correspondence between X and **N**, we can say that X is countable. All all finite sets are countable. Let's look more closely at the set N. The set of the natural numbers meets the following five axioms:

1. The number 1 is a natural number.

2. For every natural number n there is a single number that is called successor of n, which we will denote by n'.

3. For every natural number n we have $n' \neq 1$.

4. If $m' = n'$, then $m = n$.

5. The mathematical induction is valid: If a set M of natural numbers includes the number 1 and if every time a certain natural number n and all the numbers less than n can be considered as belonging to the set M, the number $n + 1$ may also be inferred to belong to M, then the set M includes all the natural numbers.

Some examples of countable sets:

- The set of all natural even numbers, using the one-to-one correspondence $n \leftrightarrow 2n$.

- The set **Z** of all the integer numbers, as we can establish the bijection:

$$\{0, -1, 1, -2, 2, ...\} \leftrightarrow \{1, 2, 3, 4, 5, ...\}$$

- The set **Q** of all rational numbers. Let's examine this set. Since our first Elementary Education courses we became accustomed to handle fractions, but now we need to give a more rigorous definition the set **Q** of the rational numbers. Consider the set consisting of all ordered pairs (a, a') as follows: the first element of the pair is an integer: $a \in \mathbf{Z}$; and the second element of each pair is an integer not null: $a' \in \{\mathbf{Z} - 0\}$.

Now we will consider the relation R which is defined in the set of those ordered pairs in the following way:

$$(a, a') R (b, b') \leftrightarrow ab' = ba'$$

That is:

$$(a, a') R (b, b') \leftrightarrow \frac{a}{a'} = \frac{b}{b'}$$

This is an equivalence relation: ($aa' = aa'$; if $ab' = ba'$ then $ba' = a'b$; and transitivity is easy to check). Therefore, the relationship will create a partition of the set of ordered pairs into classes. This set of ordered pairs

Z × **Z**' is what we call the set of rational numbers, **Q**. Each class of equivalence, created by this equivalence relation in the set **Q** is a rational number.

For example, $\frac{1}{2}$, $\frac{5}{10}$ and $\frac{12}{24}$ belong to the same equivalence class: they are equivalent fractions. Each one of them is a representative of the equivalence class $q = (1,2)$. Let's note that $\frac{-1}{3}$ and $\frac{1}{-3}$ belong to the same equivalence class, they represent the same rational number $(-1, 3)$.

How we can demonstrate that **Q** is countable? Let us take the set **Q** of all the ordered pairs (a, b) that we have-defined above. Take $b > 0$ to represent each class of equivalence. We can sort all the rational numbers as follows: we take the sum of the two elements in absolute value: $|a| + b$. We call this sum the height of the rational number. Now, we will write the rational numbers in order of increasing height, and in ascending order of the element a for those numbers who have the same height. For example:

$$\frac{0}{1}, \frac{-1}{1}, \frac{1}{1}, \frac{-2}{1}, \frac{-1}{2}, \frac{1}{2}, \frac{2}{1}, \dots$$

In this way we can establish a bijection between the sets **Q** and **N**, and therefore **Q** is countable.

The set **Q** is dense, which means that between any two rational numbers a and b, with $a < b$, we can find so many rational numbers as we want, just by calculating $a + \nu \frac{b-a}{n+1}$, para $\nu = 1, 2, \dots n$. Let us look at this with a numerical example calculated with Python.

```python
# -*- coding: utf-8 -*-
"""
Mathematics and Python Programming    www.pysamples.com
p1a.py
"""

import numpy as np

maximum = 1000
n = 30  # number of integers to calculate

num_a = np.random.randint(1, maximum)
den_ab = np.random.randint(1, maximum)
num_b = num_a + 1
print ('a = ' + str(num_a) + '/' + str(den_ab) +
       ' = ' + "%.6f" % (1.0 * num_a / den_ab))
print ('b = ' + str(num_b) + '/' + str(den_ab) +
       ' = ' + "%.6f" % (1.0 * num_b / den_ab))
print 'b-a = 1/', den_ab
print ('q = (' + str(num_a) + '*' + str(n + 1) +
       '+ i)/(' + str(den_ab) + '*' +
       str(n + 1) + ')')
print ('q = (' + str(num_a * (n + 1)) +
       '+ i)/' + str(den_ab * (n + 1)))


def qdense():
    for i in range(1, n + 1):
        num = num_a * (n + 1) + i
        den = den_ab * (n + 1)
        print ('(' + str(i) + ') ' + str(num) +
```

```
            '/' + str(den) + ' = ' +
            "%.8f" % (1.0 * num / den))
```

`qdense()`

We run the program to calculate 30 rational numbers between any two rational numbers whose numerators differ only in one unit:

```
─────────────────── 01p01.py - output of the program ───────────────────
a = 832/152 = 5.473684
b = 833/152 = 5.480263
b-a = 1/ 152
q = (832*31+ i)/(152*31)
q = (25792+ i)/4712
 (1) 25793/4712 = 5.47389643      (2) 25794/4712 = 5.47410866
 (3) 25795/4712 = 5.47432088      (4) 25796/4712 = 5.47453311
 (5) 25797/4712 = 5.47474533      (6) 25798/4712 = 5.47495756
 (7) 25799/4712 = 5.47516978      (8) 25800/4712 = 5.47538200
 (9) 25801/4712 = 5.47559423     (10) 25802/4712 = 5.47580645
(11) 25803/4712 = 5.47601868     (12) 25804/4712 = 5.47623090
(13) 25805/4712 = 5.47644312     (14) 25806/4712 = 5.47665535
(15) 25807/4712 = 5.47686757     (16) 25808/4712 = 5.47707980
(17) 25809/4712 = 5.47729202     (18) 25810/4712 = 5.47750424
(19) 25811/4712 = 5.47771647     (20) 25812/4712 = 5.47792869
(21) 25813/4712 = 5.47814092     (22) 25814/4712 = 5.47835314
(23) 25815/4712 = 5.47856537     (24) 25816/4712 = 5.47877759
(25) 25817/4712 = 5.47898981     (26) 25818/4712 = 5.47920204
(27) 25819/4712 = 5.47941426     (28) 25820/4712 = 5.47962649
(29) 25821/4712 = 5.47983871     (30) 25822/4712 = 5.48005093
```

We can also give some examples of uncountable sets, which we now just mention. The following sets of real numbers are uncountable:

- The set **R** of real numbers.
- The set $[0, 1]$
- The set of points which belong to any closed interval $[a, b]$
- The set of points which belong to any open interval (a, b)
- The set of all the points of a plane.

Finally, note that any infinite set M has a countable subset, which we can get simply by taking an element $a_1 \in M$, then another element a_2, etc until we get our countable set: $A = \{a_1, a_2, a_3, ...\} \subset M$.

1.3 Three examples of Number Theory

Theory of numbers is a branch of mathematics that deals with studying the properties of the numbers, in particular of the integer numbers. In this section, we'll look at three well known examples solved with Python programs.

1.3.1 Prime Numbers

Prime numbers are those natural numbers that are divisible only by 1 and by themselves. The set of prime numbers is an infinite subset of set **N** of natural numbers, but the set of prime numbers less than a given natural number is finite and can be calculated using a method called sieve of Eratosthenes. Let n be the given number. The method consists in drawing up a list of all the natural numbers in the interval $[2, n]$, and then delete from the list all the multiples of prime numbers between 2 y $\sqrt{n}$.

The following Python program draws up a list of all the prime numbers less than a given natural number n, following the method of the sieve of Eratosthenes. Then it shows the result for $n = 100$

and $n = 10000$, the latter summarizing the list of prime numbers. The time in milliseconds for each case is displayed after the results.

```python
# -*- coding: utf-8 -*-
"""
Mathematics and Python Programming     www.pysamples.com
p1b.py
"""

import numpy as np

n = 100000
print 'calculating for n< ' + str(n) + '. (about 2 seconds for n=50000).'
mylist = np.arange(1, n + 1, 2)
mylist[0] = 2
longitud = len(mylist)
root = np.floor(np.sqrt(n))


def f(x):
    return x != 0

for i in range(0, longitud):
    if mylist[i] > 5:
        if (mylist[i] % 3 == 0):
            mylist[i] = 0
        elif (mylist[i] % 5 == 0):
            mylist[i] = 0
mylist = filter(f, mylist)
longitud = len(mylist)
print 'Checking', str(longitud), ' numbers less than ', str(n), ':'
i = 0
j = 1
while i <= root:
    if mylist[i] > 0:
        while j < (longitud):
            if mylist[j] != 0:
                resto = mylist[j] % mylist[i]
                if resto == 0:
                    mylist[j] = 0
            j += 1
        if (longitud >= 1000) and (i % 500 == 0):
            print 'calculating for ' + str(mylist[i])
    i += 1
    j = i + 1
primos = filter(f, mylist)
print 'list of prime numbers: ', primos
print 'There are ' + str(len(primos)) + ' prime numbers <= ' + str(n)
```

―――――――――――――――――――――――――― output of the program ――――――――――――――――――――――――――
```
calculating for n< 100. (about 2 seconds for n=50000).
Checking 28  numbers less than  100 :
list of prime numbers:  [2, 3, 5, 7, 11, 13, 17, 19, 23, 29, 31, 37, 41,
43, 47, 53, 59, 61, 67, 71, 73, 79, 83, 89, 97]
There are 25 prime numbers <= 100
1.00493431091  ms

calculating for n< 1000. (about 2 seconds for n=50000).
Checking 268  numbers less than  1000 :
list of prime numbers:  [2, 3, 5, 7, 11, 13, 17, 19, 23, 29, 31, 37, 41, 43, 47, 53,
59, 61, 67, 71, 73, 79, 83, 89, 97, 101, 103, 107, 109, 113, 127, 131, 137, ... ,
```

```
9887, 9901, 9907, 9923, 9929, 9931, 9941, 9949, 9967, 9973]
There are 168 prime numbers <= 1000
32.378911972  ms
```

1.3.2 Euclidean algorithm

Euclidean algorithm calculates the greatest common divisor (GCD) and the least common multiple (lcm) of two natural numbers. The following program calculates the GCD and the lcm of two natural numbers. Let's look at an example and then we'll explain the program which performs the Euclid's algorithm:

```
──────────────────────── output of the program ────────────────────────
numbers:  256 , 60
remainder of  256 / 60  =  16
remainder of  60 / 16  =  12
remainder of  16 / 12  =  4
remainder of  12 / 4  =  0
GCD (greatest common divisor) = 4
lcm (least common multiple) = 3840
```

The program must calculate the GCD of $a = 256$ and $b = 60$. The algorithm divides a/b and calculates its remainder r. If the remainder is not zero, it carries out again a division whose dividend is the former divider, and whose divisor is the remainder previously obtained, i.e. b/r, and it calculates the new remainder; if this is not null, the program repeats the procedure until it reaches a null remainder. The greatest common divisor is the last remainder which is not zero. When we have obtained the GCD, we can immediately obtain the least common multiple because:

$$lcm = \frac{a \cdot b}{GCD}$$

The Python program code is as follows:

```python
# -*- coding: utf-8 -*-
"""
Mathematics and Python Programming      www.pysamples.com
p1c.py
"""


def gcd_lcm(x, y):
    print 'numbers: ', x, ',', y
    GCD = gcd(x, y)
    print 'GCD (greatest common divisor) = ' + str(GCD)
    if GCD > 1:
        lcm = x * y / GCD
    else:
        print str(x) + ' y ' + str(y), ' are relative primes'
        lcm = x * y
    s = 'lcm (least common multiple) = ' + str(lcm)
    return s


def gcd(a, b):  # x > y
    if b == 0:
        return 0
    else:
        r = a % b
        print 'remainder of ', a, '/', b, ' = ', r
        if r > 0:
            return gcd(b, r)
        else:
```

```
            return b

print gcd_lcm(256, 60)
```

The program is executed by calling the function $gcd_lcm(a, b)$, being $a > b$, through the statement print $gcd_lcm(256, 60)$. Let's look at some examples with other pairs of numbers:

```
―――――――――――――――――――――――― output of the program ――――――――――――――――――――――――
numbers:  13 ,  8
remainder of  13 / 8  =  5
remainder of  8 / 5  =  3
remainder of  5 / 3  =  2
remainder of  3 / 2  =  1
remainder of  2 / 1  =  0
GCD (greatest common divisor) = 1
13 y 8  are relative primes
lcm (least common multiple) = 104

numbers:  1470 ,  1155
remainder of  1470 / 1155  =  315
remainder of  1155 / 315  =  210
remainder of  315 / 210  =  105
remainder of  210 / 105  =  0
GCD (greatest common divisor) = 105
lcm (least common multiple) = 16170
```

1.3.3 Pythagorean triple

We already know since we were at primary school the Pythagorean theorem: $h^2 = a^2 + b^2$, where h is the hypotenuse, a and b are the other two sides of a right triangle. Pythagorean triples are the sets of three integers which are equivalent to the lengths of the sides of a right triangle: they provide all the right triangles whose sides are integers. The following Python program calculates some of these triangles and the Pythagorean triples, and represents them using turtle graphics:

```
# -*- coding: utf-8 -*-
"""
Mathematics and Python Programming     www.pysamples.com
p1d.py
"""

import turtle as tt

tt.mode('logo')
tt.reset()
tt.home()
screen = tt.getscreen()
screen.colormode(255)
screen.screensize(700,700)
colors = ['black', 'green', 'red', 'grey', 'blue', 'orange']
color = 0
tt.pencolor(colors[color])
tt.speed('slow')
tt.pensize(2)
tt.penup()
tt.setpos(-300, -300)
print ' m   n    a    b    h'
print '_____'
for m in range(2, 7):
    for n in range(1, m):
        x = m ** 2 - n ** 2
        y = 2 * m * n
        z = m ** 2 + n ** 2
        tt.penup()
        tt.right(90)
```

1.3. THREE EXAMPLES OF NUMBER THEORY

```
        tt.pendown()
        tt.forward(10 * x)
        tt.write(' ' + str(x))
        tt.left(90)
        tt.forward(10 * y)
        tt.write(str(y))
        tt.setpos(-300, -300)
        if color < 5:
            color += 1
        else:
            color = 0
        tt.pencolor(colors[color])
        print "%2d" % m, "%2d" % n, "%5d" % x, "%5d" % y, "%5d" % z
tt.hideturtle()
```

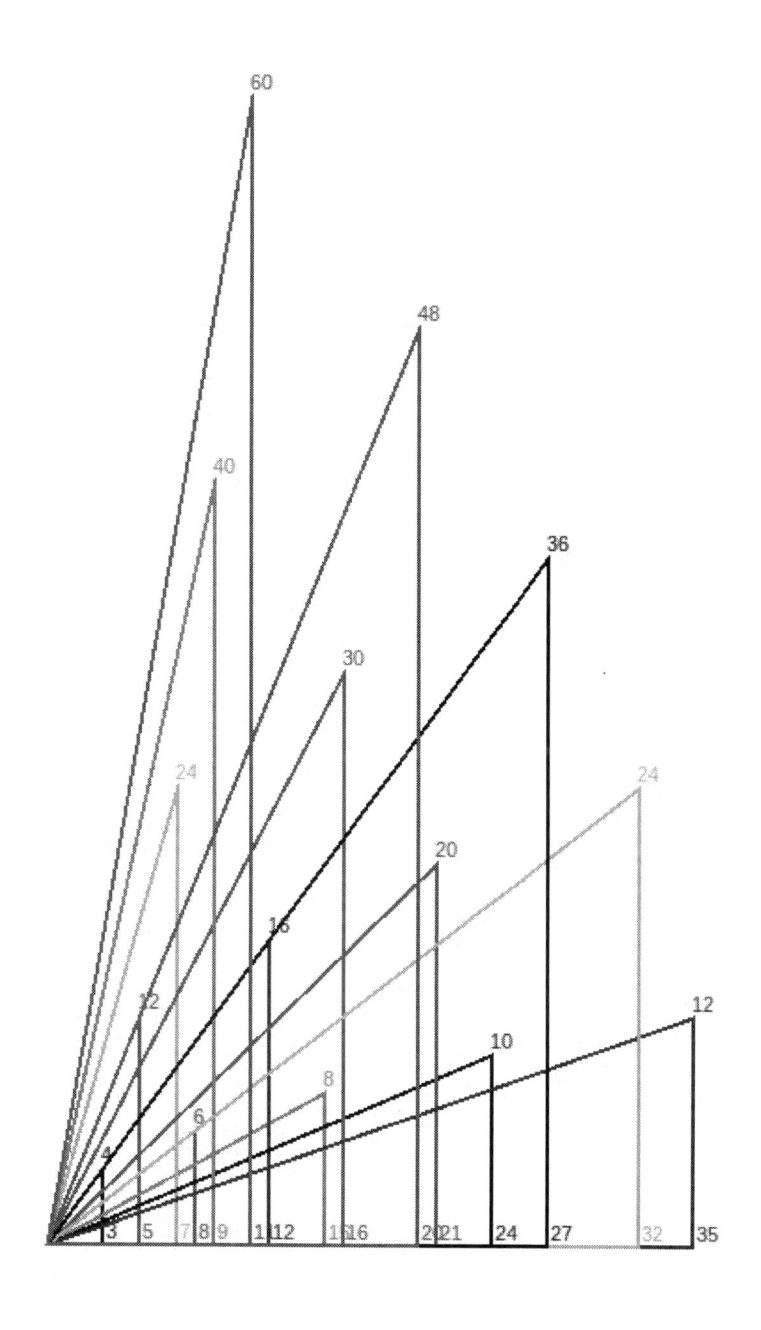

─────────────────── output of the program ───────────────────

m	n	a	b	h
2	1	3	4	5
3	1	8	6	10
3	2	5	12	13
4	1	15	8	17
4	2	12	16	20
4	3	7	24	25
5	1	24	10	26
5	2	21	20	29
5	3	16	30	34
5	4	9	40	41
6	1	35	12	37
6	2	32	24	40
6	3	27	36	45
6	4	20	48	52
6	5	11	60	61

───

1.4 Equivalent sets. Ordered sets.

In this section we are going to define two concepts that will be useful later: the equivalence of sets and ordered sets. We will say that two sets M and N are equivalent, and denote it as $M \sim N$, if there is a bijection between the elements of M and the elements of N. Two finite sets shall be equivalent if and only if both of them have the same number of elements. Given any finite set X, any subset of X itself will have less number of items that X, so that it is impossible to establish a bijection between both. On the other hand, if X is infinite, this will not happen, but that there will exist a proper subset of X such that it is possible to establish a bijection between X and its proper subset.

This property of the finite sets allows us to stablish another definition of an infinite set: an infinite set is a set which is equivalent to one of its proper subsets. At this point it is suitable for us to study the structure of the infinite set that we will use most in this book: the set $\mathbf{R}$ of the real numbers.

Given a binary relation R on a set X, we can say that this relationship is a partial order relation, or that X is partially ordered if the relation R meets the following conditions:

1. Reflexivity: $aRa, \forall a \in X$.

2. Antisymmetry: if aRb and bRa, then $a = b$.

3. Transitivity: if aRb and bRc, then aRc.

For example, the set of the integers greater than 1, if we define aRb to mean that b is divisible by a, is a partially ordered set. Given two elements a and b of a partially ordered set X, it may happen that these items are not related by a relation aRb nor bRa. In this case we can say that the elements a and b are not comparable. So that the relation R is only defined for certain pairs of numbers, and so we'll say that the set X is partially ordered. On the other hand, if X does not have any comparable elements we'll say that X is an ordered set. That is, X is ordered if, given any two elements $a, b \in X$, we have aRb or bRa. Any subset of an ordered set, is an ordered set.

Some examples of ordered sets, defining aRb to be $a \leq b$, are the set of all natural numbers and the set Q of rational numbers. It follows that an ordered set is a special and important type of a partially ordered set, in which the relationship is valid for any pairs of numbers of the set. An especially important type of ordered set is the well-ordered set. An ordered set X is said to be well-ordered if every non-empty subset $A \subset X$ has a first element (minimal element) $\mu \leq a \quad \forall a \in X$. Some examples of well-ordered sets are:

- Any finite ordered set is a well-ordered set.

- Every non empty subset of a well-ordered set, is a well-ordered set.

1.5 The set of real numbers

Many authors define axiomatically the set of real numbers. We are going to take as a starting point the sets that we already know: $\mathbf{N} \subset \mathbf{Z} \subset \mathbf{Q}$. Why there is a need for a new set? As it is well known, the operation $\sqrt{2}$, has no solution in the set $\mathbf{Q}$. Let's suppose that this is false and there is a rational number $q = \frac{a}{b}$, being $\frac{a}{b}$ an irreducible fraction such that $q = \frac{a}{b} = \sqrt{2}$. Then $\frac{a^2}{b^2} = 2$, and $a^2 = 2b^2$ should be true:

If a is odd, then a^2 will be odd and therefore $a^2 \neq 2b^2$. If a is even, then b must be odd, as well as b^2. But at the same time we have $b^2 = \frac{a^2}{2}$, which is impossible since an odd number may not be the result of dividing the square of an even number by 2:

$$b^2 = \frac{a^2}{2} = \frac{(2n)^2}{2} = \frac{4n^2}{2} = 2n^2$$

because it would contradict the former premise that b is odd. Therefore, $\sqrt{2}$ has no rational solution.

```
>>> sqrt(2)
1.4142135623730951
```

In a similar way we can prove that $\sqrt{5}$ is not rational. Suppose the contrary: there exists a rational number $\frac{a}{b} = \sqrt{5}$, and a and b, have no common factor. Then $a^2 = 5b^2$ but if a^2 is a natural number multiple of 5, then we can affirm that $a = 5k$, where k is the product of the other prime factors of a.

$$25k^2 = 5b^2; \qquad 5k = b$$

And b would have the same prime factors that a, which contradicts the initial assumption, and therefore $\sqrt{5}$ is an irrational number.

One last example: the number $\log 5$. If this number were rational, let's say $q = \frac{m}{n}$, with $m, n \neq 0$, then we would have

$$10^{\frac{m}{n}} = 5$$
$$2^{\frac{m}{n}} \cdot 5^{\frac{m}{n}} = 5$$

Which is impossible because we'd have $2^a = 5^b$, and therefore the number $\log 5$ is not rational.

```
>>> log10(5)
0.6989700043360188
```

Another important irrational number is the golden ratio, represented by the letter φ. Two numbers $a > b$ are said to be in the golden ratio if

$$\frac{a}{b} = \frac{a+b}{a}$$

If we call $\varphi = \frac{a}{b}$ we get:

$$\varphi = 1 + \varphi^{-1}; \qquad \varphi^2 = 1 + \varphi; \qquad \varphi^2 - \varphi - 1 = 0$$

We solve the equation:

$$\varphi = \frac{1 \pm \sqrt{1+4}}{2}$$

and the positive solution is

$$\varphi = \frac{1 + \sqrt{5}}{2}$$

We can calculate it using Python, for example with 20 decimal places:

```
>>> import numpy as np
>>> print "%22.20f" % ((1 + np.sqrt(5)) / 2)
1.61803398874989490253
```

The golden ratio is related to π as we can see in the following geometric proof: φ is equal to the radius of the circle circumscribed to a decagon unit of side:

$$\frac{KB}{AB} = \frac{FB}{BJ}$$

$$\frac{r}{1} = \frac{FB}{BJ}$$

$$KL = BL = AB = 1$$

$$LA = r - 1$$

$$\frac{FB}{BJ} = \frac{AB}{r-1} = \frac{1}{r-1}$$

$$r = \frac{1}{r-1}$$

$$r^2 - r - 1 = 0$$

$$r = \varphi$$

$$sen(\pi/10) = \frac{1}{2\varphi} \qquad \varphi = \frac{1}{2sen(\pi/10)}$$

```
>>> print 1 / (2*np.sin(np.pi/10))
1.61803398875
```

The golden ratio appears in the proportions of many works of art from antiquity. It also has some interesting mathematical properties:

$$\varphi = \varphi^2 - 1; \qquad \varphi - 1 = \frac{1}{\varphi}$$

```
>>> phi = ((1 + np.sqrt(5)) / 2)
>>> print phi
1.61803398875
>>> print str(phi ** 2 - 1)
1.61803398875
>>> print 1 / phi
0.61803398875
```

We have thus arrived at the evidence that the set $\mathbf{Q}$ is incomplete. Our next objective will be to complete it. An elegant and not axiomatic way to do it was proposed by the great German mathematician Dedekind (1831-1916) through the cuts that bear his name: Dedekind cuts. A Dedekind cut in Q is a couple of subsets A, B of $\mathbf{Q}$ such that:

1. A y B are a partition of $\mathbf{Q}$: $A \cup B = \mathbf{Q}$.
   $A \neq \emptyset$, $B \neq \emptyset$; $A \cap B = \emptyset$.

2. If $a \in A$ and $b \in B$, then $a < b$.

3. A does not have a last element, that it to say that $\forall a \in A \ \exists a' \in A$ such that $a < a'$.

Dedekind cuts can also be defined in another way: a Dedekind cut is a subset A of $\mathbf{Q}$ such that it meets the following three conditions:

1. $A \neq \emptyset \neq \mathbf{Q}$.

1.5. THE SET OF REAL NUMBERS

2. If $b \in A$ and $c < b$ then $c \in A$.

3. A does not have a last element, that it to say that $\forall b \in A \ \exists b' \in A$ such that $b < b'$.

We will denote this cut as $A|B$. Well, now we are in conditions to define a real number: a real number is a Dedekind cut in $\mathbf{Q}$. The set $\mathbf{Q}$ of rational numbers is incomplete, it has gaps, one of which occurs at the point $\sqrt{2}$. These gaps are actually very small, of zero amplitude and our goal now is to complete $\mathbf{Q}$ by filling these gaps. To do this we will use the order relationship that we saw in the previous paragraph. Let's look at some definitions:

- Given two cuts $x = A|B$ y $y = C|D$, we will say that $x \leq y$ if $A \subset C$.

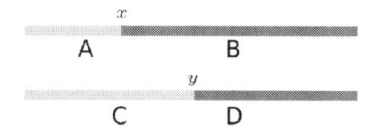

- If S is a subset of a partially ordered set P, an element $u \in P$ is called upper bound of S if $x < u$ for every $x \in S$. Similarly we define the lower bound as the element w such that $w < x \quad \forall x \in S$.

- The least bound of of the upper bounds is called the least upper bound (lub) of S. That is, if $\mu \in P$ and $\mu < u \ \forall u$ upper bound of S, and we denote it as $\mu = \text{lub}S$. Similarly we define de greatest bound of the lower bounds as the greatest lower bound (glb) of S, that is, if $\varphi > w$ $\forall w$ lower bound of S, and we denote it as $\varphi = \text{glb}S$.

It should also be noted that given a set S, it may not have any upper and/or lower bound, and if it is bounded, the $\text{lub}S$ and the $\text{glb}S$ may not belong to set S. A couple of examples: the number 3 is an upper bound for the set of negative integers; the number -1 is the lub of the set of negative integers.

The idea behind of Dedekind cuts is that we can define a real number x using a subset of $\mathbf{Q}$: the set of all integers less than x. And we define the set of real numbers as the set of all the cuts of Dedekind.

To further clarify the concept of lub, let's look at an example: let S be the set of rational numbers q such that $q \cdot q < 2$. This set is bounded from above, for example, by the rational number $\frac{2}{1}$, but it does not have a lub belonging to $\mathbf{Q}$, i.e. $\text{lub}S \notin \mathbf{Q}$, since, as we have just seen, $\sqrt{2} \notin \mathbf{Q}$. This circumstance that the lub
of a bounded subset of $\mathbf{Q}$ does not belong to $\mathbf{Q}$ is not present in the case of the real numbers: any bounded subset of $\mathbf{R}$ has a lub belonging to $\mathbf{R}$.

It is possible to find a $\text{lub}A \in \mathbf{R}$ for any subset $A \subset \mathbf{R}$ such that A is bounded from above.

This statement constitutes an important theorem in mathematical analysis and it is necessary in order to prove later, for example, that a continuous function on a closed interval reaches a maximum. Now we declare some consequences of this property:

1. For any real number x, there exists an integer n such that $x < n$.

2. For any positive real number ϵ there exists an integer n such that $\frac{1}{n} < \epsilon$.

3. For any real number x there exists an integer n such that $n \leq x < n + 1$.

4. For any real number x and any positive integer N, there exists an integer n such that $\frac{n}{N} \leq x < \frac{n+1}{N}$.

5. Any real number x can be approximated as much as we wish by a rational number. That is: if $x, \epsilon \in \mathbf{R}$, $\epsilon > 0$, then there is a rational number q such that $|x - q| < \epsilon$.

Bibliography for this chapter: [6], [8], [11], [15], [16], [18], [22], [27], [29] [32], [50], [59], [62], [63]

2 | Fields

2.1 The field of real numbers

We have advanced in the previous chapter from the general concept of a set to reach the sets of natural numbers, rational numbers and at last, the set of real numbers. There are many calculus books that practically begin with the concept of function or even with that of continuity. In this book we prefer to move forward step by step, from the most fundamental concepts, but without claiming this book to be an encyclopedia of definitions, theorems and proofs, nor a book of exercises or formulas. At this point, we need to study in more detail the set of real numbers, and we will introduce some concepts that in later chapters will prove to be essential to the ability to move forward, but in any case we will limit ourselves to the minimum necessary concepts.

We will begin with the concept of binary operation: a binary operation $*$ in a set G is a rule that assigns to each ordered pair (a, b) of elements of the set, some element of the set. We must note that this assigned element must also belong to the set G, so it is said that G is closed under the binary operation $*$. Let us add that this assigned element can be one of the members of the pair (a, b), or cannot be. A binary operation in a set G is commutative if and only if $a * b = b * a, \quad \forall a, b \in G$. The binary operation is associative if and only if $(a * b) * c = a * (b * c) \quad \forall a, b, c \in G$.

A group is a set G, together with a binary operation $*$ in G that meets the following axioms:

1. The binary operation $*$ is associative.
2. There is an element $e \in G$ such that $e * a = a * e, \forall a \in G$. This element e is called identity element (or neutral element) for the binary operation $*$ in G.
3. For each element $a \in G$ there exists an element $a' \in G$ such that $a' * a = a * a' = e$. This element a' is called symmetrical element of a with respect to the binary operation $*$ in G.

If we denote the binary operation as $+$, it is customary to denote $-a$ the symmetrical element of a, and call it the opposite of a. If we denote the binary operation as $\cdot$ it is customary to refer to the symmetrical element as the inverse of a. If the binary operation is commutative, the group is called abelian. The identity element e in a group G is unique. For each element $a \in G$ there exists only one symmetrical element a'. Let's look at some examples of groups, and examples of sets and operations that are not groups:

- The set $\mathbf{Z}$ under addition is a group. However, the set $\mathbf{Z}^+$ under addition is not a group, in the absence of an identity element belonging to $\mathbf{Z}^+$.
- The set $\mathbf{Q}^+$ under multiplication is an abelian group.
- The set $\mathbf{Z}$ under multiplication is not a group, because there is no inverse element for every element belonging to $\mathbf{Z}$. For example, any number other than 1 does not have an inverse element in $\mathbf{Z}$.

What happens if instead of one binary operation, we introduce two binary operations? A ring is a set A, together with two binary operations defined in A, which we will call $+$ (addition) y $\cdot$ (multiplication), (and we denote it as $\langle A, +, \cdot \rangle$), such that the following axioms are met:

2.1. THE FIELD OF REAL NUMBERS

1. The set A under addition is an abelian group.

2. The multiplication is associative.

3. Multiplication distributes over addition: $\forall a, b, c \in A$, that is: $a \cdot (b + c) = (a \cdot b) + (a \cdot c)$ and $(a + b) \cdot c = (a \cdot c) + (b \cdot c)$.

If we call 0 the identity element under addition in a ring A, then:

- $0 \cdot a = a \cdot 0 = 0$
- $a \cdot (-b) = (-a) \cdot b = -(a \cdot b)$
- $(-a)(-b) = ab$

Some examples of rings: $\langle \mathbf{Z}, +, \cdot \rangle$, $\langle \mathbf{Q}, +, \cdot \rangle$ y $\langle \mathbf{R}, +, \cdot \rangle$. The set $M_n\{\mathbf{R}\}$ of square matrices of order n in $\mathbf{R}$ under matrix addition and multiplication, is also a ring.

We have demanded only one condition to the multiplication for the set to be a ring: that the multiplication is associative. If this operation has a neutral element in A, then we call it a ring with unit element. That is, if there exists an element 1 which is the multiplicative identity $1 \in A$ such that $1 \cdot a = a \cdot 1 = a$, $\forall a \in A$. This multiplicative identity, if there exists, is unique.

Now let's take a ring with multiplicative identity. If multiplication also meets that for every element $a \neq 0$ there exists a symmetrical element that we denote as a^{-1}, such that $a \cdot a^{-1} = a^{-1} \cdot a = 1$, then A is called a semifield.

A field is a semifield in which the multiplication is commutative. That is, if we have a set F such that $\langle F, + \rangle$ is an abelian group, and $\langle F, \cdot \rangle$ is an abelian group for every non-zero element, and both groups are united by the distributive property, then $\langle F, +, \cdot \rangle$ has field structure.

Some examples of fields: $\langle \mathbf{Q}, +, \cdot \rangle$ y $\langle \mathbf{R}, +, \cdot \rangle$. Thus, the set of real numbers under the operations of addition and multiplication, is a field. Finally, we must note two important properties of the fields:

1. There are no dividers of zero: there are no elements a, b such that $a \cdot b = 0$ and $a \neq 0, b \neq 0$.

2. The equations $b \cdot x = a$ and $y \cdot b = a$, with $b \neq 0$, admit only one solution.

The equation $x^2 + 1 = 0$ has no solution in the field of real numbers, because there is no real number $x = \sqrt{-1}$. Girolamo Cardano, in his work Ars magna, in 1545, has the merit of having proposed a new type of numbers, although at that time their meaning was not fully understood. Cardano he said that those numbers "were as subtle as useless".

In 1702, Leibniz himself described i, the square root of -1, as "that amphibian between existence and non-existence". The root of this difficulty to understand complex numbers is perhaps due to a kind of psychological blockage to accept the existence of a number whose square is equal to -1. However, we will see below that there is a set of matrices, which has field structure, and in which there is a solution to the equation $A \cdot A = -E$, being E the unit matrix. Hopefully, this reasoning may contribute to eliminate prejudices that have historically burdened these numbers, which are fundamental in a multitude of calculations, in areas such as waves, electromagnetism, acoustics ...

2.2 Square matrices

Let's review some concepts about matrices. At this time we are interested in square matrices, for example:

$$\begin{pmatrix} a_{11} & a_{12} & a_{13} \\ a_{21} & a_{22} & a_{23} \\ a_{31} & a_{32} & a_{33} \end{pmatrix}$$

and in particular, the set $M_2\{\mathbf{R}\}$ of second order square matrices whose elements are real numbers:

$$\begin{pmatrix} a & b \\ c & d \end{pmatrix}$$

Let's take a look at the properties of this set, which henceforth we will refer simply as M_2, under the operations of addition and matrix multiplication. Let A and B be:

$$A = \begin{pmatrix} a_{11} & a_{12} \\ a_{21} & a_{22} \end{pmatrix}, \quad B = \begin{pmatrix} b_{11} & b_{12} \\ b_{21} & b_{22} \end{pmatrix}$$

then

$$A + B = \begin{pmatrix} a_{11} + b_{11} & a_{12} + b_{12} \\ a_{21} + b_{21} & a_{22} + b_{22} \end{pmatrix}$$

Since the sum of real numbers is commutative, the sum of matrices will be also commutative: $A + B = B + A$.
The associative property is also met:

$$(A + B) + C = A + (B + C)$$

There is a neutral element, the matrix

$$0 = \begin{pmatrix} 0 & 0 \\ 0 & 0 \end{pmatrix}$$

such that $A + 0 = 0 + A = A \quad \forall A \in M_2$.
For every matrix $A \in M_2$, there exists a matrix $-A \in M_2$ such that $A + (-A) = 0$.

$$-A = \begin{pmatrix} -a_{11} & -a_{12} \\ -a_{21} & -a_{22} \end{pmatrix}$$

Therefore, the set M_2 under the addition of matrices is an abelian group.
Now let's look at the matrix multiplication:

$$A \cdot B = \begin{pmatrix} a_{11}b_{11} + a_{12}b_{21} & a_{11}b_{12} + a_{12}b_{22} \\ a_{21}b_{11} + a_{22}b_{21} & a_{21}b_{12} + a_{22}b_{22} \end{pmatrix}$$

The product of square matrices of order 2 meets the associative property:

$$(A \cdot B) \cdot C = A \cdot (B \cdot C)$$

and the distributive property of multiplication with respect to the addition::

$$A \cdot (B + C) = A \cdot B + A \cdot C$$

$$(B + C) \cdot A = B \cdot A + C \cdot A$$

Note that the multiplication is not commutative:

$$A \cdot B \neq B \cdot A \quad \forall A, B \in M_2$$

The following Python program shows how to perform operations with matrices:

2.2. SQUARE MATRICES

```python
# -*- coding: utf-8 -*-
'''
Mathematics and Python Programming    www.pysamples.com
p2a.py
'''

import numpy as np

A = np.array([[1, 2], [3, 4]])
C = np.array([[5, 6], [7, 8]])
E = np.array([[1, 0], [0, 1]])
print 'A = ', str(A)
print 'A.E = ', np.dot(A, E)
print 'E.A = ', np.dot(E, A)
inv_A = np.linalg.inv(A)
print 'inverse matrix of A = ', inv_A
print 'A.inv_A = ', np.dot(A, inv_A)
print 'matrices product:'
print 'C = ', str(C)
print 'A.C = ', np.dot(A, C)
print 'C.A = ', np.dot(C, A)
```

―――――――――――――――――――――――― output of the program ――――――――――――――――――――――――
```
A =  [[1 2]
 [3 4]]
A.E =  [[1 2]
 [3 4]]
E.A =  [[1 2]
 [3 4]]
inverse matrix of A =  [[-2.   1. ]
 [ 1.5 -0.5]]
A.inv_A =  [[  1.00000000e+00   1.11022302e-16]
 [  0.00000000e+00   1.00000000e+00]]
matrices product:
C =  [[5 6]
 [7 8]]
A.C =  [[19 22]
 [43 50]]
C.A =  [[23 34]
 [31 46]]
```

Each row is shown in Python in square brackets, for example,

```
A.B =  [[19 22]
 [43 50]]
```

is equivalent to:

$$A \cdot B = \begin{pmatrix} 19 & 22 \\ 43 & 50 \end{pmatrix}$$

Therefore, the set $\langle M_2, +, \cdot \rangle$ is a ring. This ring has unit element: the matrix

$$E = \begin{pmatrix} 1 & 0 \\ 0 & 1 \end{pmatrix}$$

However, not every square matrices of order 2 has an inverse matrix. For example, the determinant of any matrix whose four components are equal, will be null, and the matrix does not have inverse matrix.

2.3 The field of complex numbers

Now we are going to consider a subset P of M_2, the set of square matrices of order two which have the form:

$$A = \begin{pmatrix} a & b \\ -b & a \end{pmatrix} \quad C = \begin{pmatrix} c & d \\ -d & c \end{pmatrix} \quad \text{etc.}$$

This set P is closed under de operations of addition and multiplication:

$$A + C = \begin{pmatrix} a+c & b+d \\ -b-d & a+c \end{pmatrix} \in M_2$$

$$A \cdot C = \begin{pmatrix} ac-bd & ad+bc \\ -bc-ad & -bd+ac \end{pmatrix}$$

$$A \cdot C = \begin{pmatrix} ac-bd & ad+bc \\ -(ad+bc) & ac-bd \end{pmatrix} \in M_2$$

the neutral element of the sum of matrices, the matrix 0 belongs to the set P:

$$0 = \begin{pmatrix} 0 & 0 \\ 0 & 0 \end{pmatrix} \in P$$

and therefore it also has structure of ring with identity element, since

$$E = \begin{pmatrix} 1 & 0 \\ 0 & 1 \end{pmatrix} \in P$$

Let's see if there is an inverse matrix for any matrix $A \in P$:

$$\left(\begin{array}{cc|cc} a & b & 1 & 0 \\ -b & a & 0 & 1 \end{array} \right)$$

By means of elementary transformations on the rows this augmented matrix:

```
# -*- coding: utf-8 -*-
"""
Mathematics and Python Programming    www.pysamples.com
p2b.py
"""

import sympy as sy

a, b, r2 = sy.symbols('a b r2')
sy.init_printing(use_unicode=True)

row1 = sy.Matrix([[a, b, 1, 0]])
row2 = sy.Matrix([[-b, a, 0, 1]])
print row1
print row2
print 'r2 = a**2 + b**2'


def printoutput():
    print '---------------------'
    print row1
    print row2


def H(row, k):
```

2.3. THE FIELD OF COMPLEX NUMBERS

```
        for col in range(0, 3):
            Haux = row * k
        return Haux

row2 = row2 + H(row1, b / a)
printoutput()
row2[1] = r2 / a
printoutput()
row2 = row2 * (a / r2)
printoutput()
row1 = row1 + H(row2, -b)
printoutput()
row1[2] = a ** 2 / r2
row1 = row1 * (1 / a)
printoutput()
print
print 'inverse matrix = '
print '(1/r2) . [ ', row1[2] * r2, ', ', row1[3] * r2, ' ]'
print '          [ ', row2[2] * r2, ', ', row2[3] * r2, ' ]'
```

─────────────────────────── output of the program ───────────────────────────

```
Matrix([[a, b, 1, 0]])
Matrix([[-b, a, 0, 1]])
r2 = a**2 + b**2
-------------
Matrix([[a, b, 1, 0]])
Matrix([[0, a + b**2/a, b/a, 1]])
-------------
Matrix([[a, b, 1, 0]])
Matrix([[0, r2/a, b/a, 1]])
-------------
Matrix([[a, b, 1, 0]])
Matrix([[0, 1, b/r2, a/r2]])
-------------
Matrix([[a, 0, -b**2/r2 + 1, -a*b/r2]])
Matrix([[0, 1, b/r2, a/r2]])
-------------
Matrix([[1, 0, a/r2, -b/r2]])
Matrix([[0, 1, b/r2, a/r2]])

inverse matrix =
(1/r2) . [ a ,  -b ]
        [ b ,   a ]
```

───

we get:
$$\left(\begin{array}{cc|cc} 1 & 0 & \frac{a}{r^2} & \frac{-b}{r^2} \\ 0 & 1 & \frac{b}{r^2} & \frac{a}{r^2} \end{array} \right)$$

where $r^2 = a^2 + b^2$, and therefore, for any matrix $A \in P$, there is a matrix $A^{-1} \in P$ such that $A \cdot A^{-1} = E$.

$$A^{-1} = \left(\begin{array}{cc} \frac{a}{r^2} & \frac{-b}{r^2} \\ \frac{b}{r^2} & \frac{a}{r^2} \end{array} \right) = \frac{1}{r^2} \left(\begin{array}{cc} a & -b \\ b & a \end{array} \right)$$

Note that the inverse of the matrix A is equal to the transpose of A, multiplied by $\frac{1}{r^2}$:

$$A^{-1} = \frac{1}{r^2} \cdot A^t$$

$$A \cdot A^{-1} == \frac{1}{r^2} [\left(\begin{array}{cc} a & b \\ -b & a \end{array} \right) \left(\begin{array}{cc} a & -b \\ b & a \end{array} \right)]$$

$$A \cdot A^{-1} == \frac{1}{r^2} \left(\begin{array}{cc} a^2 + b^2 & -ab + ba \\ -ba + ab & b^2 + a^2 \end{array} \right) = \frac{1}{r^2} \left(\begin{array}{cc} r^2 & 0 \\ 0 & r^2 \end{array} \right) = E$$

Therefore, P is a field. Let's see if the equation $x^2 + 1 = 0$ has a solution in this field, i.e. if there is a matrix J such that $J \cdot J = -E$.

$$J = \begin{pmatrix} 0 & 1 \\ -1 & 0 \end{pmatrix}$$

$$J \cdot J = \begin{pmatrix} 0 & 1 \\ -1 & 0 \end{pmatrix} \begin{pmatrix} 0 & 1 \\ -1 & 0 \end{pmatrix} = \begin{pmatrix} -1 & 0 \\ 0 & -1 \end{pmatrix} = -E$$

Now that we have come to obtain a set P with field structure, and in which the equation that historically resulted in the emergence of complex numbers, has a solution, let's see how this set P is related to the field of complex numbers. In the first place, each matrix $A \in P$ can be written as a sum:

$$A = \begin{pmatrix} a & b \\ -b & a \end{pmatrix} = a \begin{pmatrix} 1 & 0 \\ 0 & 1 \end{pmatrix} + b \begin{pmatrix} 0 & 1 \\ -1 & 0 \end{pmatrix} = aE + bJ$$

The set $\mathbf{C}$ whose elements have the form:

$$z = a + bj$$

and the set P are isomorphic. The set $\mathbf{C}$ is called the field of complex numbers. The numbers a and b are real numbers, and $j = \sqrt{-1}$, or said another way, $j^2 = -1$. The real number a is called real part of the complex number, $Re(z)$, and the real number b which is the coefficient of j, is called imaginary part of the complex number, $Im(z)$. If the real part of a complex number is zero, it is said that the complex is a pure imaginary number.

The similarity with the representation of the elements of P as a sum in which a and b are real numbers and $J \cdot J = -E$, is clear: each ordered pair of real numbers (a, b) can matched to a square matrix belonging to P:

$$(a, b) \in \mathbf{C} \mapsto \begin{pmatrix} a & b \\ -b & a \end{pmatrix} \in P$$

The field P has a subfield, the set $\{aE, \; a \in \mathbf{R}\}$ which is isomorphic to $\mathbf{R}$. The field $\mathbf{C}$ has a subfield, the set of real numbers $\mathbf{R}$.

$$\mathbf{Q} \subset \mathbf{R} \subset \mathbf{C}$$

Gauss at the end of the XVIII and early XIX century gave the name of complex numbers to this class of numbers, and also granted them a geometrical interpretation. W. R. Hamilton in 1837 and 1853 considered the complex number as an ordered pair of real numbers.

Let's look at the correlation between the two fields P y $\mathbf{C}$:

$$
\begin{aligned}
A &= aE + bJ & z &= a + bj & \in \mathbf{C} \\
A &= \begin{pmatrix} a & b \\ -b & a \end{pmatrix} \in P & z &= a + bj & \in \mathbf{C} \\
A^{-1} &= \frac{1}{r^2} \begin{pmatrix} a & -b \\ b & a \end{pmatrix} \in P & z^{-1} &= \frac{a}{r^2} - \frac{b}{r^2} j & \in \mathbf{C} \\
A^{-1} &= \frac{1}{r^2} \cdot A^t \quad \in P & z^{-1} &= \frac{1}{r^2} \bar{z} & \in \mathbf{C} \\
0 &= \begin{pmatrix} 0 & 0 \\ 0 & 0 \end{pmatrix} \in P & z &= 0 + 0j & \in \mathbf{C} \\
E &= \begin{pmatrix} 1 & 0 \\ 0 & 1 \end{pmatrix} \in P & z &= 1 + 0j & \in \mathbf{C} \\
J &= \begin{pmatrix} 0 & 1 \\ -1 & 0 \end{pmatrix} \in P & z &= 0 + 1j & \in \mathbf{C} \\
J^2 &= -E & j^2 &= -1 \\
r^2 &= a^2 + b^2 & |z| &= +\sqrt{a^2 + b^2}
\end{aligned}
$$

Compare how the operations of addition and multiplication on both fields are done:

2.3. THE FIELD OF COMPLEX NUMBERS

Addition:
$$A + C = \begin{pmatrix} a+c & b+d \\ -b-d & a+c \end{pmatrix} \in M_2$$
$$z_1 + z_2 = (a+bj) + (c+dj) = (a+c) + (b+d)j$$

Multiplication:
$$A \cdot C = \begin{pmatrix} ac-bd & ad+bc \\ -(ad+bc) & ac-bd \end{pmatrix} \in M_2$$
$$z_1 \cdot z_2 = (a+bj) + (c+dj) = (ac-bd) + (ad+bc)j$$

There are other ways to represent complex numbers, which we will see in later chapters. We'll be glad for the time being to see how operations with complex numbers can be easily done in Python:

```python
# -*- coding: utf-8 -*-
"""
Mathematics and Python Programming    www.pysamples.com
p2c.py
"""

import numpy as np


def r2(z):
    a = z.real
    b = z.imag
    rdos = a ** 2 + b ** 2
    return rdos


def length(z):
    m = np.sqrt(r2(z))
    return m


def explaincomplex(z):
    explain = (str(z) + '; Re(z) = ' + str(z.real) + '; Im(z) = ' +
               str(z.imag) + '; |z| = ' + "%6.4f" % length(z) +
               '; r2(z) = ' + str(r2(z)))
    return explain


def str_inverse(z):
    coef = '(1/' + str(r2(z)) + ')'
    conjugate = np.conjugate(z)
    strinverse = coef + '*' + str(conjugate)
    return strinverse


def checkinverse(z):
    print 'z = ' + str(z)
    print 'conjugate of z = ' + str(np.conjugate(z))
    print 'z * conjugate(z) = ', z * np.conjugate(z)
    print 'inverse of z = ' + str_inverse(z)
    print ('z * inv(z) = ' + str(z) +
           ' * [' + str_inverse(z) + '] = (1/' +
           str(r2(z)) + ') * ' + str(np.conjugate(z)) + ' = 1')
```

```python
print 'powers of j:'
j = complex(0, 1)
print 'j^1 = j'
print 'j^2 = ' + str(j * j) + ' = -1'
print 'j^3 = ' + str(-1 * j) + ' = -j'
print 'j^4 = ' + str(-j * j) + ' = 1'
print '_____'
z1 = 3 + 4j
z2 = complex(-1, 1)
print 'z1 = ' + explaincomplex(z1)
print 'z2 = ' + explaincomplex(z2)
print 'z1 + z2 = ', z1 + z2
print 'z1 - z2 = ', z1 - z2
print 'z1 * z2 = ', z1 * z2
print '_____'
print 'checking z1 inverse:'
checkinverse(z1)
print '_____'
print 'checking z2 inverse:'
checkinverse(z2)
```

When we run the program we get:

```
_____ complejos.py - ejecución _____
powers of j:
j^1 = j
j^2 = (-1+0j) = -1
j^3 = (-0-1j) = -j
j^4 = (1-0j) = 1
--------------------------------
z1 = (3+4j); Re(z) = 3.0; Im(z) = 4.0; |z| = 5.0000; r2(z) = 25.0
z2 = (-1+1j); Re(z) = -1.0; Im(z) = 1.0; |z| = 1.4142; r2(z) = 2.0
z1 + z2 =   (2+5j)
z1 - z2 =   (4+3j)
z1 * z2 =   (-7-1j)
--------------------------------
checking z1 inverse:
z = (3+4j)
conjugate of z = (3-4j)
z * conjugate(z) =   (25+0j)
inverse of z = (1/25.0)*(3-4j)
z * inv(z) = (3+4j) * [(1/25.0)*(3-4j)] = (1/25.0) * (3-4j) = 1
--------------------------------
checking z2 inverse:
z = (-1+1j)
conjugate of z = (-1-1j)
z * conjugate(z) =   (2+0j)
inverse of z = (1/2.0)*(-1-1j)
z * inv(z) = (-1+1j) * [(1/2.0)*(-1-1j)] = (1/2.0) * (-1-1j) = 1
```

Complex numbers can be represented as points in the complex plane. The real part corresponds to the X-coordinate, and the imaginary part corresponds to the Y-coordinate. The following Python program represents two complex z_1 and z_2, as well as the opposite, the conjugate and the inverse of z_1; the sum $z_1 + z_2$, and checks that the product of a complex by its inverse is equal to the unit.

```python
# -*- coding: utf-8 -*-
"""
Mathematics and Python Programming    www.pysamples.com
p2d.py
"""

import matplotlib.pyplot as plt
import numpy as np

j = complex(0, 1)
z1 = complex(2, 2)
```

2.3. THE FIELD OF COMPLEX NUMBERS

```python
z2 = complex(-1.5, -0.5)


def inverse(z):
    ro2 = z.real ** 2 + z.imag ** 2
    inversoz = (np.conj(z) / ro2)
    return inversoz

plt.figure()
plt.ylabel('Im')
plt.xlabel('Re')
plt.axhline(color='black', lw=1)
plt.axvline(color='black', lw=1)
plt.grid(b=None, which='major')
plt.ylim(-5, 3)
plt.xlim(-3, 3)


def arrow(z, texto):
    dx = z.real
    dy = z.imag
    plt.arrow(0, 0, dx, dy, width=0.02, fc='b',
              ec='none', length_includes_head=True, lw=0.5, head_width=0.05, head_length=0.1 )
    if dx == 0:
        xtexto = dx + 0.05
    else:
        xtexto = np.sign(dx) * (abs(dx) * 1.2)
    if dy == 0:
        ytexto = dy + 0.05
    else:
        ytexto = np.sign(dy) * (abs(dy) * 1.2)
    plt.text(xtexto, ytexto, texto,
             horizontalalignment='center')

arrow(j, 'j')
arrow(z1, 'z1')
arrow(z2, 'z2')
arrow(np.conj(z1), 'conjugated(z1)')
arrow(- z1, 'opposite(z1)')
arrow(inverse(z1), 'inv(z1)')
arrow(z1 + z2, 'z1+z2')
arrow(z1 * j, 'z1*j')
arrow(z1 * z2, 'z1*z2')
arrow(z1 * inverse(z1), 'z1*inv(z1)')
plt.show()
```

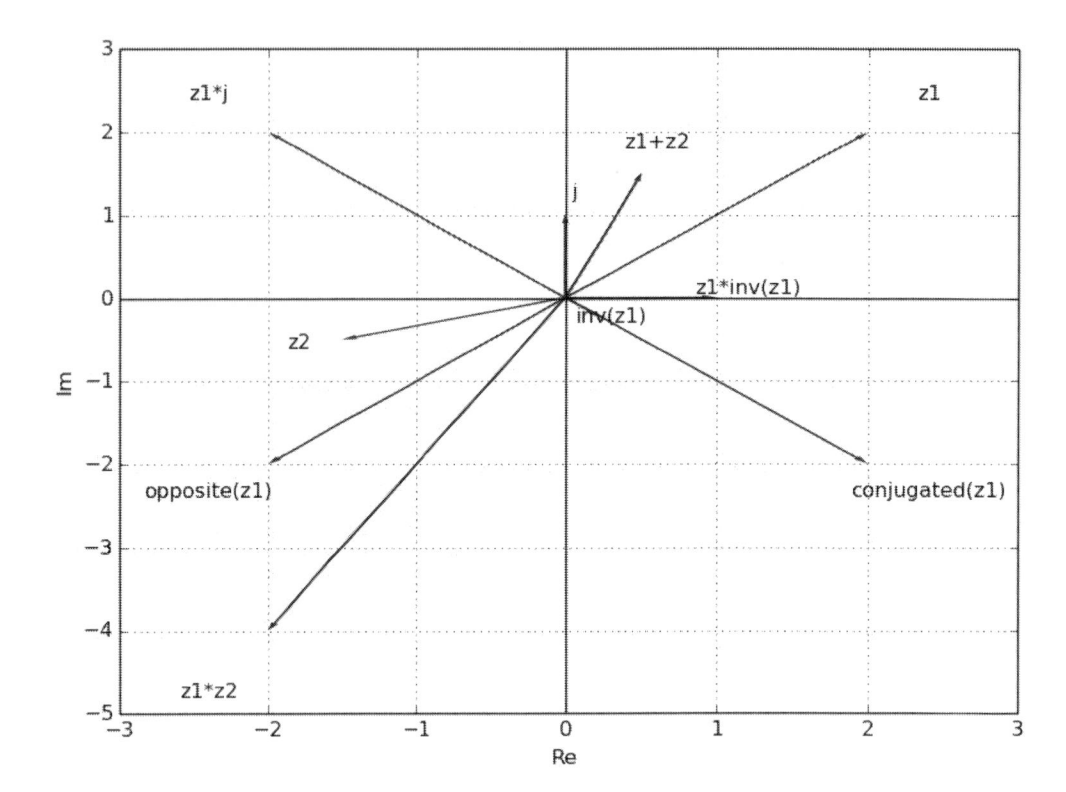

The following program solves and represents graphically the solutions of the quadratic equation $az^2 + bz + c = 0$:

```python
# -*- coding: utf-8 -*-
"""
Mathematics and Python Programming    www.pysamples.com
p2e.py
"""

import numpy as np
import matplotlib.pyplot as plt
from matplotlib import rc

# solves equation a z +bz + c = 0
a = 1.0
b = 2.0
c = 3.0


def equation():
    if b >= 0:
        strb = '+ ' + str(b)
    else:
        strb = str(b)
    if c >= 0:
        strc = '+ ' + str(c)
    else:
        strc = str(c)
    strec = '$' + str(a) + 'z^{2} ' + strb + 'z ' + strc + '=0$\n'
    strec = strec + '$z_{1}=' + str(z1) + '$\n$z_{2}=' + str(z2) + '$'
    return strec
```

2.3. THE FIELD OF COMPLEX NUMBERS

```python
plt.figure()
rc('text', usetex=True)
rc('font', family='serif')
plt.ylabel('Im')
plt.xlabel('Re')
plt.axhline(color='black', lw=1)
plt.axvline(color='black', lw=1)
plt.grid(b=None, which='major')


def pointsR(t, sol1, sol2):
    x = []
    y = [0, 0, 0]
    plt.ylim(-1, 1)
    #-b/2a
    x.append(t)
    x.append(float(sol1))
    x.append(float(sol2))
    xmin = np.floor(x[np.argmin(x)])
    xmin = xmin - abs((xmin / 10))
    xmax = np.ceil(x[np.argmax(x)])
    xmax = xmax + abs((xmax / 10))
    plt.xlim(xmin, xmax)
    plt.plot(x, y, 'ko')
    if sol1 == sol2:
        plt.text(1.01 * x[1], -0.25, '$z_{1}$',
                 horizontalalignment='center', color='blue', fontsize=20)
        plt.text(0.99 * x[2], -0.25, '$z_{2}$',
                 horizontalalignment='center', color='red', fontsize=20)
        plt.text(x[0], 0.15, r"$\displaystyle\frac{-b}{2a}$",
                 horizontalalignment='center', color='green', fontsize=16)
    else:
        plt.text(x[1], 0.15, '$z_{1}$',
                 horizontalalignment='center', color='blue', fontsize=20)
        plt.text(x[2], 0.15, '$z_{2}$',
                 horizontalalignment='center', color='red', fontsize=20)
        plt.text(x[0], 0.15, r"$\displaystyle\frac{-b}{2a}$",
                 horizontalalignment='center', color='green', fontsize=16)
    textstr = equation()
    props = dict(boxstyle='round', facecolor='#FCE945',
                 alpha=0.9)
    plt.text(0.9 * xmin, 0.90,
             textstr, fontsize=14, verticalalignment='top', bbox=props)


def pointsC(t, re, im):
    x = []
    y = [0]
    #-b/2a
    x.append(t)
    x.append(float(re))
    x.append(float(re))
    #xmin = np.floor(x[np.argmin(x)])
    xmin = x[0] - abs((x[0] / 5))
    xmax = np.ceil(x[np.argmax(x)])
    xmax = xmax + abs((xmax / 10))
```

```
        if xmax <= 0:
            xmax = abs(xmin / 3)
        if xmin >= 0:
            xmin = -1 * xmax / 3
        plt.xlim(xmin, xmax)
        y.append(float(im))
        y.append(float(-im))
        ymin = np.floor(y[np.argmin(y)])
        ymin = ymin - abs((ymin / 10))
        ymax = np.ceil(y[np.argmax(y)])
        ymax = ymax + abs((ymax / 10))
        plt.ylim(ymin, ymax)
        plt.plot(x, y, 'ko')
        plt.text(1.05 * re, 1.15 * y[1], '$z_{1}$',
                 horizontalalignment='center', color='blue',
                 fontsize=20)
        plt.text(1.05 * re, 1.05 * y[2], '$z_{2}$',
                 horizontalalignment='center', color='red', fontsize=20)
        if x[0] == 0:
            plt.text(xmax, 0.25, r"$\displaystyle\frac{-b}{2a}$",
                     horizontalalignment='left', color='green', fontsize=16)
        else:
            plt.text(x[0], 0.25, r"$\displaystyle\frac{-b}{2a}$",
                     horizontalalignment='center', color='green', fontsize=16)
        textstr = equation()
        props = dict(boxstyle='round', facecolor='#FCE945',
                     alpha=0.9)
        if re >= 0:
            plt.text(0.9 * xmin, 0.9 * ymax,
                     textstr, fontsize=14,
                     verticalalignment='top', bbox=props)
        else:
            plt.text(0.3 * xmin, 0.90 * ymax,
                     textstr, fontsize=14, verticalalignment='top', bbox=props)

if a != 0:
    radicando = b ** 2 - 4 * a * c
    t1 = -b / (2 * a)
    t2 = radicando / (2 * a)
    if radicando > 0:
        z1 = "%6.4f" % (t1 + t2)
        z2 = "%6.4f" % (t1 - t2)
        pointsR(t1, z1, z2)
    elif radicando == 0:
        z1 = "%6.4f" % t1
        z2 = z1
        pointsR(t1, z1, z1)
    else:
        aa = float("%6.4f" % t1)
        bb = float("%6.4f" % (np.sqrt(-radicando) / (2 * a)))
        z1 = complex(aa, bb)
        z2 = complex(aa, -bb)
        plt.plot([0, aa], [0, bb], 'b', lw=2)
        plt.plot([0, aa], [0, -bb], 'r', lw=2)
        pointsC(t1, aa, bb)
    print 'z1 = ', z1
```

2.3. THE FIELD OF COMPLEX NUMBERS

```
        print 'z2 = ', z2
else:
        print 'coefficient a cannot be null'

plt.show()
```

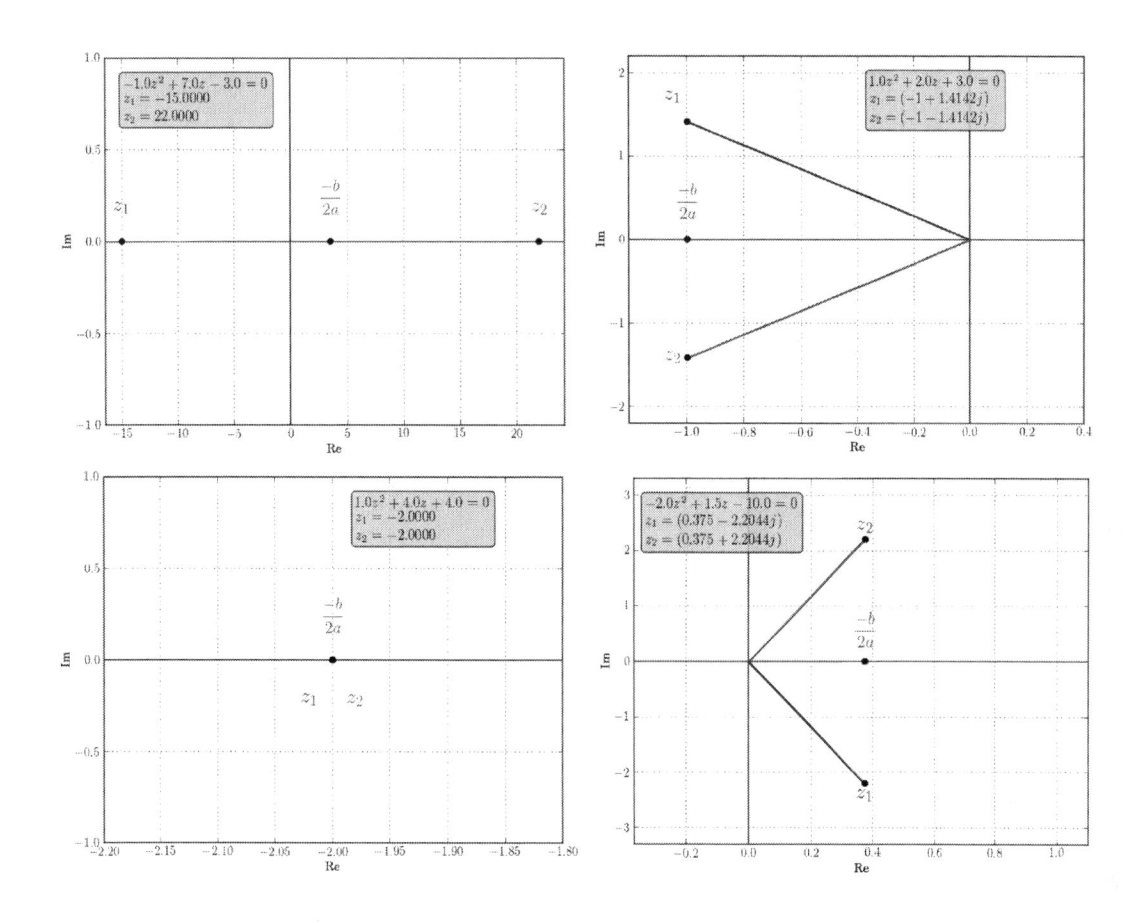

Bibliography for this chapter: [14], [17], [20], [33], [42], [43], [48] [52]

3 | Metric spaces

3.1 Metric spaces

The concept of limit is of fundamental importance in mathematical analysis. We cannot get to the concept of limit without understanding another concept: the distance. Until this time we have spoken of numeric sets: **N, Z, Q, R**... One of the objectives of this book is to present mathematical analysis in a useful way, so that it can serve as a bridge between calculus that is studied at the end of secondary education, and the mathematics studied in the first courses of any science college. And we can achieve this objective using a modern programming language, accessible to any student, as it is Python.

For this reason, we're not going to limit to study the concept of limit on **R**, but we are going to talk about a generic set E. This extra effort will provide us with a better overall picture and will make it easier for the reader to have the possibility to access the texts of mathematical analysis of university level. If we are satisfied at this time with a vision of the analysis restricted to the set of real numbers, in the future the student will find a real barrier to understand mathematical texts that in fact do not present concepts which are very different from those that the student had already studied in secondary education, but those text do it with a level of abstraction such that for many students it is impossible to find a point of connection with what they know.

A set E is called space metric if in that set there is a rule that assigns to each pair (x, y) of elements, a real number $d(x, y) \geq 0$ called distance, in such a way that it will meet the following properties, which are called axioms of metric:

1. $d(x, y) \geq 0 \ \forall x, y \in E$

2. $d(x, y) = 0$ if and only if $x = y$

3. $d(x, y) = d(y, x)$ (symmetry axiom)

4. $d(x, z) \leq d(x, y) + d(y, z)$ (triangle inequality)

Let's have a look at some examples of metric spaces:

- Any set E of elements in which a relationship of equality has been defined, can be converted into a metric space if we define the distance as

$$d(x, y) = \begin{cases} 0 & \text{si } x = y \\ 1 & \text{si } x \neq y \end{cases}$$

- The set $E = \mathbf{R}$ with the distance $d(x, y) = |x - y|$

- If E is a metric space with a distance d, and E_1 is a subset of E, then E_1 with the distance d has also the structure of a metric space, and it is called subspace of E.

- The set of all ordered n-tuples

$$x = (x_1, x_2, x_3, ..., x_n)$$

with $x_i \in \mathbf{R}$, and the distance

$$d(x,y) = \sqrt{\sum_{i=1}^{n}(x_i - y_i)^2}$$

is called euclidean space of dimension n, $\mathbf{R}^n$.

- The set of all ordered n-tuples
$$x = (x_1, x_2, x_3, ..., x_n)$$
with $x_i \in \mathbf{R}$ and the distance
$$d_1(x,y) = \sum_{i=1}^{n}|x_i - y_i|$$
is denoted by $\mathbf{R}_1^n$.

- The set of all ordered n-tuples
$$x = (x_1, x_2, x_3, ..., x_n)$$
and the distance
$$d_0(x,y) = \max_{1 \leq i \leq n}|x_i - y_i|$$
is denoted by $\mathbf{R}_0^n$.

A consequence of the triangular inequality is as follows: if x, y, z are three points of the metric space E then we have:

$$d(x,z) \leq d(x,y) + d(y,z) \qquad d(y,z) \leq d(y,x) + d(x,z)$$
$$d(x,z) - d(y,z) \leq d(x,y) \qquad d(y,z) - d(x,z) \leq d(y,x)$$

and uniting both inequalities we get

$$|d(x,z) - d(y,z)| \leq d(x,y)$$

In geometry that corresponds to the fact that the difference between two sides of a triangle is less than the third side.

3.2 Open sets, closed sets, bounded sets

High school calculus examines the concept of real interval. Let's take $a, b \in \mathbf{R}$:

- The following subset of $\mathbf{R}$ is called open interval with endpoints a y b:
$$(a,b) = \{x \in \mathbf{R} : a < x < b\}$$

- The following subset of $\mathbf{R}$ is called closed interval with endpoints a y b:
$$[a,b] = \{x \in \mathbf{R} : a \leq x \leq b\}$$

Now we are going to expand this concept in a metric space E. Let x_0 be a point, and $r > 0$ a real number:

- The subset of E:
$$\{x \in E : d(x_0, x) < r\}$$
is called open ball of radius r centered at x_0.

- The subset of E:
$$\{x \in E : d(x_0, x) \leq r\}$$
is called closed ball of radius r centered at x_0.

- A subset S of a metric space E is open if for any point $p \in S$, S contains an open ball centered at p.

- A subset S of a metric space E is closed if its complement is open.

Examples of open subsets of E are the empty subset $\emptyset$; the subset E; the union of any collection of open subsets of E; the intersection of a finite number of open subsets of E. In any metric space E, an open ball is an open subset. We can also define an open set as the union of all the open balls that it contains.

Examples of closed subsets of E are the subset E; the empty subset $\emptyset$; the intersection of any collection of closed subsets of E; the union of a finite number of closed subsets of E. In any metric space E, a closed ball is a closed subset.

- It is said that a subset S of a metric space E is bounded if it is contained in some ball.

Suppose that S is a subset of $\mathbf{R}$ bounded superiorly. Then there is an element $a = \text{lub} S$. We are going to determine if a belongs to the subset S. If $a \notin S$, then a belongs to the complementary subset of S, which we call cS. Since cS is open, it must contain some ball centered at a contained in cS. Let ϵ be the radius of this open ball that is contained in cS, then there is no element of S greater than $a - \epsilon$, and $a - \epsilon$ will be an upper bound of S, but this contradicts the assumption that a was the $\text{lub} S$, and hence the assertion $a \notin S$ must be false, and we conclude that $a \in S$.

- A point $p \in S$ is said to be interior to the set S if there is an open ball of centered ad p, contained in S. The interior of a set S is the subset of S which contains all the points in S which are interior points of S. The interior of a set S is the larger open set that is contained in S.

- A point $p \in S$ is said to be outside the set S if there is an open ball centered at p, in which every point does not belong to S.

- A point $p \in S$ is said to be a boundary point of S if it is a point that is not inside or out of S, and therefore every ball centered at p contains at least one point of S and at least one point not belonging to S.

- The collection of all the boundary points of a set S is called the boundary of the set S. The boundary of a set S is always a closed set.

- The closure of a set S, which is denoted as $\bar{S}$ is the union of S and its boundary: $\bar{S} = S \cup$ boundary(S). The closure of a set S is the smallest closed set containing S.

- The border of a set can also be defined as the intersection of the closure of S and the closure of the complementary of S.

- A point $p \in S$ is said to be isolated if there is a neighbourhood of p that contains no other point of S.

- Two disjoint non-empty sets are said to be mutually separated if none contains a boundary point of the other.

- A set is said to be disconnected if it is the union of separated subsets.

- A set is said to be connected if it is not disconnected.

While the structure of open and closed sets in a general metric space E can be quite complicated, we can say for the particular case of $\mathbf{R}$:

- Every open subset of $\mathbf{R}$ is the union of a finite collection of countable or disjoint open intervals of $\mathbf{R}$.

- Every closed subset of **R** can be obtained by eliminating a finite or countable collection of disjoint intervals of **R**.

- Every subset of a single element $\{x_0\}$ is closed.

3.3 Sequences

If we match a real number x_n to every number n of the series of natural numbers $1, 2, ..., n, ...$, the set of real numbers $x_n = x_1, x_2, ..., x_n, ...$ is called a sequence of real numbers.

Some examples of well-known sequences already in secondary education are the arithmetic progression, the geometric progression and the Fibonacci sequence. Let us briefly review its features without the demonstrations, which are simple and well known since the secondary education courses, which explained the concept of general term of the sequence and sum S_n of its n first terms. We will see also some numerical examples with the help of Python.

- Arithmetic progression: its n-th term is defined as follows:

$$x_n = x_{n-1} + d$$

or

$$x_n = x_1 + (n-1)d$$

The arithmetical progression has the property that the sums of the terms which equidistant from both endpoints, are equal, that is to say that if $m + n = k + l$ we will have $x_m + x_n = x_k + x_l$, and the sum of n first terms is

$$S_n = \frac{(x_1 + x_n) \cdot n}{2}$$

Let's see a concrete example calculated with a Python program. At the end of the results it displays the execution time in milliseconds, to give an idea of the calculation speed of Python.

```python
# -*- coding: utf-8 -*-
"""
Mathematics and Python Programming    www.pysamples.com
p3a.py
"""

import numpy as np


def arithmetical(x1, n, d):
    x = np.zeros(n + 1, int)
    x[1] = x1
    suma = x[1]
    sequence = str(x[1]) + ', '
    for i in range(2, n + 1):
        x[i] = x[i - 1] + d
        sequence = sequence + str(x[i]) + ', '
        suma = suma + x[i]
    sequence = sequence[0:len(sequence) - 2]
    print 'x1 =', x[1], '; d =', d, '; n =', n
    print ('The ' + str(n) + ' first terms of the sequence:')
    print sequence
```

```
        sn = (x[1] + x[n]) * n / 2
        print
        print ('S(' + str(n) + ') : (' + str(x[1]) +
                ' + ' + str(x[n]) + ')' + str(n) + '/ 2 = ' + str(sn))
        print 'S(', n, ') added one by one:', suma
        print
        print ('Check: xm + xn = ' + 'xk+ xl si m+n = k+l: ')
        m = 1
        ene = np.random.random_integers(2, n)
        k = np.random.random_integers(2, ene)
        l = m + ene - k
        print 'm + n = k + l:'
        print m, ' + ', ene, ' = ', k, ' + ', l, ': '
        print (str(x[m]) + ' + ' + str(x[ene]) +
                ' = ' + str(x[k]) + ' + ' + str(x[l]))
        print x[m] + x[ene], ' = ', x[k] + x[l]

arithmetical(10, 100, -3)   # arithmetical(x1,n,d)
```

when we run the program, we get:

──────────────── output of the program ────────────────
```
x1 = 10 ; d = -3 ; n = 100
The 100  first terms of the sequence:
10, 7, 4, 1, -2, -5, -8, -11, -14, -17, -20, -23, -26, -29, -32, -35, -38, -41, -44, -47,
-50, -53, -56, -59, -62, -65, -68, -71, -74, -77, -80, -83, -86, -89, -92, -95, -98, -101,
-104, -107, -110, -113, -116, -119, -122, -125, -128, -131, -134, -137, -140, -143, -146,
-149, -152, -155, -158, -161, -164, -167, -170, -173, -176, -179, -182, -185, -188, -191,
-194, -197, -200, -203, -206, -209, -212, -215, -218, -221, -224, -227, -230, -233, -236,
-239, -242, -245, -248, -251, -254, -257, -260, -263, -266, -269, -272, -275, -278, -281, -284, -287

S(100) : (10 + -287)100/ 2 = -13850
S( 100 ) added one by one: -13850

Check: xm + xn = xk+ xl si m+n = k+l:
m + n = k + l:
1 + 86 = 32 + 55 :
10 + -245 = -83 + -152
-235 = -235
1.20401382446  ms
```
──

- Geometric progression: Its n-th term is defined as follows:

$$x_n = x_{n-1} \cdot r$$

where r is a constant number, $r \neq 1$ which is called the common ratio of the progression. The above formula is equivalent to:

$$x_n = x_1 \cdot r^{n-1}$$

The sum of its n first terms is

$$S_n = \frac{x_1(1 - r^n)}{1 - r}$$

The following Python program calculates the terms of a geometric progression.

```
# -*- coding: utf-8 -*-
"""
Mathematics and Python Programming    www.pysamples.com
p3b.py
"""

import numpy as np
```

3.3. SEQUENCES

```python
def geometric(x1, n, r):
    x = np.zeros(n + 1, float)
    x[1] = x1
    suma = x[1]
    sequence = str(x[1]) + ', '
    for i in range(2, n + 1):
        x[i] = x[i - 1] * r
        sequence = sequence + "%.8f" % x[i] + ', '
        suma = suma + x[i]
    sequence = sequence[0:len(sequence) - 2]
    print ('x1 =' + str(x[1]) + '; r =' +
        "%.2f" % r + '; n =' + str(n))
    print ('The ' + str(n) + ' first terms of the sequence:')
    print sequence
    sn = x[1] * (1 - r ** n) / (1 - r)
    print
    strsuma = ('S(' + str(n) + ') : ' + str(x1) + '(1 - ' + "%.2f" % r)
    strsuma = (strsuma + '^' + str(n) + ') / (1-' + "%.2f" % r)
    strsuma = strsuma + ') = ' + "%.8f" % sn
    print strsuma
    print ('S(' + str(n) + ') added one by one:' + "%.8f" % suma)

geometric(100.0, 20, 0.5)   # geometric(x1,n,r)
```

We run the program to calculate 20 terms:

```
———————————————————————— output of the program ————————————————————————
x1 = 100.0 ; r = 0.50 ; n = 20
The 20 first terms of the sequence:
100.0, 50.00000000, 25.00000000, 12.50000000,
6.25000000, 3.12500000, 1.56250000, 0.78125000,
0.39062500, 0.19531250, 0.09765625, 0.04882812,
0.02441406, 0.01220703, 0.00610352, 0.00305176,
0.00152588, 0.00076294, 0.00038147, 0.00019073

S(20) : 100.0(1 - 0.50^20) / (1-0.50) = 199.99980927
S( 20 ) added one by one: 199.99980927
0.225067138672   ms
```

If $r > 1$, the terms of the progression grow very quickly (shown below in columns for clarity):

```
———————————————————————— output of the program ————————————————————————
x1 = 2.0 ; r = 7.50 ; n = 20
The 20 first terms of the sequence::
1                       2.0
2                      15.0000
3                     112.5000
4                     843.7500
5                    6328.1250
6                   47460.9375
7                  355957.0312
8                 2669677.7344
9                20022583.0078
10              150169372.5586
11             1126270294.1895
12             8447027206.4209
13            63352704048.1567
14           475145280361.1755
15          3563589602708.8164
16         26726922020316.1250
17        200451915152370.9375
18       1503389363642782.0000
19      11275420227320864.0000
20      84565651704906480.0000

S(20) : 2.0(1 - 7.50^20) / (1-7.50) = 97575751967199776.00000000
S( 20 ) added one by one: 97575751967199776.00000000
0.220060348511   ms
```

- Fibonacci sequence: its first two terms are: $x_1 = 1$, $x_2 = 1$. Each term starting from the third one is obtained by adding the two previous terms, that is to say: $x_n = x_{n-1} + x_{n-2}, \forall n \geq 3$. The terms of this progression grow very rapidly and have been formatted in the Python program result with commas every three digits to facilitate your reading; in addition, the result of the program has been distributed in columns to fit the page:

```
# -*- coding: utf-8 -*-
"""
Mathematics and Python Programming     www.pysamples.com
p3c.py
"""

dictionaryiter = {1: 1, 2: 1}


def fibiter(n):
    i = 3
    while i <= n:
        dictionaryiter[i] = (dictionaryiter[i - 1] +
                             dictionaryiter[i - 2])
        i += 1


fibiter(100)


def comma(n):
    s = str(n)
    pos = len(s)
    while pos > 3:
        pos = pos - 3
        s = s[:pos] + ',' + s[pos:]
    return s

for i, value in dictionaryiter.iteritems():
    term = comma(value)
    print '%3d %28s' % (i, term)
```

———————————— output of the program ————————————

```
  1                        1       51               20,365,011,074
  2                        1       52               32,951,280,099
  3                        2       53               53,316,291,173
  4                        3       54               86,267,571,272
  5                        5       55              139,583,862,445
  6                        8       56              225,851,433,717
  7                       13       57              365,435,296,162
  8                       21       58              591,286,729,879
  9                       34       59              956,722,026,041
 10                       55       60            1,548,008,755,920
 11                       89       61            2,504,730,781,961
 12                      144       62            4,052,739,537,881
 13                      233       63            6,557,470,319,842
 14                      377       64           10,610,209,857,723
 15                      610       65           17,167,680,177,565
 16                      987       66           27,777,890,035,288
 17                    1,597       67           44,945,570,212,853
 18                    2,584       68           72,723,460,248,141
 19                    4,181       69          117,669,030,460,994
 20                    6,765       70          190,392,490,709,135
 21                   10,946       71          308,061,521,170,129
 22                   17,711       72          498,454,011,879,264
 23                   28,657       73          806,515,533,049,393
 24                   46,368       74        1,304,969,544,928,657
 25                   75,025       75        2,111,485,077,978,050
 26                  121,393       76        3,416,454,622,906,707
 27                  196,418       77        5,527,939,700,884,757
 28                  317,811       78        8,944,394,323,791,464
```

3.3. SEQUENCES

29	514,229	79	14,472,334,024,676,221
30	832,040	80	23,416,728,348,467,685
31	1,346,269	81	37,889,062,373,143,906
32	2,178,309	82	61,305,790,721,611,591
33	3,524,578	83	99,194,853,094,755,497
34	5,702,887	84	160,500,643,816,367,088
35	9,227,465	85	259,695,496,911,122,585
36	14,930,352	86	420,196,140,727,489,673
37	24,157,817	87	679,891,637,638,612,258
38	39,088,169	88	1,100,087,778,366,101,931
39	63,245,986	89	1,779,979,416,004,714,189
40	102,334,155	90	2,880,067,194,370,816,120
41	165,580,141	91	4,660,046,610,375,530,309
42	267,914,296	92	7,540,113,804,746,346,429
43	433,494,437	93	12,200,160,415,121,876,738
44	701,408,733	94	19,740,274,219,868,223,167
45	1,134,903,170	95	31,940,434,634,990,099,905
46	1,836,311,903	96	51,680,708,854,858,323,072
47	2,971,215,073	97	83,621,143,489,848,422,977
48	4,807,526,976	98	135,301,852,344,706,746,049
49	7,778,742,049	99	218,922,995,834,555,169,026
50	12,586,269,025	100	354,224,848,179,261,915,075

0.711917877197 ms

As a curiosity, if we write the terms of the Fibonacci sequence in binary notation, for example using the following Python program:

```python
# -*- coding: utf-8 -*-
"""
Mathematics and Python Programming    www.pysamples.com
p3d.py
"""

import numpy as np

f = []
f.append(0)
f.append(1)
terminos = 10    # number of terms >0
i = 2
while i <= terminos:
    termino = f[i - 1] + f[i - 2]
    f.append(termino)
    i += 1
j = 1
while j <= terminos:
    print str(j) + ': ' + str(f[j]) + ' = ' + np.binary_repr(f[j])
    j += 1
```

─────────────────── output of the program ───────────────────
```
1: 1 = 1
2: 1 = 1
3: 2 = 10
4: 3 = 11
5: 5 = 101
6: 8 = 1000
7: 13 = 1101
8: 21 = 10101
9: 34 = 100010
10: 55 = 110111
```

and if you render the Fibonacci sequence in binary notation, you can get images as the next one, which represents the first hundred terms. You can find more examples on our website www.pysamples.com:

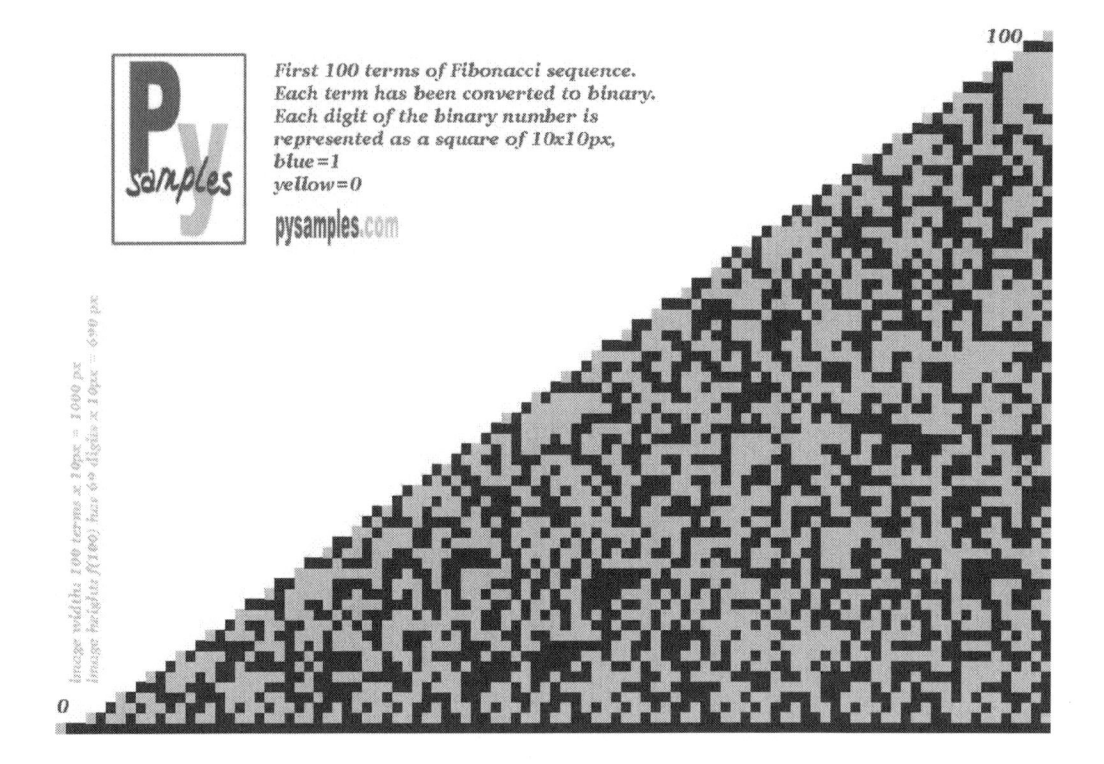

First 100 terms of Fibonacci sequence. Each term has been converted to binary. Each digit of the binary number is represented as a square of 10x10px, blue=1 yellow=0

As we have seen in the previous examples, some sequences grow indefinitely, while others decrease. If we look at the examples of geometric progression, we see that in the first case, with $r = 0.5$ the terms are becoming ever smaller, and the sum of the 20 first terms was a number very close to 200, 199.999809. We can intuit that if we calculate more terms, these terms will be increasingly small and their sum will be a number very close to 200. In contrast, in the second example we had $r = 7.5$ the terms grew very quickly, so much so that the tenth term was already a quantity of the order of hundreds of millions: 150169372.558594.

We can define two types of sequences: the infinitely large and the infinitely small.

- A sequence $\{x_n\}$ is said to be infinitely large if for all positive real number A, there exists a natural number N such that for $n > N$ satisfies $|x_n| > A$.
  It would be, for example, the case of our geometric progression of $r = 7.5$. We can choose any positive real number A, no matter how large it is, and we can calculate a sufficiently large number of terms to reach a term x_n which will be greater than A in absolute value.

- A sequence $\{\alpha_n\}$ is said to be infinitely small if for all positive real number ε, however small it may be, there is a natural number N such that for $n > N$ satisfies $|\alpha_n| < \varepsilon$. It would be, for example, the case of our geometric progression of $r = 0.5$. We can choose any positive real number ε, as small as we want, and we will just calculate a sufficiently large number of terms to reach a term x_n which will be less than ε in absolute value. We will see below a numerical example calculated with Python.

- A sequence $\{x_n\}$ is said to be bounded from above (below) if there is a number M (a number m) such that every term x_n of this sequence satisfies the inequality $x_n \leq M$, $(x_n \geq m)$.

- A sequence x_n is said to be bounded, if it is bounded from above and from below: there exists M, and m such that every element of the sequence satisfies the inequalities: $m \leq x_n \leq M$. That is, if there exists $|x_n| \leq K$ for all element x_n of the sequence.

Let's take the example of a infinitely small sequence, calculated using the program that we wrote before for geometric sequences, but now slightly modified to show the terms in scientific notation. We run it to calculate 50 terms:

3.3. SEQUENCES

```
─────────────────────── 03p04.py - ejecución ───────────────────────
x1 = 100.0 ; r = 0.5 ; n = 50
First 50 terms of the sequence:
    1      100.0              26    2.9802e-06
    2      50                 27    1.4901e-06
    3      25                 28    7.4506e-07
    4      12.5               29    3.7253e-07
    5       6.25              30    1.8626e-07
    6       3.125             31    9.3132e-08
    7       1.5625            32    4.6566e-08
    8       0.78125           33    2.3283e-08
    9       0.39062           34    1.1642e-08
   10       0.19531           35    5.8208e-09
   11       0.097656          36    2.9104e-09
   12       0.048828          37    1.4552e-09
   13       0.024414          38    7.276e-10
   14       0.012207          39    3.638e-10
   15       0.0061035         40    1.819e-10
   16       0.0030518         41    9.0949e-11
   17       0.0015259         42    4.5475e-11
   18       0.00076294        43    2.2737e-11
   19       0.00038147        44    1.1369e-11
   20       0.00019073        45    5.6843e-12
   21       9.5367e-05        46    2.8422e-12
   22       4.7684e-05        47    1.4211e-12
   23       2.3842e-05        48    7.1054e-13
   24       1.1921e-05        49    3.5527e-13
   25       5.9605e-06        50    1.7764e-13

S(50) : 100.0(1 - 0.5^50) / (1-0.5) = 200
S( 50 ) added one by one: 200
0.591039657593  ms
```

In effect, we see that the terms become very small: if we run the program to calculate 100 terms, we would get that for $n = 100$ the terms are the order of 10^{-28}, and their sum is very close to 200, as we saw before. In fact, we will demonstrate later that the sum of the terms of a geometric progression of $|r| < 1$ is $S = \frac{x_1}{1-r}$. In our example: $S = \frac{100}{1-0.5} = \frac{100}{0.5} = 200$.

We need to make an observation: if a sequence is infinitely large, it will not be bounded. But the reverse statement is false: an unbounded sequence may not be infinitely large. For example the unbounded sequence $1, 0, 2, 0, 3, 0, 4, 0, 5, \ldots$ is not infinitely large.

What relationship is there between the infinitely large sequence and the infinitely small sequence? The following: if $\{x_n\}$ is an infinitely large succession, and $x_n \neq 0, \forall n \in \mathbf{N}$, then the sequence $\{\alpha_n\} = \{\frac{1}{x_n}\}$ is an infinitely small sequence. The reverse statement is also true: if $\{\alpha_n\}$ is an infinitely small sequence and $\alpha_n \neq 0, \forall n \in \mathbf{N}$, then the sequence $\{x_n\} = \{\frac{1}{\alpha_n}\}$ is infinitely large.

We are approaching a fundamental concept as is the limit of a sequence, but before we get there, we are going to list some properties of the infinitely small sequences:

1. The difference of two infinitely small sequences is a infinitely small succession.

2. The sum of a finite number of infinitely small sequences is an infinitely small sequence.

3. The product of a finite number of infinitely small sequences is an infinitely small sequence.

4. The product of a bounded sequence by another infinitely small sequence is an infinitely small sequence.

5. The product of a number by an infinitely small sequence is an infinitely small sequence.

Bibliography for this chapter: [9], [24], [32], [47], [50], [55], [59], [61]

4 | Limit

4.1 Convergent sequences. Limit

We will begin by giving the definition of a convergent sequence in a metric space E, and then we will define it for the metric space $\mathbf{R}$.

1. Let $p_1, p_2, p_3, \ldots$ be a sequence of points in a metric space E. It is said that the sequence is convergent if there is a point $p \in E$, such that for any real number $\varepsilon > 0$ there exists a natural number N and starting at the N-th term, all the terms of the sequence are at a distance from p less than ε. That is, if for all p_n with $n > N$, $d(p, p_n) < \varepsilon$. The point p is called the limit of the sequence, and it is said that the sequence of points converges to p.

$$\lim_{n \to \infty} d(p, p_n) = 0$$

2. The number l is called limit of a sequence $\{x_n\}$ of real numbers if for every positive number ε there exists a number N such that for $n > N$ the following inequality is met:

$$|x_n - l| < \varepsilon$$

and then we say that the sequence $\{x_n\}$ is convergent:

$$\{x_n\} \to l \text{ for } n \to \infty.$$

$$\lim_{n \to \infty} x_n = l$$

To put it another way: the number l is called limit of the sequence x_n if for all ε-neighbourhood of the point l there exists a number N such that all the elements x_n, with $n > N$, are in this ε-neighbourhood.

Let's take as example the sequence $x_n = a^n$, $a \in \mathbf{R}$, $|a| < 1$. Suppose that the sequence is convergent to the limit l. Then there is a number N, such that starting with it $|a^n - l| < \varepsilon$ for all $\varepsilon \in \mathbf{R}$. This example is in fact a geometric progression as we saw previously, with $x_1 = a = r$.

In the following Python program we will calculate that number N for the sequence $x_n = a^n$, with $a = 0.15$, such that starting with that number all the terms are at a distance from the limit $l = 0$, less than $\varepsilon = 10^{-10}$.

```
# -*- coding: utf-8 -*-
"""
Mathematics and Python Programming    www.pysamples.com
p4a.py
"""

import numpy as np


def geometric(epsilon, a):
```

4.1. CONVERGENT SEQUENCES. LIMIT

```
    print ('a = ' + str(a) +
           '; required epsilon =' +
           "%8.4g" % epsilon)
    n = (int(round(np.log10(epsilon) / np.log10(abs(a))))) + 1
    x = np.zeros(n + 1, float)
    x[1] = a * 1.0
    print ('It is necessary to calculate ' + str(n) +
           ' terms of the sequence:')
    print "%3.0f" % 1.0, ': ', "%8.2g" % a
    for i in range(2, n + 1):
        x[i] = x[i - 1] * a
        print "%3.0f" % i, ': ', "%8.4g" % x[i]

geometric(1e-10, 0.15)   # geometric(epsilon, a)
```

———————————————————————— output of the program ————————————————————————
```
a = 0.15; required epsilon =    1e-10
It is necessary to calculate 13 terms of the sequence:
  1 :    0.15
  2 :    0.0225
  3 :    0.003375
  4 :    0.0005062
  5 :    7.594e-05
  6 :    1.139e-05
  7 :    1.709e-06
  8 :    2.563e-07
  9 :    3.844e-08
 10 :    5.767e-09
 11 :    8.65e-10
 12 :    1.297e-10
 13 :    1.946e-11
0.248908996582  ms
```
———

If instead of $a = 0.15$, we take a negative value, less than 1, this new sequence also complies with $|a| < 1$ and it converges to zero. For example, we take $a = -0.08$:

———————————————————————— output of the program ————————————————————————
```
a = -0.08 ; required epsilo =    1e-10
It is necessary to calculate 10 terms of the sequence:
  1 :   -0.08
  2 :    0.0064
  3 :   -0.000512
  4 :    4.096e-05
  5 :   -3.277e-06
  6 :    2.621e-07
  7 :   -2.097e-08
  8 :    1.678e-09
  9 :   -1.342e-10
 10 :    1.074e-11
0.217914581299  ms
```
———

We can see that this sequence is coming ever closer to the limit value $l = 0$. In effect, if

$$\lim_{n \to \infty} a^n = 0$$

According to the definition of convergent sequence this means that there is a number N from which $|a^n - 0| = |a^n| < \varepsilon$

$$|a^n| < \varepsilon$$

$$\left|\frac{1}{a}\right|^n > \frac{1}{\varepsilon}$$

$$n \cdot \log \frac{1}{|a|} > \log \frac{1}{\varepsilon}$$

$$n > \frac{-\log \varepsilon}{-\log |a|}$$

$$n > \frac{\log \varepsilon}{\log |a|}$$

Therefore, for any positive ε that we choose, it will be that starting with the term $N > \frac{\log \varepsilon}{\log |a|}$, the terms of the sequence will be at a distance from the limit $l = 0$ less than ε. For example, with $a = 0.15$, if we choose $\varepsilon = 10^{-10}$, we will get

$$N = \frac{\log 10^{-10}}{\log 0.15} = \frac{-10}{-0.8239} \simeq 12.13$$

─────────── Cálculo de N ───────────
```
>>> log10(1e-10)/log10(0.15)
12.137266547917129
```
────────────────────────────────────

And from the term number 13, all the terms of the sequence will be at a distance $l = 0$ less than $\varepsilon = 10^{-10}$, as you can see from the results of the previous program, and the limit of the sequence $\{a^n\}$, $|a| < 1$, is zero.

4.2 Properties of the sequences

Now let's look at some properties of the sequences.

1. Let $\{x_n\}$ be a convergent sequence whose limit is l, then every element x_n can be represented as
$$x_n = l + \alpha_n$$
where α_n is an element of the infinitely small sequence $\{\alpha_n\}$.

2. If all the elements of a infinitely small sequence $\{\alpha_n\}$ are equal to the same number c, then $c = 0$.

3. The limit of a convergent sequence us unique. Let's suppose that sequence x_n has two limits, a y b. Then, according to the first property:
$$x_n = a + \alpha_n$$
$$x_n = b + \beta_n$$
where α_n and β_n are elements of the infinitely small sequences $\{\alpha_n\}$ y $\{\beta_n\}$. If we subtract both equations:
$$\{x_n\} - \{x_n\} = 0 = a - b + \{\alpha_n\} - \{\beta_n\}$$
we get $a - b = \{\alpha_n\} - \{\beta_n\}$, but the difference of two infinitely small sequences is another infinitely small sequence$\{\alpha_n - \beta_n\}$, and $a - b$ is a constant c. Thus we have an infinitely small sequence, all whose term are equal to a constant. By the previous property, this constant should be zero and then: $a - b = 0$, and hence $a = b$, and the limit of the convergent sequence is unique.

4. Any convergent sequence is bounded, since from the term number N, we will have $|x_n| < |l| + \varepsilon$. However, an unbounded sequence can be not convergent, for example the succession $1, -1, 1, -1, ...$.

5. The product of two convergent sequences is another convergent, whose limit is equal to the product of the limits of the two convergent sequences given The sum (difference) of two convergent sequences is another convergent sequence, whose limit is the sum (difference) of the limits of the two convergent sequences given. We can apply this property to calculate the sum of a decreasing geometric progression. We know that the sum of n first terms of a geometric progression of reason R, with $|r| < 1$ is
$$S_n = \frac{x_1(1 - r^n)}{1 - r}$$
$$\lim_{n \to \infty} S_n = \lim_{n \to \infty} \frac{x_1}{1 - r} - \lim_{n \to \infty} \frac{x_1 \cdot r^n}{1 - r}$$

$$\lim_{n\to\infty} S_n = \frac{x_1}{1-r} - \frac{x_1}{1-r} \cdot \lim_{n\to\infty} r^n$$

$$\lim_{n\to\infty} S_n = \frac{x_1}{1-r} - \frac{x_1}{1-r} \cdot 0$$

$$\lim_{n\to\infty} S_n = \frac{x_1}{1-r}$$

6. The product of two convergent sequences is another sequence converged, whose limit is equal to the product of the limits of the two convergent sequences given.

7. If a sequence $\{x_n\}$, is convergent to a non-zero limit, then the sequence $\{\frac{1}{x_n}\}$ is bounded.

8. The quotient of two convergent sequences x_n and and_n, the latter with limit non-zero, is another convergent sequence, whose limit is equal to the quotient of the limits of the two sequences given.

9. A set is closed if and only if it contains the limit of any convergent sequence whose terms belong to S.

10. If after a certain number of order N, the elements of two convergent sequences $\{x_n\}$ and $\{y_n\}$ satisfy the inequality $x_n \leq y_n$, then their limits verify the same inequality:

$$\lim_{n\to\infty} x_n \leq \lim_{n\to\infty} y_n$$

11. If all the elements of a convergent sequence $\{x_n\}$ are in an interval $[a, b]$, then the limit of this sequence is also in the same interval.

12. Let $\{x_n\}$ and $\{z_n\}$ be two convergent sequences, which converge to the same limit l. Let $\{y_n\}$ be a third sequence which meets from a certain term the inequality $x_n \leq y_n \leq z_n$, then the sequence $\{y_n\}$ also converges to the limit l.

13. If l is the limit of the sequence $\{x_n\}$, then you can extract from this sequence a subsequence which converges to the number l.

14. The infinitely large sequences have no limit in the sense that we have defined it, and thus it is considered that the infinitely large sequences have a limit equal to ∞, and we write:

$$\lim_{n\to\infty} x_n = \infty$$

4.3 Monotonic sequences

Let's take the sequence of real numbers $\{x_n\} = x_1, x_2, x_3, ...$

- If $x_1 < x_2 < x_3 < ...$, the sequence is said to be increasing.
- If $x_1 \leq x_2 \leq x_3 \leq ...$, the sequence is said to be not decreasing.
- If $x_1 \geq x_2 \geq x_3 \geq ...$, the sequence is said to be not increasing.
- If $x_1 > x_2 > x_3 > ...$, the sequence is said to be decreasing.

In any of these four cases, the sequence is said to be monotone. Remember that in section 3.3 we defined a bounded sequence as that for which there exists K such that $|x_n| \leq K$ for all x_n of the sequence. Well, a fundamental property of real numbers is as follows:

Every monotone bounded sequence is convergent.

We are going to prove it. Let us take a real numbers sequence $x_1 \leq x_2 \leq x_3 \leq ....$ Suppose it is bounded. Then there must be for this set $\{x_n\}$ of real numbers a real number $a = \text{lub}(\{x_n\})$. Since a is upper bound:

$$a \geq x_n \qquad \forall n$$

And at the same time, since a is the lowest upper bound, we must find that for any real number $\varepsilon > 0$ there will be elements of $\{x_n\}$ greater that $a-\varepsilon$, since otherwise a would not be the LOWEST of the upper bounds. This means that there will be for our increasing sequence a number of order N from which

$$x_n > a - \varepsilon \qquad \forall n > N$$

If we unite both inequalities to the following obvious fact $a < a + \varepsilon$:

$$a - \varepsilon < x_n \leq a < a + \varepsilon \qquad \forall n > N$$

$$|x_n - a| < \varepsilon \qquad \forall n > N$$

But this last expression is precisely the definition of limit, and therefore the sequence $\{x_n\}$ converges, and its limit is precisely $a = \text{lub}(\{x_n\})$.

$$\lim_{n \to \infty} x_n = a$$

Similarly it is shown for the case of a decreasing monotone sequence, by changing the inequalities and by replacing the lub for the glb. Let's look at an example with Python. We take the sequence

$$x_n = \frac{1}{2}\left(x_{n-1} + \frac{A}{x_{n-1}}\right)$$

with $A > 0$, $x_1 > 0$ y $x_n > \sqrt{A}$. We will proof that this sequence is monotone. If we subtract

$$x_n - x_{n+1} = x_n - \frac{1}{2}\left(x_n + \frac{A}{x_n}\right)$$

$$x_n - x_{n+1} = \frac{1}{2}\left(x_n - \frac{A}{x_n}\right) = \frac{x_n^2 - A}{2x_n} > 0$$

Therefore the sequence is decreasing, and bounded, since all their terms are positive. As it is monotone and bounded, it must be convergent. Let $a \geq A$ be its limit:

$$\lim_{n \to \infty} x_n = a$$

$$\lim_{n \to \infty} x_{n+1} = a$$

$$\lim_{n \to \infty} x_{n+1} = \lim_{n \to \infty} \frac{1}{2}\left(x_n + \frac{A}{x_n}\right)$$

$$\lim_{n \to \infty} x_{n+1} = \frac{1}{2}\left(a + \frac{A}{a}\right) = a$$

$$a = \frac{a}{2} + \frac{A}{2a}$$

$$2a^2 = a^2 + A$$

$$a = \sqrt{A}$$

Therefore, it is proved that this bounded monotone sequence and therefore convergent: its limit is $\sqrt{A}$. This sequence provides a method for the calculation of the square root of any real number $A > 0$. Let's see a numerical example of this sequence calculated with Python.

```
# -*- coding: utf-8 -*-
"""
Mathematics and Python Programming      www.pysamples.com
p4b.py
"""

import numpy as np
```

```python
def root(n, A):
    print ('Calculate the square root of ' + "%3.1f" % A +
           ' with ' + str(n) + ' terms.')
    y = np.zeros(n + 1, float)
    y[1] = A / 2.0        # begin with A/2:
    for i in range(2, n + 1):
        y[i] = 0.5 * (y[i - 1] + (A / y[i - 1]))
        s = ('y[' + "%.0f" % i + '] = 0.5(' +
             "%12.8f" % y[i - 1] + '+(' + "%3.1f" % A +
             '/' + "%12.8f" % y[i - 1] + ')) = ' +
             "%12.8f" % y[i])
        print s
    print ('Direct calculation: raiz(A) = raiz(' +
           "%3.1f" % A + ') = ' +
           "%.15f" % np.sqrt(A))
#n: number of terms to calculate; A: number whose root we want to calculate
root(10, 3000)   # root(n, A)
```

```
──────────────────────────────── output of the program ────────────────────────────────
Calculate the square root of 3000.0 with 10 terms.
y[2]  = 0.5(1500.00000000+(3000.0/1500.00000000)) = 751.00000000
y[3]  = 0.5(751.00000000+(3000.0/751.00000000)) = 377.49733688
y[4]  = 0.5(377.49733688+(3000.0/377.49733688)) = 192.72220641
y[5]  = 0.5(192.72220641+(3000.0/192.72220641)) = 104.14432667
y[6]  = 0.5(104.14432667+(3000.0/104.14432667)) =  66.47525228
y[7]  = 0.5( 66.47525228+(3000.0/ 66.47525228)) =  55.80241452
y[8]  = 0.5( 55.80241452+(3000.0/ 55.80241452)) =  54.78176454
y[9]  = 0.5( 54.78176454+(3000.0/ 54.78176454)) =  54.77225658
y[10] = 0.5( 54.77225658+(3000.0/ 54.77225658)) =  54.77225575
Direct calculation: sqrt(A) = sqrt(3000.0) = 54.772255750516614
0.210046768188  ms
```

4.4 Cauchy sequence

Often we are not concerned to calculate the limit of a sequence, but only to know if it is convergent. In this section we introduce the important concept of Cauchy sequence. As we shall see below, all convergent sequences are Cauchy sequences, and although the reciprocal assertion is not true in general, it is true for certain metric spaces.

A sequence of points $p_1, p_2, p_3, ...$ in a metric space is a Cauchy sequence if, given any real number $\varepsilon > 0$, there exists a positive integer N such that the distance $d(p_n, p_m) < \varepsilon$ when $n, m > N$.

This condition of Cauchy is a necessary condition for the convergence of a sequence, because if $p_1, p_2, p_3, ...$ converges to p, then for any $\varepsilon > 0$ there exists a positive integer N such that $d(p, p_n) < \frac{\varepsilon}{2}$ when $n > N$. Therefore, from a number of order $n, m > N$ we have:

$$d(p_n, p_m) \leq d(p_n, p) + d(p, p_m) < \frac{\varepsilon}{2} + \frac{\varepsilon}{2} = \varepsilon$$

In this way, we have shown that the condition of Cauchy is a necessary condition for the convergence of a sequence. Let's make some comments:

1. Not any Cauchy sequence in a metric space is convergent. For example, the sequence $x_n = 1, \frac{1}{2}, \frac{1}{3}, \frac{1}{4}, ...$, in the space $E = \mathbf{R} - \{0\}$. This sequence does not converge to any point of E, in spite of be a Cauchy sequence.

2. Any subsequence of a Cauchy sequence, is itself a Cauchy sequence.

3. Any Cauchy sequence having a convergent subsequence, is itself convergent.

4. Any Cauchy sequence of rational numbers converges to a real number (which can be rational or irrational).

5. A metric space E is said to be complete if every Cauchy sequence of points of E converges to a point of E.

6. **R** is complete.

This means that in **R**, not only any convergent sequence must be a Cauchy sequence, but conversely, that any Cauchy sequence will be convergent. We have already shown the first part, that any convergent sequence of real numbers is a Cauchy sequence. We are going to demonstrate the reciprocal statement: that any Cauchy sequence of real numbers will be convergent. Let $\{a_n\} = a_1, a_2, a_3, ...$ be a Cauchy sequence of real numbers. We must prove that this sequence converges to a real number. Consider a sequence $\{\varepsilon_n\}$. We assign to each real number a_p of the sequence $\{a_n\}$ a rational number r_p such that

$$|a_p - r_p| < \varepsilon_p$$
$$|r_p - r_q| = |r_p - a_p + a_p - a_q + a_q - r_q|$$
$$|r_p - r_q| \leq |r_p - a_p| + |a_p - a_q| + |a_q - r_p|$$
$$|r_p - r_q| \leq \varepsilon_p + \varepsilon + \varepsilon_q$$

This assures us that the sequence of rational numbers r_n is a Cauchy sequence, and as we have seen in the fourth property, every Cauchy sequence of rational numbers converges to a real number, and the sequence r_n will define a real number a, (as we saw in the first chapter (if $a, \epsilon \in \mathbf{R}$, $\epsilon > 0$, then there exists a rational number r such that $|a - r| < \epsilon$).

$$|a_n - a| \leq |a_n - r_n| + |r_n - a| \qquad n > N$$
$$|a_n - a| \leq \varepsilon_n + \varepsilon$$

It follows that the sequence $\{a_n\}$ also converges to the point a, as we wanted to show.

Now let's look at an numeric example with Python. Consider the sequence

$$x_n = \left(1 + \frac{1}{n}\right)^n$$

we are going to calculate some terms with a Python program:

```
# -*- coding: utf-8 -*-
"""
Mathematics and Python Programming     www.pysamples.com
p4c.py
"""

import numpy as np

n = 5000
y = np.zeros(n + 1, float)
y[1] = np.power((1 + (1.0 / 1)), 1)


def esequence(n):
    i = 2
    while i <= n:
        y[i] = np.power((1 + (1.0 / i)), i)
        i += 1

esequence(n)
print 'i          x[i]         x[i+1]-x[i]'
```

4.4. CAUCHY SEQUENCE

```
print '[   1]:', "%10.8f" % y[1], '       ; '
for i in range(2, n + 1):
    if i > 99:
        print ('[' + str(i) + ']: ' +
               "%10.8f" % y[i] + '   ; ' +
               "%10.4g" % (y[i] - y[i - 1]))
    elif i > 9:
        print ('[ ' + str(i) + ']: ' +
               "%10.8f" % y[i] + '   ; ' +
               "%10.4g" % (y[i] - y[i - 1]))
    else:
        print ('[  ' + str(i) + ']: ' +
               "%10.8f" % y[i] + '   ; ' +
               "%10.4g" % (y[i] - y[i - 1]))
```

In the following results many terms have been omitted:

```
────────────────────────────── output of the program ──────────────────────────
     i        x[i]              x[i+1]-x[i]
[   1]: 2.00000000       ;
[   2]: 2.25000000       ;       0.25
[   3]: 2.37037037       ;       0.1204
[   4]: 2.44140625       ;       0.07104
[   5]: 2.48832000       ;       0.04691
[   6]: 2.52162637       ;       0.03331
[   7]: 2.54649970       ;       0.02487
[   8]: 2.56578451       ;       0.01928
[   9]: 2.58117479       ;       0.01539
[  10]: 2.59374246       ;       0.01257
...
[  99]: 2.70467904       ;       0.0001375
[100]: 2.70481383        ;       0.0001348
[101]: 2.70494598        ;       0.0001321
...
[999]:  2.71692257       ;       1.361e-06
[1000]: 2.71692393       ;       1.358e-06
...
[2999]: 2.71782877       ;       1.511e-07
[3000]: 2.71782892       ;       1.510e-07
...
[4999]: 2.71801000       ;       5.438e-08
[5000]: 2.71801005       ;       5.436e-08
64.7749900818  ms
───────────────────────────────────────────────────────────────────────────────
```

It can be shown that this sequence of real numbers is increasing and bounded, and therefore convergent in **R**. As we have just seen, the difference between two consecutive terms can be made as small as we like, simply by calculating terms of a number of order sufficiently great. Well, the limit of this sequence is a number called number e.

```
>>> print "%25.22f" % math.e
 2.7182818284590450907956
```

It is algebraically cumbersome to determine the value of the difference of any two terms of this sequence, however it is easy to write a Python program that calculates the number of order N from which the difference between two successive terms is less than any real number ε we choose:

```
# -*- coding: utf-8 -*-
"""
Mathematics and Python Programming     www.pysamples.com
p4d.py
difference between two successive terms is less than epsilon
"""

import numpy as np


def esequence2(epsilon):
```

```
    i = 1
    incremento = 1000
    while incremento > epsilon:
        y = np.power((1 + (1.0 / i)), i)
        y1 = np.power((1 + (1.0 / (i + 1))), (i + 1))
        incremento = abs(y1 - y)
        i += 1
    print 'required epsilon: ', epsilon
    print 'number of terms: N = ', i - 1
    y = np.power((1 + (1.0 / (i - 1))), (i - 1))
    y1 = np.power((1 + (1.0 / i)), i)
    print 'x[', i, '] - x[', i - 1, '] = '
    print "%30.27f" % y1, ' -', "%30.27f" % y, ' = '
    print ' = ', y1 - y

esequence2(1e-3)    # must be >1e-16
```

We ran the program for some values of ε and show the times of timed execution of the program in each case:

```
─────────────────────────────────── output of the program ───────────────────────────────────
required epsilon:   0.001
number of terms: N =    36
x[ 37 ] - x[ 36 ] =
 2.682435477308525495487856460 -  2.681464420300858630952234307 =
=   0.000971057007667
0.818967819214  ms

required epsilon:   1e-06
number of terms: N =    1165
x[ 1166 ] - x[ 1165 ] =
 2.717117100087719006040742897 -  2.717116101106833880862723163 =
=   9.98980885125e-07
20.3380584717  ms

required epsilon:   1e-15
number of terms: N =    223071
x[ 223072 ] - x[ 223071 ] =
 2.718275735651982660812109316 -  2.718275735651982216722899466 =
=   4.4408920985e-16
2770.68591118  ms
```

4.5 Nested intervals

Let I_n be a sequence of closed intervals of real numbers, such that the intervals are nested within one another:
$$I_1 \supset I_2 \supset I_3 ... = [a_1, b_1] \supset [a_2, b_2] \supset [a_3, b_3], ...$$
So that the differences $d_n = b_n - a_n$ form a null sequence.

We shall call it a sequence of nested intervals. Using this concept, we can prove a criterion for the completeness of a metric space, in this case $\mathbf{R}$. A metric space is complete if and only if any sequence of nested intervals of the metric space is such that the intersection of all the intervals when $n \to \infty$ is not an empty set.
$$\bigcap_1^\infty I_n \neq \emptyset$$
The fact that these intervals are nested implies:
$$a_1 \leq a_2 \leq ... \leq a_k \leq a_{k+1} \leq b_{k+1} \leq b_k \leq ... \leq b_2 \leq b_1 \qquad \forall k$$

Thus, $\{a_n\}$ is a increasing sequence and bounded from above by b_1, and $\{b_n\}$ is a decreasing sequence and bounded from below by a_1. Therefore, these sequences will be both convergent:
$$\lim_{n \to \infty} a_n = a$$

4.5. NESTED INTERVALS

$$\lim_{n\to\infty} b_n = b$$

and a and b are such that:

$$a_n \le a \le b \le b_n \quad \forall n$$

This means that the interval $[a, b]$ is a subset that belongs to each of the intervals I_n, and it is precisely the intersection of all of them. Since the differences $d_n = b_n - a_n$ form a null sequence null, we will have:

$$\lim_{n\to\infty} d_n = \lim_{n\to\infty} (b_n - a_n) = 0$$

$$\lim_{n\to\infty} (b_n - a_n) = \lim_{n\to\infty} b_n - \lim_{n\to\infty} a_n = b - a = 0$$

Then, $a = b$, and the intersection of all intervals is a single point. We do a couple of observations:

- The theorem of the nested intervals is not met if we take open intervals instead of closed. For example, in the sequence of open intervals:

$$(0,1) \supset \left(0, \frac{1}{2}\right) \supset \left(0, \frac{1}{4}\right), ... \supset \left(0, \frac{1s}{2^n}\right) \supset ...$$

there exists no point belonging to all the intervals.

- The number e can be defined as the intersection of all the intervals of the following form:

$$[a, b] = \left[\left(1 + \frac{1}{n}\right)^n, \left(1 + \frac{1}{n}\right)^{n+1}\right]$$

The following program calculates some terms of the sequence of nested intervals whose limit is the number e. The commented instruction print shows the intervals as decimal values, instead of in their fraction form. First it has been run to display the first 10 intervals in fraction form, and then it has been run to display the first 100 intervals in decimal format:

```
# -*- coding: utf-8 -*-
"""
Mathematics and Python Programming      www.pysamples.com
p4e.py
"""

import fractions


def nested(n):
    i = 1
    while i <= n:
        anum = (i + 1) ** i
        bnum = anum * (i + 1)
        aden = i ** i
        bden = aden * i
        # fractions
        fraca = fractions.Fraction(anum, aden)
        fracb = fractions.Fraction(bnum, bden)
        print (str(i) + ': ' + '%s < e < %s' % (fraca, fracb))
#        print (str(i) + ': ' + "%15.12f" % (1.0 * anum / aden) +
#                ' < e < ' + "%15.12f" % (1.0 * bnum / bden))
        i += 1

nested(10)   # must be <=140 to calculate in decimal format
```

```
------- output of the program -------
1: 2 < e < 4
2: 9/4 < e < 27/8
3: 64/27 < e < 256/81
4: 625/256 < e < 3125/1024
5: 7776/3125 < e < 46656/15625
6: 117649/46656 < e < 823543/279936
7: 2097152/823543 < e < 16777216/5764801
8: 43046721/16777216 < e < 387420489/134217728
9: 1000000000/387420489 < e < 10000000000/3486784401
10: 25937424601/10000000000 < e < 285311670611/100000000000
7.47680664062   ms
```

$$\left[\frac{2}{1}, \frac{4}{1}\right] \supset \left[\frac{9}{4}, \frac{27}{8}\right] \supset \left[\frac{64}{27}, \frac{256}{81}\right] \supset \left[\frac{625}{256}, \frac{3125}{1024}\right] \supset \left[\frac{7776}{3125}, \frac{46656}{15625}\right] \supset \left[\frac{117649}{46656}, \frac{823543}{279936}\right] \supset \left[\frac{2097152}{823543}, \frac{16777216}{5764801}\right]$$

$$\supset \left[\frac{43046721}{16777216}, \frac{387420489}{134217728}\right] \supset \left[\frac{1000000000}{387420489}, \frac{10000000000}{3486784401}\right] \supset \left[\frac{25937424601}{10000000000}, \frac{285311670611}{100000000000}\right] \supset \ldots$$

This sequence of nested intervals corresponds to the following real values, represented in the figure:

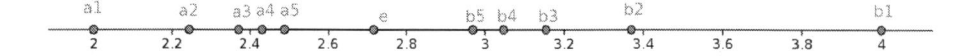

```
------- output of the program -------
1:   2.000000000000 < e <   4.000000000000
2:   2.250000000000 < e <   3.375000000000
3:   2.370370370370 < e <   3.160493827160
4:   2.441406250000 < e <   3.051757812500
5:   2.488320000000 < e <   2.985984000000
...
99:  2.704679036165 < e <   2.731999026429
100: 2.704813829422 < e <   2.731861967716
9.09495353699   ms
```

The following program represents the first 50 nested intervals. The horizontal red line represents the actual value of the number e. The blue line marks the middle point of each interval. This blue line approaches the red line representing number e.

```python
# -*- coding: utf-8 -*-
"""
Mathematics and Python Programming
    www.pysamples.com
p4f.py
"""

import numpy as np
import matplotlib.pyplot as plt

terms = 50
x = []
yerror = []
ycenter = []
ye = []


def nested(n):
    i = 1.0
    while i <= n:
        start = ((i + 1) ** i) / (i ** i)
        end = (i + 1) ** (i + 1) / (i ** (i + 1))
        print start, end
        x.append(i)
```

4.6. SERIES

```
            yerror.append((end - start) / 2)
            ycenter.append(start + (end - start)/2)
            ye.append(np.e)
            i += 1

nested(terms)
print yerror
print ycenter
x0 = np.zeros(terms, float)
xerror = np.zeros(terms, float)
fig = plt.figure()
ax = fig.add_subplot(111)
ax.set_ylim(1.9, 4.1)
ax.errorbar(x, ycenter, xerr=xerror, yerr=yerror)
ax.plot(x,ye, lw=1.0, color='r')
plt.xlabel('n')
plt.ylabel('length of the intervals')
plt.show()
```

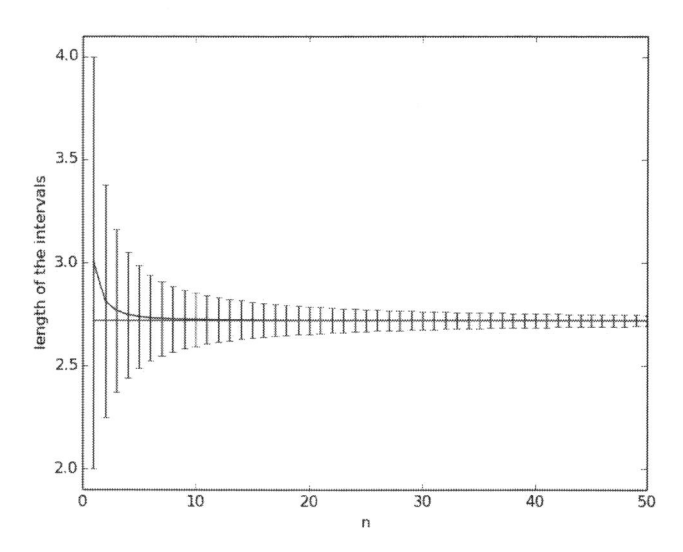

4.6 Series

Let's take the sequence of real numbers $\{x_n\} = x_1, x_2, ..., x_n, ...$ The sum of an infinite number of terms of the sequence is called numerical series.

$$x_1 + x_2 + x_3 + ... + x_k + ... = \sum_{i=1}^{\infty} x_i$$

The sum of n terms of the sequence is called partial sum.

$$S_n = x_1 + x_2 + x_3 + ... + x_n = \sum_{i=1}^{n} x_i$$

If the succession $s_1, s_2, ..., s_n, ...$ is convergent, it is said that the series is convergent, and its sum is equal to the limit of the sequence $\{s_n\}$. It is said that a numerical series satisfies the condition of Cauchy if given any real number $\varepsilon > 0$, there is a positive integer N such that the following inequality is met

$$|x_{n+1} + x_{n+2} + ... + x_m| < \varepsilon \quad \forall \ m > n > N$$

In fact, we have
$$|s_m - s_n| = |x_{n+1} + x_{n+2} + ... + x_m| < \varepsilon$$

And the Cauchy convergence condition for the series is fully equivalent to the condition of convergence for the sequence.

If we take the sums of order n and $n - 1$, the following inequality is met:
$$|s_n - s_{n-1}| = |x_n| < \varepsilon$$

And therefore we will have:
$$\lim_{n \to \infty} x_n = 0$$

And this is a necessary condition for the convergence of a series. We are going to study the limit of a sequence which later will be useful:
$$\lim_{n \to \infty} \frac{a^n}{n!} = 0$$

for any constant $a > 0$. To prove it, let $k \geq 0$ be smallest integer such that $a < k + 1$. We call $\gamma = \frac{a}{k+1} < 1$. Suppose that $n \geq k$, so that $n = k + l, \quad l \geq 0$.

$$\frac{a^n}{n!} = \frac{a^{k+l}}{(k+l)!} = \frac{a^k}{k!} \cdot \frac{a^l}{(k+1)...(k+l)} \leq \frac{a^k}{k!} \frac{a^l}{(k+1)^l}$$

$$\frac{a^n}{n!} \leq \frac{a^k}{k!} \gamma^l = \frac{a^k}{k!\gamma^k} \gamma^{k+l}$$

$$\frac{a^n}{n!} \leq \frac{a^k}{k!\gamma^k} \gamma^n = b\gamma^n$$

Thus, we have $\frac{a^n}{n!} \leq b\gamma^n$ for $n \geq k$. In the other case $(n < k)$, we will find that there only exists a finite amount of terms for which the previous equality is not met, and from a certain term, the previous inequality will be met, and there will exists a constant c such that $\frac{a^n}{n!} \leq c\gamma^n$. If we take the larger of the two numbers, $M = max_{b,c}$, the two inequalities will be met at the same time, and we will have
$$\frac{a^n}{n!} \leq M\gamma^n$$

And since
$$\lim_{n \to \infty} \gamma^n = 0$$

In fact it is a geometric progression of ratio $\gamma < 1$, the given sequence also meets:
$$\lim_{n \to \infty} \frac{a^n}{n!} = 0$$

Let's look at an example calculated using Python, with $a = 3.5$:

4.6. SERIES

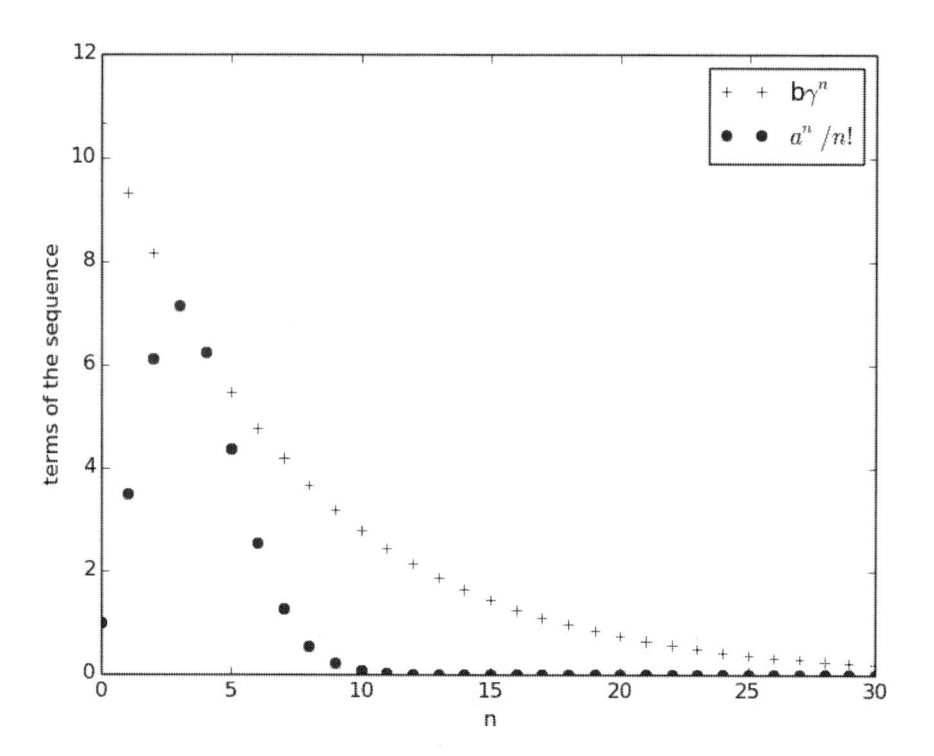

```
──────────────────── output of the program ────────────────────
a = 3.5; k = 3
gamma = 0.875000
b = 10.666667; n = 30
0:   1              10.666667       16: 2.424e-05    1.259382
1:   3.5            9.333333        17: 4.99e-06     1.101959
2:   6.125          8.166667        18: 9.703e-07    0.964215
3:   7.146          7.145833        19: 1.787e-07    0.843688
4:   6.253          6.252604        20: 3.128e-08    0.738227
5:   4.377          5.471029        21: 5.213e-09    0.645948
6:   2.553          4.787150        22: 8.293e-10    0.565205
7:   1.277          4.188756        23: 1.262e-10    0.494554
8:   0.5585         3.665162        24: 1.84e-11     0.432735
9:   0.2172         3.207017        25: 2.577e-12    0.378643
10:  0.07602        2.806139        26: 3.469e-13    0.331313
11:  0.02419        2.455372        27: 4.496e-14    0.289899
12:  0.007055       2.148451        28: 5.62e-15     0.253661
13:  0.001899       1.879894        29: 6.783e-16    0.221954
14:  0.0004748      1.644907        30: 7.914e-17    0.194209
15:  0.0001108      1.439294
```

```python
# -*- coding: utf-8 -*-
"""
Mathematics and Python Programming    www.pysamples.com
p4g.py
"""

import numpy as np
import matplotlib.pyplot as plt


def fact(x):
    if x == 0:
        return 1
    else:
        return x * fact(x - 1)

a = 3.5
k = int(np.trunc(a))
```

```
gamma = a / (k + 1)
b = np.power(a, k) / (fact(k) * (np.power(gamma, k)))
n = 10 * k
sequence = np.zeros(n + 1, float)
M = np.zeros(n + 1, float)
points = np.zeros(n + 1, float)
for i in range(0, n + 1):
    sequence[i] = np.power(a, i) / fact(i)
    M[i] = b * np.power(gamma, i)
    points[i] = i
p1, = plt.plot(points, M, 'k+')
p2, = plt.plot(points, sequence, 'bo')
plt.ylabel('terms of the sequence')
plt.xlabel('n')
plt.legend(('b$\gamma ^{n}$', '$ a^n /n!$'), loc='best')
print 'a = ' + str(a) + '; k = ' + str(k)
print 'gamma = ' + "%8.6f" % gamma
print 'b = ' + "%8.6f" % b + '; n = ' + str(n)
for i in range(0, n + 1):
    print (str(i) + ': ' + "%9.4g" % sequence[i] + "%12.6f" % M[i])
plt.show()
```

Let's look at some important types of series:

1. Let $x_1 + x_2 + x_3 + ... + x_n + ...$ and $y_1 + y_2 + y_3 + ... + y_n + ...$ be two convergent series, whose sums are s and t respectively, the series can be added together term by term, and the sum of both series converges to the number $s + t$.

2. Let $x_1 + x_2 + x_3 + ... + x_n + ...$ and $y_1 + y_2 + y_3 + ... + y_n + ...$ be two convergent series, whose sums are s and t respectively, the series can be subtracted from term to term and the difference of both series converges to the number $s - t$.

3. A series $x_1 + x_2 + x_3 + ... + x_n + ...$ is said to be absolutely convergent series if $|x_1| + |x_2| + |x_3| + ... + |x_n| + ...$ is convergent. If a series is absolutely convergent, is already convergent because:
$$|x_{n+1}| + |x_{n+2}| + ... + |x_m| < \varepsilon$$
$$|x_{n+1} + x_{n+2} + ... + x_m| \leq |x_{n+1}| + |x_{n+2}| + ... + |x_m| < \varepsilon$$

   The sum of an absolutely convergent series does not depend on the order of its terms.

4. Given a series $x_1 + x_2 + x_3 + ... + x_n + ...$ with non-negative terms, and the series $\nu_1 + \nu_2 + \nu_3 + ... + \nu_n + ...$ which is obtained by rearranging arbitrarily the terms of the original series, then if the original series was convergent, the new series also converges and it has the same sum as the original series.

5. Let's take the series $x_1 + x_2 + x_3 + ... + x_n + ...$, and an arbitrary number c other than zero. Then the series
$$cx_1 + cx_2 + cx_3 + ... + cx_n + ...$$
will converge if and only if the original series was convergent, and if the sum of this was s, the sum of the series multiplied by c will be cs.

6. If we remove a finite number of the terms of a series, its convergence does not change: if the initial series was convergent, the sum of the series obtained will be less than the initial series in an amount equal to the sum of the terms deleted.

We shall return later to the theme of the series. Up to now the concept of function has not been used in the text: the time has come to do so.

Bibliography for this chapter: [9], [24], [32], [29], [44], [47], [50], [55], [60]

5 | Function. Continuity

5.1 Function

Let's start by defining the concept of function. To try to understand more clearly the concept of function, we set out below three definitions of different authors that can be consulted in the bibliography listed at the end of the chapter:

Definition 1: We can define a function ϕ of a set X in a set Y as a rule that assigns to each element $x \in X$ exactly one element $y \in Y$, and then it is said that ϕ transforms x into y, and ϕ transforms X into Y, and this is denoted:

$$\phi : X \to Y$$
$$\phi(x) = y$$

or also:

$$x\phi = y$$

and we say that element y is the image of x under ϕ.

Definition 2: Let X be an arbitrary set of points $x \in X$, then x is called a variable, and it is said that X is the domain of x. If there exists a rule which matches a definite number y to each point x of X, it is said that y is a function of the variable x, and is denoted:

$$y = f(x)$$

where f represents the rule prescribed. X is called domain of definition, and x is called argument of the function. The entire range of values of y that correspond to points $x \in X$ is called domain of values (or range) of the function on X.

Along with the term "function", it is also used the equivalent term "application", and it is written $f : x \to y$, and it is said that the application f transforms the number x into the number y, or what is the same, that the number y is the image of the number x under the application f.

Definition 3: Let E and E' be two metric spaces with distances d and d' respectively. The function f of the metric space E in the metric space E' assigns each point $p \in E$ a point $f(p) \in E'$.

$$F : E \to E'$$

A function $f(x)$ is said to be bounded in an interval $[a, b]$ if there is a number $M > 0$ such that for all the points $x \in [a, b]$, $|f(x)| \leq M$.

It is not necessary that the function to be given in an explicit way, but only that a value y will match in a not ambiguous manner to each value x. This concept of function is so large that we could hardly cover it with a few theorems. Fortunately, the features that are most useful in science and mathematics are of a type to which we can limit ourselves merely demanding that our functions are differentiable. This property of differentiability of a function, which we will discuss later, has important consequences for the nature of the function. The differentiability presupposes another requirement: the continuity of the function.

5.2 Limit of a function

The number l it is called the limit of a function $f(x)$ as x approaches the value x_0, if for every sequence of values of x
$$x_1, x_2, x_3, ..., x_n, ...$$
which converges to x_0, the corresponding sequence of values of $f(x)$
$$f(x_1), f(x_2), f(x_3), ..., f(x_n), ...$$
converges to the number l, and it is written:
$$\lim_{x \to x_0} f(x) = l$$
Since the limit of a sequence is unique, a function can have a single limit in a point x_0.

We define the concept of one-sided limit of a function: the number l is called right (left) limit of the function $f(x)$ in the point x_0 if for any sequence of values of the argument x which converges to x_0 and whose elements are greater (lesser) than x_0, the sequence of corresponding values of $f(x)$ converges to the number l. The limit on the right side is denoted as
$$\lim_{x \to x_0^+} f(x)$$
and the left side limit is denoted as
$$\lim_{x \to x_0^-} f(x)$$
For the function $f(x)$ to have limit in the point x_0, it is necessary and sufficient that both side limits exist and are equal.

The function $f(x)$ is called infinitely small (or infinitesimal) at the point x_0 if
$$\lim_{x \to x_0} f(x) = 0$$
. Another equivalent definition is as follows: the function $\alpha(x) = f(x) - l$ is said to be an infinitesimal when x approaches x_0 if the limit
$$\lim_{x \to x_0} f(x) = l$$

The function $f(x)$ is called infinitely large (or infinite) at the point x_0 if
$$\lim_{x \to x_0} f(x) = \infty$$
Let $\alpha(x)$ and $\beta(x)$ be two infinitesimal functions:.

1. If $\lim_{x \to x_0} \frac{\alpha(x)}{\beta(x)} = 0$, it is said that $\alpha(x)$ is a an infinitesimal of higher order than $\beta(x)$, and it is denoted $\alpha(x) = o(\beta(x))$.

2. If $\lim_{x \to x_0} \frac{\alpha(x)}{\beta(x)} = A \neq 0$, it is said that $\alpha(x)$ and $\beta(x)$ are infinitesimals of the same order.

3. If $\lim_{x \to x_0} \frac{\alpha(x)}{\beta(x)} = 1$, it is said that $\alpha(x)$ and $\beta(x)$ are equivalent infinitesimals. If in the expression of a function, an infinitesimal factor or divider is replaced by another equivalent one, the limit of the function does not vary. For example, $\lim_{x \to 0} \frac{sen x}{x} = 1$.

4. If $\lim_{x \to x_0} \frac{\alpha(x)}{\beta^n(x)} = A \neq 0$, it is said that $\alpha(x)$ is an infinitesimal of order n with respect to $\beta(x)$. For example, $\lim_{x \to 0} \frac{1 - \cos x}{x^2} = \frac{1}{2}$, and it is said that the function $\alpha(x) = 1 - \cos x$ is a second order infinitesimal with respect to $\beta(x) = x$.

5.3 Continuity

Definition 1: A function $f(x)$ is said to be continuous at the point a if, for any sequence $x_1, x_2, ..., x_n, ...$ of values of its argument, which converges to a, the corresponding sequence $f(x_1), f(x_2), ..., f(x_n), ...$ of values of the function converges to the number $f(a)$.

Definition 2: A function $y = f(x)$ defined in x is said to be continuous if the limit $\lim_{x \to x_0} f(x)$ exists and is equal to $f(x_0)$. That is to say: if we choose any real number $\varepsilon > 0$, it is always possible to assign a number $\delta = \delta(\varepsilon) > 0$, such that $|f(x) - f(x_0)| < \varepsilon$ for all the points x that meet $|x - x_0| < \delta$.

Definition 3: Let E and E be two metric spaces with distances d and d' respectively, and $f: E \to E'$, a function, and $p_0 \in E$. It is said that f is continuous in p_0 if, given any real number $\varepsilon > 0$, there is a real number $\delta > 0$ such that if $p \in E$ y $d(p, p_0) < \delta$, then $d'(f(p), f(p_0)) < \varepsilon$.

$$f : E \to E'$$

We can summarize the definitions by saying that a function is continuous if its values $f(x)$ differ from $f(x_0)$ in arbitrarily small amounts when x is at a sufficiently small distance of x_0.

The function $f(x)$ is called right-continuous (left-continuous) at the point a if the right (left) limit of this function exists and is equal to the particular value $f(a)$. It is denoted: Right-continuous function:

$$\lim_{x \to a^+} f(x) = f(a)$$

Left-continuous function:

$$\lim_{x \to a^-} f(x) = f(a)$$

If the function is right-continuous and left-continuous at the point a, then it is continuous at that point. Potential functions $f(x) = x^n$, with $n \in N$, as well as polynomials are found among the types of continuous functions on **R**.

Let's take as an example the function $f : \mathbf{R} \to \mathbf{R}$ given by

$$f(x) = x^2$$

Its graph has been drawn up with Python program p5i.py, which is shown at the end of the chapter.

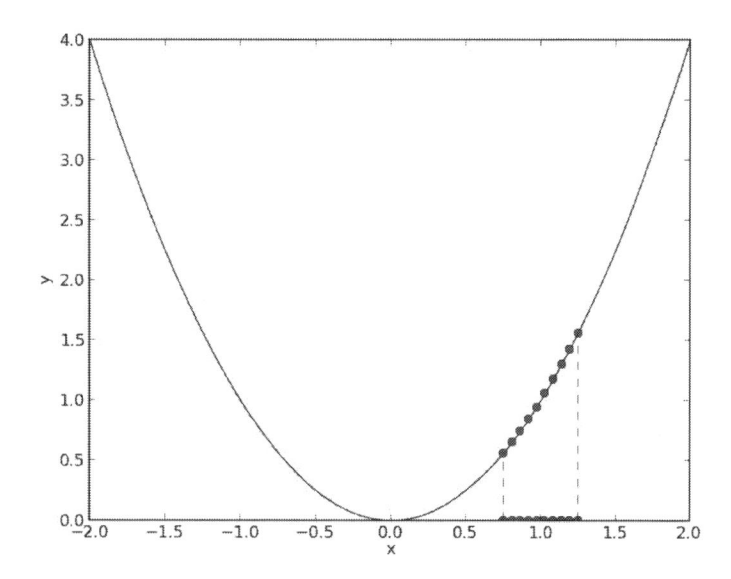

This is a continuous functions, as we are going to prove:

$$|f(x) - f(x_0)| = |x^2 - x_0^2| < \varepsilon$$

$$|x^2 - x_0^2| = |(x + x_0)(x - x_0)| < \varepsilon$$

$$|x^2 - x_0^2| = |(x - x_0 + 2x_0)(x - x_0)| < \varepsilon$$

$$|x^2 - x_0^2| \leq (|x - x_0| + 2|x_0|)|x - x_0| < \varepsilon$$

we take $|x - x_0| = \delta$:

$$|x^2 - x_0^2| \leq (\delta + 2|x_0|)\delta$$

if we choose a real number $\varepsilon > 0$ such that $|x^2 - x_0^2| = \varepsilon$, then

$$\varepsilon \leq (\delta + 2|x_0|)\delta = \delta^2 + 2\delta|x_0| < 2\delta|x_0|$$

$$\varepsilon < 2\delta|x_0|$$

$$\delta > \frac{\varepsilon}{|2x_0|}$$

Thus, for each $\varepsilon > 0$ we choose, we can find a value of $\delta > \frac{\varepsilon}{|2x_0|}$. For example, we take the point $x_0 = 1$ and we want the distance between two points of the function $f(x) = x^2$ is $\varepsilon = 0.01$. We would need to take a distance δ between x and x_0 greater than $\frac{0.01}{|2 \cdot 1|} = 0.005$. That is to say, that for values of x that are less than 0.005 away from x_0, the difference $|f(x) - f(x_0)|$ will be less than 0.01. Let's verify it with a Python program:

```
# -*- coding: utf-8 -*-
"""
Mathematics and Python Programming    www.pysamples.com
p5a.py
"""

import numpy as np

x0 = 1
epsilon = 1e-3  # desired value of |f(x)-f(x0)|
delta = abs(epsilon / 2 * x0)
print 'x0 = ', x0, '; required epsilon: ', epsilon, ' delta = ', delta
print 'Points x which are at a distance from x0 less than ', delta
print 'meet |f(x)-f(x0)|<', delta, ' and are shown as #'
print
numpoints = 20
x = np.linspace(x0 - 1.5 * delta, x0 + 1.5 * delta, numpoints + 1)

print ('     x' + '          ' + '|x - x0|' + '   ' + ' |f(x) - f(x0)|')
print '_____'

for i in range(0, numpoints + 1):
    difX = abs(x[i] - x0)
    difY = abs(x[i] ** 2 - x0 ** 2)
    if difY < epsilon:
        mark = '#'
    else:
        mark = ''
    print ("%8.5f" % x[i] + '     ' + "%8.5f" % difX + '     ' + "%8.5f" % difY + mark)
```

5.3. CONTINUITY

```
------------------------------------- output of the program -------
x0 =  1 ; required epsilon:  0.001   delta =  0.0005
Points x which are at a distance from x0 less than  0.0005
meet  |f(x)-f(x0)|< 0.0005   and are shown as #

    x         |x - x0|    |f(x) - f(x0)|
   ---------------------------------------
   0.99925    0.00075     0.00150
   0.99933    0.00067     0.00135
   0.99940    0.00060     0.00120
   0.99948    0.00052     0.00105
   0.99955    0.00045     0.00090#
   0.99962    0.00038     0.00075#
   0.99970    0.00030     0.00060#
   0.99977    0.00023     0.00045#
   0.99985    0.00015     0.00030#
   0.99992    0.00008     0.00015#
   1.00000    0.00000     0.00000#
   1.00008    0.00008     0.00015#
   1.00015    0.00015     0.00030#
   1.00022    0.00022     0.00045#
   1.00030    0.00030     0.00060#
   1.00038    0.00038     0.00075#
   1.00045    0.00045     0.00090#
   1.00053    0.00053     0.00105
   1.00060    0.00060     0.00120
   1.00067    0.00067     0.00135
   1.00075    0.00075     0.00150
0.404119491577   ms
```

Let's see an example of a discontinuous function taken from the physical sciences: the amount of heat we need to provide to a gram of water to raise its temperature by one degree. The function $y = f(x)$ in this case $q = f(t)$, with q measured in $J/(g°C)$ and t in Celsius degrees. Experimentally, the following plot is is found (previously experimental data have been adjusted, and we obtained three straight lines). The plot has been made using the following Python program:

```python
# -*- coding: utf-8 -*-
"""
Mathematics and Python Programming    www.pysamples.com
p5b.py
"""

import matplotlib.pyplot as plt
import numpy as np

cpice = 2.072773; cpliquid = 4.187409; cpvapor = 1.836533
print 'cpice = ' + "%6.4f" % cpice
print 'cpliquid = ' + "%6.4f" % cpliquid
print 'cpvapor = ' + "%6.4f" % cpvapor
qmelting = 334   # J g-1
qvaporization = 2260   # J g-1
x = [-100, 0, 0, 100, 100, 200]
q = np.zeros(6, float)
q[0] = 0
q[1] = cpice * 100
q[2] = q[1] + qmelting
q[3] = q[2] + cpliquid * 100
q[4] = q[3] + qvaporization
q[5] = q[4] + cpvapor * 100
for i in range(0, 6):
    print str(x[i]) + "%6.0f" % q[i]
plt.plot([-100, 0], [0, q[1]], 'b', lw=1.5)
plt.plot([0, 100], [q[2], q[3]], 'b', lw=1.5)
plt.plot([100, 200], [q[4], q[5]], 'b', lw=1.5)
plt.ylabel('q')
plt.xlabel('t')
pointsX = [80, 85, 90, 95, 105, 110, 115, 120]
```

```
ceros = np.zeros(8, float)
pointsY = [
    q[2] + cpliquid * 80, q[2] + cpliquid * 85,
    q[2] + cpliquid * 90, q[2] + cpliquid * 95,
    q[4] + cpvapor * 5, q[4] + cpvapor * 10,
    q[4] + cpvapor * 15, q[4] + cpvapor * 20]
plt.plot([0, 0], [q[1], q[2]], 'k--', lw=0.5)
plt.plot([100, 100], [q[3], q[4]], 'k--', lw=0.5)
#points
plt.plot(pointsX, ceros, 'ro')
plt.plot(pointsX, pointsY, 'ro')
plt.show()
```

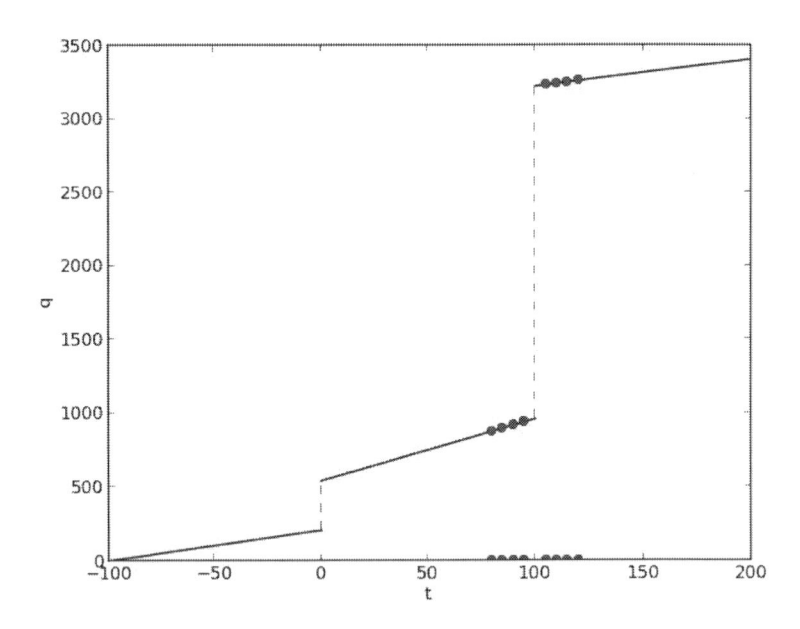

We can consider that the graph is composed of three segments:

- the first one corresponds to the warming of the ice from $-150°C$ up to $0°C$ and it follows continues the equation of a straight line: $q = 2.073t$

- the second segment corresponds to the heating of the liquid water from $0°C$ up to $100°C$ and it follows the equation of the straight line: $q = 4.187t$

- the third segment corresponds to the warming of the water vapour from $100°C$ up to $150°C$ and it follows the equation of the straight line: $q = 1.836t$

The two jumps correspond to the latent heat of melting and evaporation, whose values are $334\frac{J}{g}$ and $2260\frac{J}{g}$ respectively.

Let's look at the function in a neighbourhood of the point $t = 100°C$ by taking a sequence of values of the argument t which converges from the left side of the point $t = 100$:

t	q
99	955.813
99.9	959.581
99.99	959.958
99.999	959.996
99.9999	959.999

5.3. CONTINUITY

We calculated the corresponding values of the function taking

$$2.073 * 100 + 334 = 541.3$$

$$q = f(t) = 541.3 + 4.187t$$

the sequence of values of $f(t)$ converges to

$$q = 541.3 + 4.187 \cdot 100 = 960.0$$

If we now repeat this procedure to the right of the point $t = 100$ and calculate the corresponding values of $960 + 2260 = 3220$.

$$q = f(t) = 3220 + 1.836t$$

we get the following sequence of values which converges to

$$q = 3220 + 1.836 \cdot 0 = 3220$$

t	q
101	3221.836
100.1	3220.1836
100.01	3220.0184
100.001	3220.0018
100.0001	3220.0002

The left-side limit and the right-sided limit are not equal, and therefore, the function is not continuous at the point $t = 100$. We could check this also at the point $t = 0$, where the function is also discontinuous.

- If f and g are real functions in a metric space E. If f and g are continuous at a point $p_0 \in E$, so too will the functions $f + g$, $f - g$, fg and f/g, the latter if in addition $g(p_0) \neq 0$.

- Let g be a continuous function on a set X, and f a continuous function on a set Y. We take $p \in X$, such that $g(p) \in Y$. Then the function composition $f(g(p))$ is continuous at the point p.

- The points where a function is not continuous, are called points of discontinuity.

- Let x_0 be a point of discontinuity. That point x_0 of discontinuity is called a removable discontinuity if there is a limit of the function at that point, but either $f(x_0)$ is not defined, or the value of the function at that point does not match the limit of the function at that point.

- If at that point of discontinuity x_0 the function has finite but different left-side and right-side limits:
$$\lim_{x \to x_0^+} f(x) \neq \lim_{x \to x_0^-} f(x)$$
it is said that this point x_0 is a point of discontinuity of first class.

- A function is said to be continuous at the left side of a point a if

$$\lim_{x \to a^-} f(x) = f(a)$$

A function is said to be continuous at the right side of a point a if

$$\lim_{x \to a^+} f(x) = f(a)$$

- Functions $f(x) = \sin x$ and $g(x) = \cos x$ are continuous[1]:

$$\lim_{x \to 0} \sin x = \sin 0 = 0$$

---
[1] Python code at the end of this chapter, program p5e.py

$$\lim_{x \to 0} cos x = cos 0 = 1$$

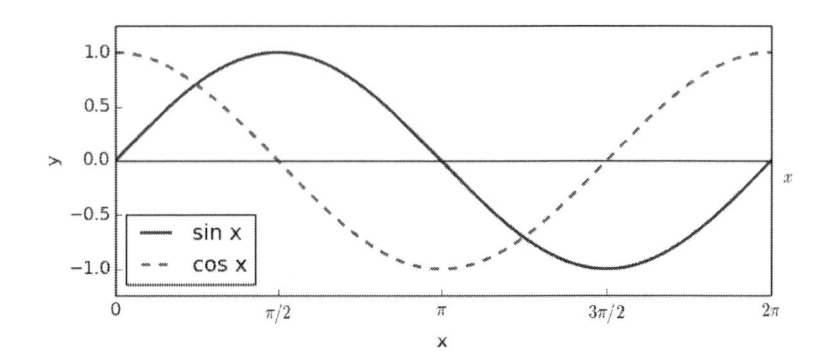

- If at a point of discontinuity x_0 the function does not have at least one side-limit, or at least one of the side-limits is infinite, it is said that this point x_0 is a point of discontinuity of second class[2].

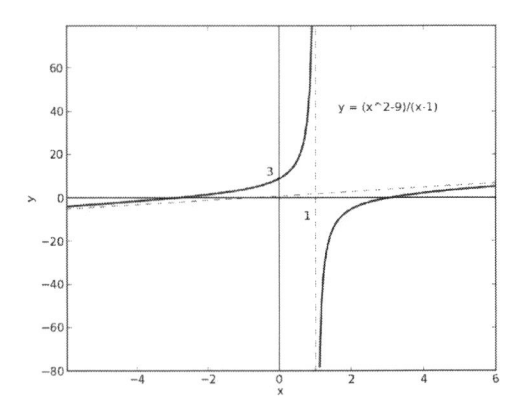

- A function is said to be piecewise continuous in the interval $[a, b]$ if it is continuous at every point of the open interval (a, b), perhaps with the exception of a finite number of points of discontinuity of first class, and in addition the function has one-side limits at the points a and b. For example, the function $f(x) = [x]$ (integer part of x) is piecewise continuous to pieces in the real line [3].

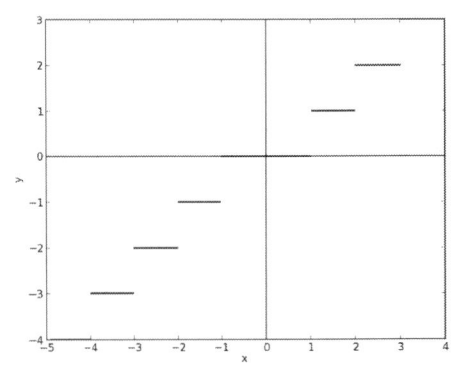

---

[2]Python code at the end of the chapter, program p5d.py
[3]Python code at the end of the chapter, program p5f.py

5.4 Continuous Functions

In this section we'll look at some properties of the continuous functions, as well as some important continuous functions.

1. Squeeze theorem.

    Let's take three functions f, g, h, a real number $\delta > 0$ and a point x_0. If for every point x such that

    $$|x - x_0| < \delta$$

    we can verify that

    $$h(x) \leq f(x) \leq g(x)$$

    Then if

    $$\lim_{x \to x_0} h(x) = \lim_{x \to x_0} g(x) = l$$

    the function f converges to the same limit l:

    $$\lim_{x \to x_0} f(x) = l$$

    To prove it, we must not that as h y g converges to l, then

    $$|h(x) - l| < \varepsilon$$

    $$|g(x) - l| < \varepsilon$$

    When the distances to the point x_0 are respectively less than δ_h and δ_g. If we take the smallest of the three distances: $\delta_{min} = min(\delta, \delta_h, \delta_g)$, we will find that when

    $$|x - x_0| < \delta_{min}$$

    the following statement will be true:

    $$l - \varepsilon < h(x) \leq f(x) \leq g(x) < l + \varepsilon$$

    and therefore $|f(x) - l| < \varepsilon$, that is to say that $f(x)$ converges to the limit l.

2. For small angles, the function $\sin x$ can be approximated through x, i.e. :

    $$\lim_{x \to 0} \frac{\sin x}{x} = 1$$

    we are going to prove it for $0 < x < \frac{\pi}{2}$. Similarly it is proven for $x \to 0^-$.

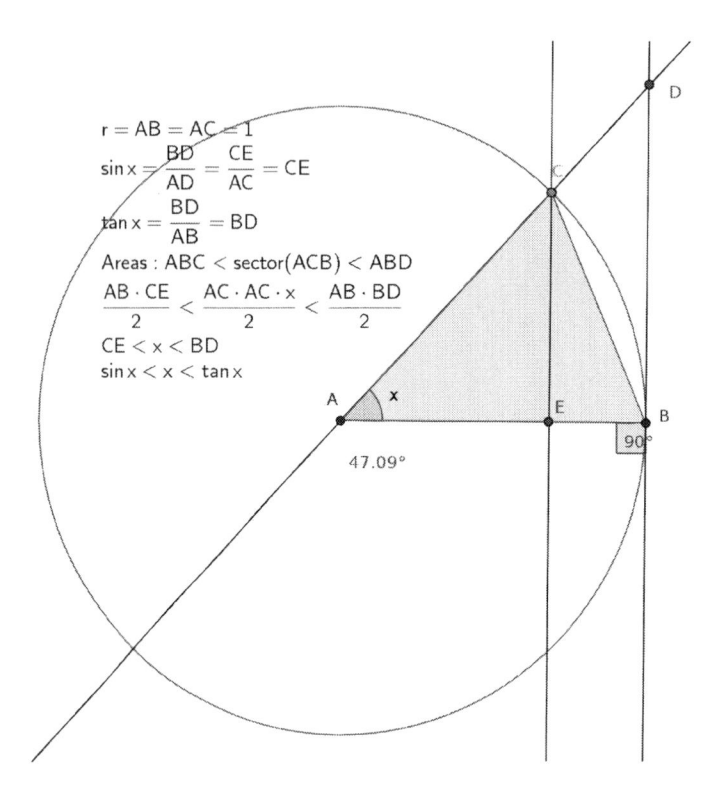

So, we have

$$\sin x < x < \tan x$$

We divide by $\sin x$:

$$1 < \frac{x}{\sin x} < \frac{1}{\cos x}$$

$$1 > \frac{\sin x}{x} > \cos x$$

and we have three functions: $g(x) = 1$; $f(x) = \frac{\sin x}{x}$; $h(x) = \cos x$, and

$$\lim_{x \to x_0} g(x) = \lim_{x \to x_0} h(x) = l$$

$$\lim_{x \to 0} 1 = \lim_{x \to x_0} cos x = 1$$

Therefore, in virtue of the Squeeze Theorem:

$$\lim_{x \to 0} f(x) = \lim_{x \to 0} \frac{\sin x}{x} = 1$$

The graph has been made using program p5g.py, which is shown at the end of the chapter:

5.4. CONTINUOUS FUNCTIONS

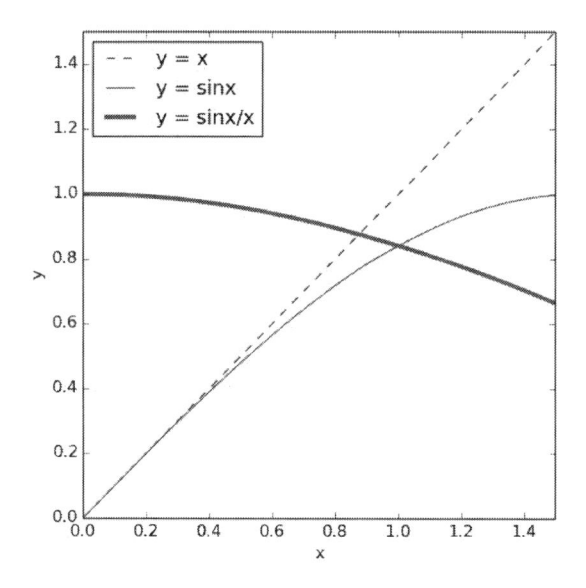

3. Theorem on the sign of a continuous function.
   Let's take a function $f(x)$ continuous at the point x_0 and such that $f(x_0) \neq 0$. Then there is a real number δ such that if $|x - x_0| < \delta$, the function $f(x)$ has the same sign that x_0.

4. First theorem of Bolzano-Cauchy.
   Let's take the function $f(x)$ continuous in the interval $[a, b]$, and such that at the endpoints of the interval, the function takes opposite signs. Then there is a point $c \in (a, b)$ where $f(c) = 0$.

5. Second theorem of Bolzano-Cauchy .
   Let's take the function $f(x)$ continues in the interval $[a, b]$, and such that at the endpoints of the interval, the values of the function are $f(a) = A$, and $f(b) = B$. Then, for any number C between A and B there is a point $c \in [a, b]$ such that $f(c) = C$.

Let's see, for example, the function $y = \frac{x \cdot \sin x}{5}$. This function is continuous on the interval $x \in [0, 10]$.

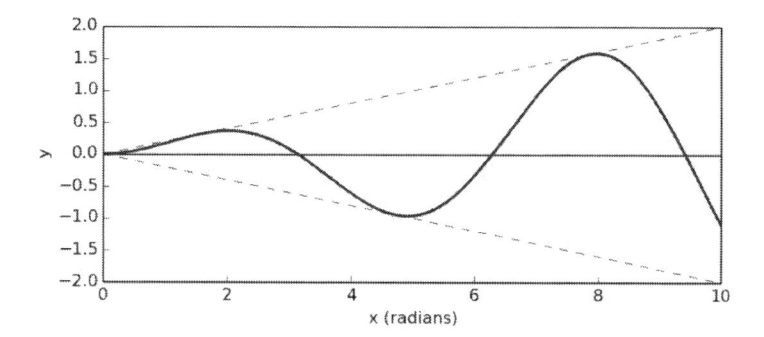

```
# -*- coding: utf-8 -*-
"""
Mathematics and Python Programming    www.pysamples.com
p5h.py
"""

import numpy as np
import matplotlib.pyplot as plt
```

```python
numpoints = 360
b = 10   # radians
x = np.linspace(0, b, numpoints)    # x * sin(x)/ 5
y = np.zeros(numpoints, float)
for i in range(0, numpoints):
    y[i] = x[i] * np.sin(x[i]) / 5
fig = plt.figure(facecolor='white')
ax = fig.add_subplot(1, 1, 1, aspect='equal')
ax.xaxis.set_ticks_position('bottom')
ax.yaxis.set_ticks_position('left')
plt.axhline(color='black', lw=1)
p1, = plt.plot(x, y, 'b', lw=2, label='y = (x sinx)/5')
plt.plot([0, b], [0, b / 5], 'k--', lw=0.5)
plt.plot([0, b], [0, -b / 5], 'k--', lw=0.5)
plt.ylabel('y')
plt.xlabel('x (radians)')
plt.show()
```

If we take $a = 5$ and $b = 8$, we see that the function takes the value $f(x) = 1 = C$ at some point of the interval $[a, b]$. The theorem says that this point exists, but it does not say how we can find that point. We can easily do this with Python. If we chose a and b conveniently, the following program finds the point c such that $f(c) = C$:

```python
# -*- coding: utf-8 -*-
"""
Mathematics and Python Programming    www.pysamples.com
p5c.py
Second theorem of Bolzano-Cauchy,
for any given C finds c such that f(c) = C
f(a) and f(b) must have opposite signs
"""

import scipy.optimize as optimize
import numpy as np

C = 1.0
#C = 0.0
a = 5
b = 8
print 'f(x) = x * sin(x) / 5'
print 'finds the point c in the interval [', a, ', ', b, ']'
print 'such that f(c) = ', C


def function(x):
    return (x * np.sin(x) / 5) - C
    #return x ** 2 - 25   # a=0 b=15
    #return np.cos(x) ** 2 + 6 - x

signo = np.sign(function(a)) * np.sign(function(b))
if signo < 0:
    c = optimize.bisect(function, a, b, xtol=1e-6)
    print 'c = ' + "%7.5f" % c
    print 'f(c) = ' + "%7.5f" % (function(c) + C)
else:
    print 'f(a) and f(b) must have opposite signs'
```

```
                                          output of the program
f(x) = x * sin(x) / 5
finds the point c in the interval [ 5 , 8 ]
such that f(c) =  1.0
c = 7.06889
f(c) = 1.00000
```

We should note that the program uses the function $y = \frac{x \cdot sen x}{5} - C$. To see an example of the first theorem, it is enough to make $C = 0$ in the previous program:

```
                                          output of the program
f(x) = x * sin(x) / 5
finds the point c in the interval [ 5 , 8 ]
such that f(c) =  0.0
c = 6.28319
f(c) = -0.00000
```

6. A function $f(x)$ continuous at a point x_0, is bounded in an neighbourhood of that point.

7. First Weierstrass theorem: If the function $f(x)$ is defined and is continuous on an interval $[a, b]$, it is bounded on that interval. Note that if we replace the closed interval by the open one, the theorem is not true. For example, the function $f(x) = \frac{1}{x}$ is continuous on the interval $(0, 1)$, but it is not bounded because $\lim_{x \to 0^+} \frac{1}{x} = +\infty$.

8. Second Weierstrass theorem: If the function $f(x)$ is defined and is continuous on an interval $[a, b]$, then there exists the points x_1, x_2 belonging to that interval such that:

$$f(x_1) = M = sup_{[a.b]}(f(x))$$

$$f(x_2) = m = inf_{[a.b]}(f(x))$$

We call M and m, the maximum and the minimum value, respectively, of the function $f(x)$ in that interval.

9. If the function $x = x(t)$ is continuous at the point t_0, and the function $y = f(x)$ is continuous at the corresponding point $x_0 = x(t_0)$, then the composite function $y = f(x(t))$ is continuous at the point t_0.

5.5 Graphics programs in this chapter

$$y = \frac{x^2 - 9}{x - 1}$$

```python
# -*- coding: utf-8 -*-
"""
Mathematics and Python Programming    www.pysamples.com
p5d.py
y = (x^2-9)/(x-1)
"""

import numpy as np
import matplotlib.pyplot as plt

numpoints = 100
asymptote = 1
epsilon = 0.1
x1 = np.linspace(-6, asymptote - epsilon, numpoints)
y1 = np.zeros(numpoints, float)
x2 = np.linspace(asymptote + epsilon, 6, numpoints)
y2 = np.zeros(numpoints, float)
#f(x) = (x^2-9)/(x-1)
```

```python
# asymptotes: y=1; y=x+1
y1[0] = (x1[0] ** 2 - 9) / (x1[0] - 1)
y2[0] = (x2[0] ** 2 - 9) / (x2[0] - 1)
for i in range(1, numpoints):
    y1[i] = (x1[i] ** 2 - 9) / (x1[i] - 1)
    y2[i] = (x2[i] ** 2 - 9) / (x2[i] - 1)
plt.plot(x1, y1, 'b', lw=2)
plt.plot(x2, y2, 'b', lw=2)
plt.plot([1, 1], [-80, 80], 'k--', lw=0.5)
plt.plot([-6, 6], [-5, 7], 'k--', lw=0.5)
plt.text(3, 40, 'y = (x^2-9)/(x-1)', horizontalalignment='center')
plt.text(-0.25, 10, '3', horizontalalignment='center')
plt.text(0.75, -10, '1', horizontalalignment='center')
plt.ylabel('y')
plt.xlabel('x')
plt.axhline(color='black', lw=1)
plt.axvline(color='black', lw=1)
plt.show()
```

$$\boxed{f(x) = \sin x \quad g(x) = \cos x}$$

```python
# -*- coding: utf-8 -*-
"""
Mathematics and Python Programming     www.pysamples.com
p5e.py
"""

import numpy as np
import matplotlib.pyplot as plt

numpuntos = 100
pi = np.pi
fig = plt.figure(facecolor='white')
x = np.linspace(0, 2 * pi, numpuntos)
ysin = np.zeros(numpuntos, float)
ycos = np.zeros(numpuntos, float)
for i in range(0, numpuntos):
    ysin[i] = np.sin(x[i])
    ycos[i] = np.cos(x[i])
ax = fig.add_subplot(1, 1, 1, aspect='equal')
p1, = plt.plot(x, ysin, 'b', lw=2, label='sin x')
p2, = plt.plot(x, ycos, 'r--', lw=2, label='cos x')
plt.legend(('sin x', 'cos x'), loc='best')
ax.xaxis.set_ticks_position('bottom')
ax.yaxis.set_ticks_position('left')
ax.autoscale_view(tight=True)
ax.set_ylim(-1.25, 1.25)
ax.set_xlim((0, 2 * pi))
ax.set_xticks([0, pi / 2, pi, 3 * pi / 2, 2 * pi])
ax.set_xticklabels(['0', r'$\pi / 2$', r'$\pi$', r'$3 \pi / 2$', r'$2\pi$'])
ax.text(2 * np.pi + .1, -.2, r'$x$')
plt.ylabel('y')
plt.xlabel('x')
plt.axhline(color='black', lw=1)
plt.show()
```

5.5. GRAPHICS PROGRAMS IN THIS CHAPTER

$$y = [x]$$

```python
# -*- coding: utf-8 -*-
"""
Mathematics and Python Programming   www.pysamples.com
p5f.py
"""

import numpy as np
import matplotlib.pyplot as plt

numpoints = 300
x = np.linspace(-4.9, 4.9, numpoints)
y = np.zeros(numpoints, int)
discontinuities = []
y[0] = np.trunc(x[0])
for i in range(1, numpoints):
    y[i] = np.trunc(x[i])
    if y[i] != y[i - 1]:
        discontinuities.append(i)
print y
print ('discontinuities:' + str(discontinuities))
j = 0
start = 0
end = 0
while (j < (numpoints - 1)):
    if (y[j + 1] != y[j]):
        end = j
        print ('from ' + str(start) + ' to ' + str(end) + ': y = ' + str(y[j]))
        plt.plot([x[start], x[end]], [y[j], y[j]], 'b', lw=2)
        start = j + 1
    j += 1
plt.ylabel('y')
plt.xlabel('x')
plt.axhline(color='black', lw=1)
plt.axvline(color='black', lw=1)
plt.show()
```

$$y = \frac{sinx}{x}$$

```python
# -*- coding: utf-8 -*-
"""
Mathematics and Python Programming   www.pysamples.com
p5g.py
sin(x)/x
"""

import numpy as np
import matplotlib.pyplot as plt

numpoints = 100
x = np.linspace(0.001, 1.5, numpoints)
y = np.zeros(numpoints, float)
sin = np.zeros(numpoints, float)
for i in range(0, numpoints):
```

```python
    sin[i] = np.sin(x[i])
    y[i] = sin[i] / x[i]
    #print x[i], y[i]
fig = plt.figure(facecolor='white')
ax = fig.add_subplot(1, 1, 1, aspect='equal')
ax.autoscale_view(tight=True)
ax.set_ylim(0, 1.5)
ax.set_xlim((0, 1.5))
p1, = plt.plot(x, x, 'b--', lw=1, label='y = x')
p2, = plt.plot(x, sin, 'g', lw=1, label='y = sinx')
p3, = plt.plot(x, y, 'r', lw=3, label='y = sinx/x')
plt.legend(('y = x', 'y = sinx', 'y = sinx/x'), loc='best')
plt.ylabel('y')
plt.xlabel('x')
plt.show()
```

$$y = x^2$$

```python
# -*- coding: utf-8 -*-
"""
Mathematics and Python Programming
    www.pysamples.com
p5i.py
y = x**2
"""

import numpy as np
import matplotlib.pyplot as plt

numpoints = 50
x = np.linspace(-2.0, 2.0, numpoints)
y = np.zeros(numpoints, float)
for i in range(0, numpoints):
    y[i] = x[i] ** 2
plt.plot(x, y)
plt.ylabel('y')
plt.xlabel('x')
x = np.linspace(0.75, 1.25, 10)
y = np.zeros(10, float)
ceros = np.zeros(10, float)
for i in range(0, 10):
    y[i] = x[i] ** 2
plt.plot(x, y, 'ro')
plt.plot(x, ceros, 'ro')
plt.ylabel('y')
plt.xlabel('x')
# vertical line
plt.plot([x[0], x[0]], [ceros[0], y[0]], 'k--', lw=0.5)
plt.plot([x[9], x[9]], [ceros[9], y[9]], 'k--', lw=0.5)
plt.show()
```

Bibliography for this chapter: [9], [17], [24], [31], [42], [47], [50], [53], [55], [56]

6 | Conics

Conics are the curves obtained as the intersection of a plane with a cone. Algebraically, the equation of a conic is as follows:

$$a_{20}x^2 + a_{11}xy + a_{02}y^2 + a_{10}x + a_{01}y + a_{00} = 0$$

The following expressions are called invariants of the conics, and they do not change when referring the equation of the conic to different axes, i.e. they do not vary by rotating the axes or moving them in parallel. There are three invariants:

- $I = a_{20} + a_{02}$

- $A_{33} = \begin{vmatrix} a_{20} & a_{11}/2 \\ a_{11}/2 & a_{02} \end{vmatrix} = \frac{1}{4} \begin{vmatrix} 2a_{20} & a_{11} \\ a_{11} & 2a_{02} \end{vmatrix} = a_{20}a_{02} - \frac{a_{11}^2}{4}$

- $\Delta = \begin{vmatrix} a_{20} & a_{11}/2 & a_{10}/2 \\ a_{11}/2 & a_{02} & a_{01}/2 \\ a_{10}/2 & a_{01}/2 & a_{00} \end{vmatrix} = \frac{1}{8} \begin{vmatrix} 2a_{20} & a_{11} & a_{10} \\ a_{11} & 2a_{02} & a_{01} \\ a_{10} & a_{01} & 2a_{00} \end{vmatrix}$

6.1 Degenerate conics

We'll start with the case that the discriminant $\Delta = 0$. In this case the conic consists of straight lines, and it is called degenerate conic. The following diagram shows the different possibilities:

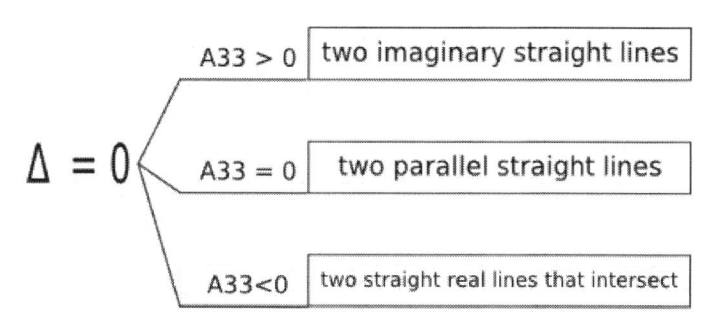

$\Delta = 0$
- A33 > 0 : two imaginary straight lines
- A33 = 0 : two parallel straight lines
- A33 < 0 : two straight real lines that intersect

The following Python program determines the type of conic if its discriminant $\Delta = 0$:

```
# -*- coding: utf-8 -*-
"""
Mathematics and Python Programming    www.pysamples.com
p6a.py
"""

import numpy as np

# example
c20 = 0.25
c11 = 1
```

```python
c02 = 1
c10 = -0.5
c01 = -1
c00 = -0.75


print c20, 'x^2 + ', c11, 'xy + ', c02, 'y^2 + ', c10, 'x + ', c01, ' y + ', c00, ' = 0'
A = 0.5 * np.array([[2 * c20, c11, c10],
                    [c11, 2 * c02, c01],
                    [c10, c01, 2 * c00]])
print A
disc = np.linalg.det(A)
print '|A| = ', "%.4f" % disc
rangoA = np.linalg.matrix_rank(A)
print 'rank(A) = ', rangoA
A33 = 0.5 * np.array([[2 * c20, c11], [c11, 2 * c02]])
print 'A33 = ', A33
det33 = np.linalg.det(A33)
print '|A33| = ', det33
rank33 = np.linalg.matrix_rank(A33)

# ------------ delta = 0 ------------
if rangoA != 3:  # delta = 0
    if rank33 == 2:  # det33 != 0
        if det33 > 0:  # 4 -2 1 -14 2 13
            print 'the equation represents two complex conjugated straight lines'
        if det33 < 0:  # -3 -2 1 7 -1 -2
            print 'the equation represets two intersecting straight lines'
    if rank33 != 2:  # det33 = 0
        print 'the equation represents two parallel or coincident straight lines,'
        print 'which can be real or complex'
    m = np.roots([1, c11, c20])
    if (m[0] - m[1]) != 0:
        print 'slopes of the lines: ', m
        n2 = (c10 + m[1] * c01) / (m[0] - m[1])
        n1 = -(n2 + c01)
        print 'n1, n2 = ', n1, ', ', n2
        print 'the two straight lines are:'
        print 'y = ', m[0], 'x + ', n1
        print 'y = ', m[1], 'x + ', n2
else:
    print 'The discriminant is not null'
```

--------------------- several outputs of the program ---------------------
```
4 x^2 +  -2 xy +  1 y^2 +  -14 x +  2  y +  13  = 0
[[  4.  -1.  -7.]
 [ -1.   1.   1.]
 [ -7.   1.  13.]]
|A| =  -0.0000
rank(A) =  2
A33 =  [[ 4. -1.]
 [-1.  1.]]
|A33| =  3.0
the equation represents two complex conjugated straight lines
slopes of the lines:  [ 1.+1.73205081j  1.-1.73205081j]
n1, n2 =  (-1-3.46410161514j) ,  (-1+3.46410161514j)
the two straight lines are:
y =  (1+1.73205080757j) x +  (-1-3.46410161514j)
y =  (1-1.73205080757j) x +  (-1+3.46410161514j)


0.25 x^2 +  1 xy +  1 y^2 +  -0.5 x +  -1  y +  -0.75  = 0
[[ 0.25  0.5  -0.25]
 [ 0.5   1.   -0.5 ]
```

```
    [-0.25 -0.5  -0.75]]
|A| =  0.0000
rank(A) =  2
A33 =  [[ 0.25  0.5 ]
 [ 0.5   1.  ]]
|A33| =  0.0
the equation represents two parallel or coincident straight lines,
which can be real or complex

0 x^2 +   2 xy +  1 y^2 +  -6 x +  -8 y +  15  = 0
[[  0.   1.  -3.]
 [  1.   1.  -4.]
 [ -3.  -4.  15.]]
|A| =  -0.0000
rank(A) =  2
A33 =  [[ 0.  1.]
 [ 1.  1.]]
|A33| =  -1.0
the equation represets two intersecting straight lines
slopes of the lines:  [-2.  0.]
n1, n2 =  5.0 ,  3.0
the two straight lines are:
y =  -2.0 x +   5.0
y =   0.0 x +   3.0
```

6.2 Ellipses and Hyperbolas

We turn now to the conics whose discriminant is not null. In this case, the type of conic is determined by the value of the invariant

$$A_{33} = \begin{vmatrix} a_{20} & a_{11}/2 \\ a_{11}/2 & a_{02} \end{vmatrix} = \frac{1}{4}\begin{vmatrix} 2a_{20} & a_{11} \\ a_{11} & 2a_{02} \end{vmatrix} = a_{20}a_{02} - \frac{a_{11}^2}{4}$$

The general equation

$$a_{20}x^2 + a_{11}xy + a_{02}y^2 + a_{10}x + a_{01}y + a_{00} = 0$$

will have its coefficient $a_{11} \neq 0$ if the axes of the conical do not match the coordinates axes, but it can be converted into another more simple equation, by rotating the axes of the conic so that they match the coordinates axes. To do this it is necessary to rotate the axes an angle:

$$tan 2\theta = \frac{a_{11}}{a_{20} - a_{02}}$$

And by replacing x and e by:

$$x = x'\cos\theta - y'\sin\theta \qquad y = x'\sin\theta + y'\cos\theta$$

so that now we will represent the conic so that their axes are parallel to the coordinates axes, and its equation will be as follows:

$$\lambda_1 x'^2 + \lambda_2 y'^2 + c_{10}x' + c_{01}y' + c_{00} = 0$$

coefficients λ_1 and λ_2 can be obtained using the equation:

$$(a_{20} - \lambda)(a_{02} - \lambda) - \frac{a_{11}^2}{4} = 0$$

$$a_{20}a_{02} - \lambda(a_{20} + a_{02}) + \lambda^2 - \frac{a_{11}^2}{4} = 0$$

$$\lambda^2 - I\lambda + A_{33} = 0$$

$$\lambda = \frac{I \pm \sqrt{I^2 - 4A_{33}}}{2}$$

Where it is clear that the product $\lambda_1 \lambda_2 = A_{33}$. There can be several cases: if $A_{33} \neq 0$, we can replace the values of λ_1 and λ_2 in the equation and this is reduced to the form

$$\lambda_1 x^2 + \lambda_2 y^2 + k = 0$$

where $k = a_{00} - \frac{a_{10}^2}{2\lambda_1^2} - \frac{a_{01}^2}{2\lambda_2^2}$, the values of x and y are referred to the coordinates axes and the curve center is at the origin of coordinates. As we shall see, the equation will represent an ellipse if $A_{33} > 0$, or a hyperbola if $A_{33} < 0$. The case that $A_{33} = 0$ corresponds to a parabola, and it will be addressed in the next section. In this section, we will see therefore the cases in which $\Delta \neq 0$ and $A_{33} \neq 0$:

1. $A_{33} > 0$, therefore λ_1 and λ_2 have equal signs.

   (a) $k < 0$.
   In this case the equation represents an ellipse

   $$\frac{x^2}{a^2} + \frac{y^2}{b^2} = 1$$

   where the coefficients a and b correspond to the major and minor axes, and are equivalent to:

   $$a = \sqrt{\frac{-k}{\lambda_1}} \qquad b = \sqrt{\frac{-k}{\lambda_2}}$$

   If $a = b$ the ellipse is in reality a circle of radius a. The parameter $c^2 = a^2 - b^2$ is the distance between each focus and the center of the ellipse. The eccentricity e of the ellipse is defined as $e = \frac{c}{a}$, and must to be $0 < e < 1$, and in the case of a circle, $e = 0$. When the eccentricity is very small, the foci are very near each other, in comparison with the dimension of the major axis.

   For example, the Moon describes an ellipse in one of the foci is the center of mass of the Earth-moon system, which is located approximately 4800 km from the center of the Earth. The following program represents the ellipse that describes the moon around the Earth: it is almost circular as its eccentricity is $e = 0.0549$. A small blue circle represents Earth.

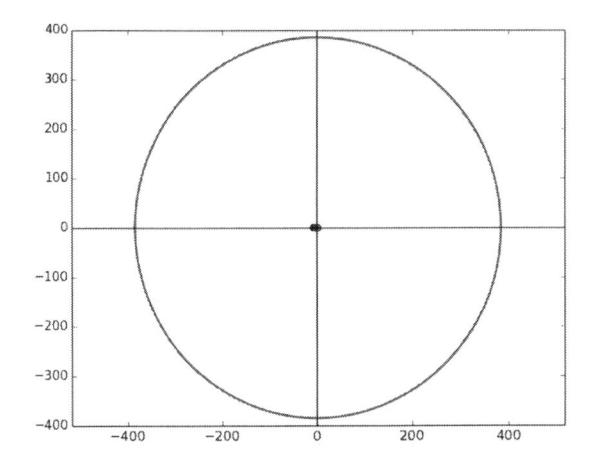

---------- output of the program ----------
```
a = 384.748
b = 384.746
c = 1.160
e = 0.0549
```

6.2. ELLIPSES AND HYPERBOLAS

```
# -*- coding: utf-8 -*-
"""
Mathematics and Python Programming    www.pysamples.com
p6b.py
"""

import numpy as np
import matplotlib.pyplot as plt
from matplotlib.patches import Ellipse

fig = plt.figure()
ax = plt.gca()
#Moon orbit (distances in thousands of kilometers)
a = 384.748
e = 0.0549
# e2 = (a2 - b2) / a; b= sqrt(a2 - ae2)
b = np.sqrt(a ** 2 - a * e ** 2)
c = a ** 2 - b ** 2
f = [[-c, 0], [c, 0]]
print 'a = ', "%.3f" % a
print 'b = ', "%.3f" % b
print 'c = ', "%.3f" % c
print 'e = ', "%.4f" % e
#graphic
elipse = Ellipse(xy=(0, 0), width=2 * a, height=2 * b, edgecolor='r', fc='None', lw=2)
ax.add_patch(elipse)
circle = plt.Circle((-6.371, 0), radius=6.371, fc='b')
plt.gca().add_patch(circle)
plt.plot(f[0][0], f[0][1], 'ro')
plt.plot(f[1][0], f[1][1], 'ro')
plt.axhline(color='black', lw=1)
plt.axvline(color='black', lw=1)
plt.axis('equal')
plt.show()
```

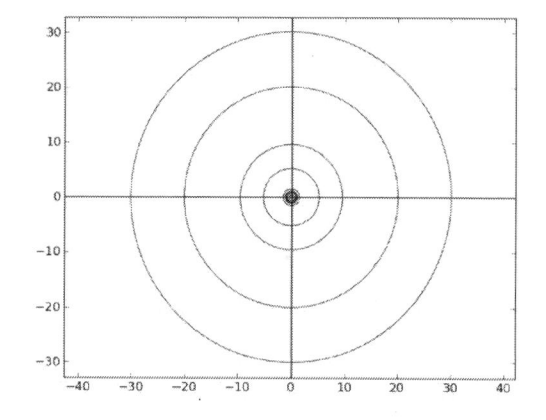

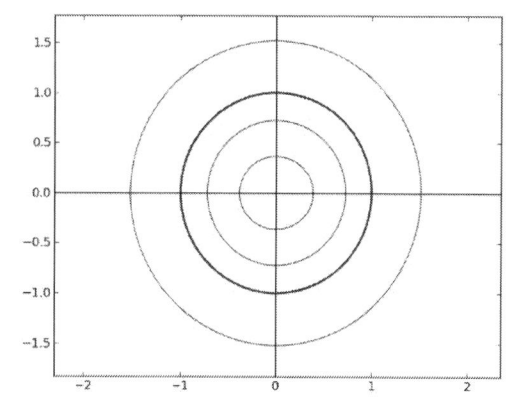

All the planets in the solar system have small eccentricities, between 0.007 for Venus, and 0,206 for Mercury. If you slightly alters the previous program and the orbits data of the planets, the graphs shown above are obtained. The right image is an extension to display the four planets closest to the Sun. The orbit of the Earth is in blue thicker stroke. Distances are in A.U. (Astronomical Units, 1 UA is equal to the average distance between the Earth and the Sun).

Sommerfeld modified Bohr atomic structure model by introducing the hypothesis that the orbits of the electrons were elliptical rather than circular. Each orbit is given by two quantum numbers: n, which determines the value of the mayor semi-axis a; and the quantum number $l = 0, 1, ..., n-1$, that determines the value of the minor semi-axis b. The following Python program represents the orbits of the model of Sommerfeld for $n = 4$:

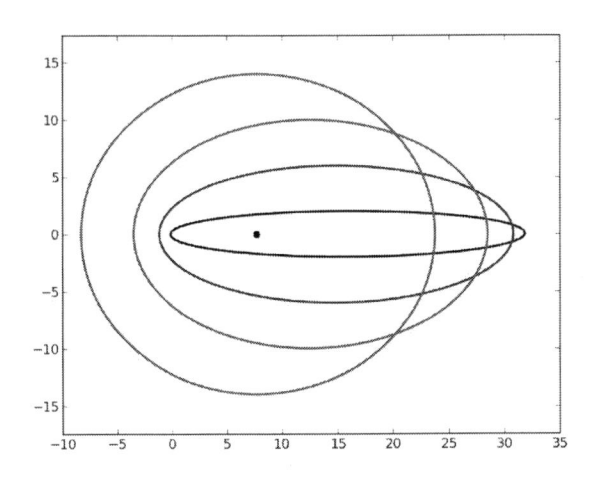

```
# -*- coding: utf-8 -*-
"""
Mathematics and Python Programming   www.pysamples.com
p6c.py
Sommerfeld atomic model orbits for n=4
"""

import numpy as np
import matplotlib.pyplot as plt
from matplotlib.patches import Ellipse

fig = plt.figure()
ax = plt.gca()

#orbits for level n=4, as a0 multiples [s p d f]
a4 = [16, 16, 16, 16]
b4 = [2, 6, 10, 14]
colors = ['k', 'b', 'g', 'r']

for i in range(0, 4):
    a = a4[i]
    b = b4[i]
    c = np.sqrt(a ** 2 - b ** 2)
    e = c / a
    f = [[-c, 0], [c, 0]]
    color = colors[i]
    elipse = Ellipse(xy=(c, 0), width=2 * a, height=2 * b,
                     edgecolor=color, fc='None', lw=2)
    ax.add_patch(elipse)

plt.plot(c, 0, 'ko')
plt.axis('equal')
plt.show()
```

(b) $k > 0$.
In this case the equation represents an imaginary ellipse whose equation is

$$\frac{x^2}{a^2} + \frac{y^2}{b^2} = -1$$

as there is no real number that satisfies this equation. The coefficients a and b correspond to the major and minor axis, and are equivalent to:

$$a = \sqrt{\frac{k}{\lambda_1}} \qquad b = \sqrt{\frac{k}{\lambda_2}}$$

(c) $k = 0$.
In this case the equation represents a degenerate ellipse, which is reduced to a single point: the origin of coordinates, because

$$\frac{x^2}{a^2} + \frac{y^2}{b^2} = 0$$

where the coefficients a and b are equivalent to:

$$a = \sqrt{\frac{1}{|\lambda_1|}} \qquad b = \sqrt{\frac{1}{|\lambda_2|}}$$

The following Python program solves the three cases in which $A_{33} > 0$:

```
# -*- coding: utf-8 -*-
"""
Mathematics and Python Programming    www.pysamples.com
p6d.py
"""

import numpy as np
import matplotlib.pyplot as plt
from matplotlib.patches import Ellipse

fig = plt.figure()
ax = plt.gca()
print 'A x  + B y  + C = 0'
print 'write A, B y C separated by one blank space:'
strdata = raw_input()
data = map(float, strdata.split())
print data
signs = np.sign(data)
if signs[0] == -1 and signs[1] == -1:
    data = np.multiply(-1.0, data)
    print data
t1 = data[0]
t2 = data[1]
k = data[2]
D = t1 * t2
print 'A33 = ', D
if D > 0:
    if np.sign([k]) == -1:
        a = np.sqrt(-k / t1)
        b = np.sqrt(-k / t2)
        c = np.sqrt(abs(a ** 2 - b ** 2))
        e = c / max(a, b)
```

```
    print 'a = ', "%.3f" % a, '; b = ', "%.3f" % b, '; c = ', "%.3f" % c
    print 'excentricity: e = ', "%.4f" % e
    J = np.pi * (3 * (a + b)) - np.sqrt((3 * a + b) * (a + 3 * b))  # Ramanujan
    area = np.pi * a * b
    if a > b:
        f = [[-c, 0], [c, 0]]
        curve = 'horizontal ellipse'
        perimeter = J
    if a == b:
        f = [[0, 0], [0, 0]]
        curve = 'circumference or radius R = ' + "%.3f" % a
        perimeter = 2 * np.pi * a
    if a < b:
        f = [[0, -c], [0, c]]
        curve = 'vertical ellipse'
        perimeter = J
    print curve
    print 'perimeter: ', "%.3f" % perimeter
    print 'inside area: ', "%.3f" % area
    print 'x ** / ', "%.3f" % a ** 2, ') + (y ** 2 / ', "%.3f" % b ** 2, ') = 1'
    ellipse = Ellipse(xy=(0, 0), width=2 * a, height=2 * b,
                      edgecolor='r', fc='None', lw=2)
    ax.add_patch(ellipse)
    plt.plot(f[0][0], f[0][1], 'ro')
    plt.plot(f[1][0], f[1][1], 'ro')
    plt.axhline(color='black', lw=1)
    plt.axvline(color='black', lw=1)
    plt.axis('equal')
    plt.show()
    if np.sign([k]) == 1:
        a = np.sqrt(k / t1)
        b = np.sqrt(k / t2)
        c = np.sqrt(abs(a ** 2 - b ** 2))
        print 'a, b, c: ', "%.3f" % a, "%.3f" % b, "%.3f" % c
        print 'imaginary ellipse'
    if np.sign([k]) == 0:
        print 'degenerated ellipse: one point'
else:
    print 'A33 <= 0'
```

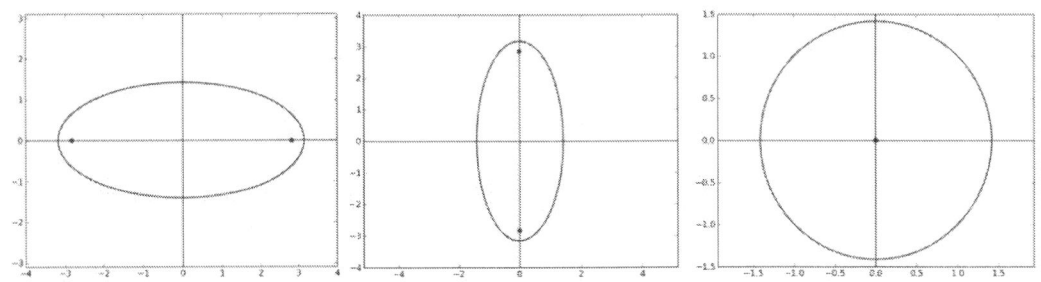

———————————— several outputs of the program ————————————

```
A x² + By² + C = 0
write A, B y C separated by one blank space:
1 5 -10
[1.0, 5.0, -10.0]
A33 =  5.0
a =  3.162 ; b =  1.414 ; c =  2.828
excentricity: e =  0.8944
horizontal ellipse
```

6.2. ELLIPSES AND HYPERBOLAS

```
            perimeter: 34.148
            inside area: 14.050
            x ** / 10.000 ) + (y ** 2 / 2.000 ) = 1

            A x² + By² + C = 0
            write A, B y C separated by one blank space:
            -1 -5 10
            [-1.0, -5.0, 10.0]
            [ 1.   5.  -10.]
            A33 = 5.0
            a = 3.162 ; b = 1.414 ; c = 2.828
            excentricity: e = 0.8944
            horizontal ellipse
            perimeter: 34.148
            inside area: 14.050
            x ** / 10.000 ) + (y ** 2 / 2.000 ) = 1

            A x² + By² + C = 0
            write A, B y C separated by one blank space:
            5 5 -10
            [5.0, 5.0, -10.0]
            A33 = 25.0
            a = 1.414 ; b = 1.414 ; c = 0.000
            excentricity: e = 0.0000
            circunference of radius R = 1.414
            perimeter: 8.886
            inside area: 6.283
            x ** / 2.000 ) + (y ** 2 / 2.000 ) = 1

            A x² + By² + C = 0
            write A, B y C separated by one blank space:
            1 5 10
            [1.0, 5.0, 10.0]
            A33 = 5.0
            a, b, c: 3.162 1.414 2.828
            imaginary ellipse

            A x² + By² + C = 0
            write A, B y C separated by one blank space:
            5 10 0
            [5.0, 10.0, 0.0]
            A33 = 50.0
            degenerated ellipse: one point
```

2. $A_{33} < 0$ and $k \neq 0$. In this case the equation represents a hyperbola whose equation is

$$\frac{x^2}{a^2} - \frac{y^2}{b^2} = 1$$

As was the case with the ellipse, the symmetry axes coincide with the coordinate axes, and the center of symmetry is the origin of coordinates. The axis OX cuts the hyperbola in two points that are called vertices, and this axis is called real axis. The axis OY is called imaginary axis. If we call now: $c^2 = a^2 + b^2$, the points located on the real axis at a distance $\pm c$ of the origin are called foci of the hyperbola. If k was null, the discriminant would be $\Delta = 0$ and this would be in the case we have already seen consisting of two intersecting straight lines. The eccentricity is defined as $e = \frac{c}{a} > 1$; if the eccentricity is large, the branches of the hyperbola are almost flat; if the eccentricity is close to 1, the branches are more pointy.

For the first quadrant, the equation of the hyperbola can be rewritten as

$$y = \frac{b}{a}\sqrt{x^2 - a^2} = \frac{b}{a}x - \frac{ab}{x + \sqrt{x^2 - a^2}}$$

and when $x \to \infty$ the second term tends to zero, and the hyperbola approaches asymptotically to the straight line $y = \frac{b}{a}x$. The same fact is found similarly in the other quadrants, so that the straight lines

$$y = \frac{b}{a}x \qquad y = \frac{-b}{a}x$$

are the asymptotes of the hyperbola. The equation of the hyperbola taking coordinates axes as their asymptotes. If we make the following change of coordinates:

$$x' = \frac{x}{a} - \frac{y}{b}, \qquad y' = \frac{x}{a} + \frac{y}{b}$$

we will have

$$\frac{x^2}{a^2} - \frac{y^2}{b^2} = 1$$

$$\left(\frac{x}{a} + \frac{y}{b}\right)\left(\frac{x}{a} + \frac{y}{b}\right) = 1$$

$$y'x' = 1 \qquad y' = \frac{1}{x'}$$

where the coordinates x' and y', are referred to the asymptotes of the parabola, i.e. that the coordinates axes are the asymptotes. The following Python program represents the hyperbola from its expression as $\lambda_1 x^2 + \lambda_2 y^2 + k = 0$, situated so that its axis coincides with the OX axis.

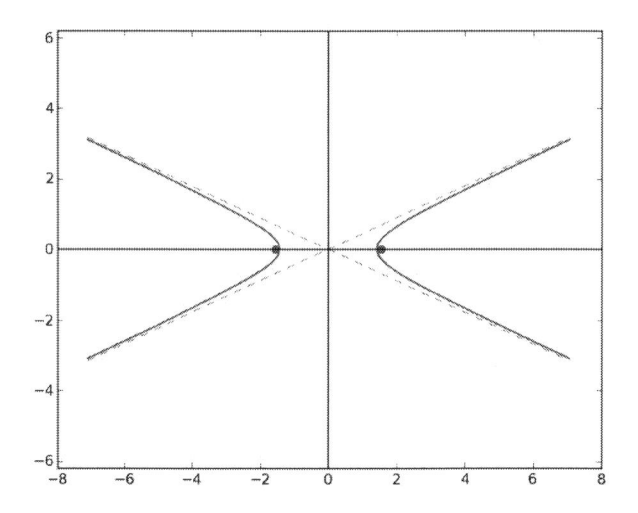

---------- output of the program ----------
```
A x² + By² + C = 0
write A, B y C separated by one blank space::
-1 5 2
[-1.0, 5.0, 2.0]
[ 1. -5. -2.]
A33 =  -5.0
[ 1. -5. -2.]
a =   1.414 ; b =   0.632 ; c =   1.549
asimptotes:
y =   0.447 x; y =  -0.447 x
vertex: ( -1.414 , 0), ( 1.414 , 0)
excentricity:  1.095
```

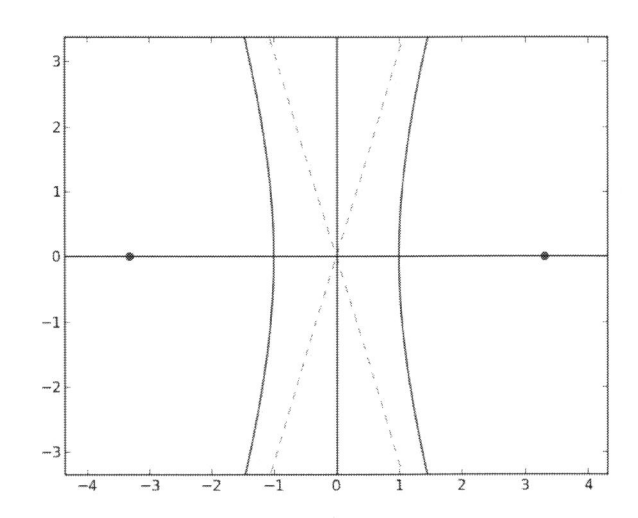

6.2. ELLIPSES AND HYPERBOLAS

─────────────── output of the program ───────────────
```
A x² + By² + C = 0
write A, B y C separated by one blank space::
5 -0.5 5
[5.0, -0.5, 5.0]
A33 =  -2.5
[5.0, -0.5, 5.0]
a =  1.000 ; b =  3.162 ; c =  3.317
asimptotes:
y =  3.162 x; y =  -3.162 x
vertex: ( -1.000 , 0), ( 1.000 , 0)
excentricidad:  3.317
```
──

```python
"""
Mathematics and Python Programming    www.pysamples.com
p6e.py
Determines the type of hyperbola and plots it if the conic has A33<0
"""

import numpy as np
import matplotlib.pyplot as plt

fig = plt.figure()
ax = plt.gca()

print 'A x   + By   + C = 0'
print 'write A, B y C separated by one blank space:'
strdata = raw_input()
data = map(float, strdata.split())
print data
signos = np.sign(data)

if signos[0] == -1:
    data = np.multiply(-1.0, data)
    print data
t1 = data[0]
t2 = data[1]
k = data[2]
D = t1 * t2
print 'A33 = ', D
if D < 0 and k != 0:
    t1 = data[0]
    t2 = data[1]
    k = data[2]
    print data
    a = np.sqrt(abs(k / t1))
    b = np.sqrt(abs(k / t2))
    c = np.sqrt(a ** 2 + b ** 2)
    print 'a = ', "%.3f" % a, '; b = ', "%.3f" % b, '; c = ', "%.3f" % c
    asymptote = b / a
    print 'asimptotes: '
    print 'y = ', "%.3f" % asymptote, 'x; y = ', "%.3f" % -asymptote, 'x'
    vertex = a
    print 'vertex: (', "%.3f" % (-1 * vertex), ', 0), (', "%.3f" % vertex, ', 0)'
    e = c / a
    print 'excentricity: ', "%.3f" % e
    y = np.zeros(200, float)
    vertex = a
    if e < 2:
        rank = 5 * vertex
```

```
        else:
            rank = 1.5 * vertex
        x = np.linspace(vertex, rank, 200)
        for i in range(0, 200):
            y[i] = asymptote * np.sqrt(x[i] ** 2 - a ** 2)
        plt.plot(-c, 0, 'ro')
        plt.plot(c, 0, 'ro')

        #graphic
        plt.plot(x, y, 'r-', lw=1.5)
        plt.plot(x, -y, 'r-', lw=1.5)
        plt.plot(-x, y, 'r-', lw=1.5)
        plt.plot(-x, -y, 'r-', lw=1.5)
        plt.plot([-rank, rank], [-rank * asymptote, rank * asymptote], color='grey', ls='-', lw=0.8)
        plt.plot([-rank, rank], [rank * asymptote, -rank * asymptote], color='grey', ls='-', lw=0.8)
        plt.axhline(color='black', lw=1)
        plt.axvline(color='black', lw=1)
        plt.axis('equal')
        plt.show()
    else:
        print 'A33 >= 0 o k=0'
```

6.3 Parabolas

If $\Delta \neq 0$ y $A_{33} = 0$, it is a real parabola. In this case, we cannot follow the path we followed for the cases of the ellipse and the hyperbola, but if we make one rotation of the axes so that the axis of the parabola is parallel to the axis X, and then we take as the center of coordinates, the vertex of the parable, the equation becomes reduced to:

$$Iy^2 \pm 2\sqrt{\frac{-\Delta}{I}}x = 0$$

$$y^2 = \pm 2\sqrt{\frac{-\Delta}{I^3}}x = 2px$$

being

$$p = \pm\sqrt{\frac{-\Delta}{I^3}} = \pm\sqrt{\frac{-\Delta}{(a_{20} + a_{02})^3}}$$

the coordinates of the focus of the parabola $\left(\frac{p}{2}, 0\right)$. The following Python program represents the conic in the case that it is a parabola:

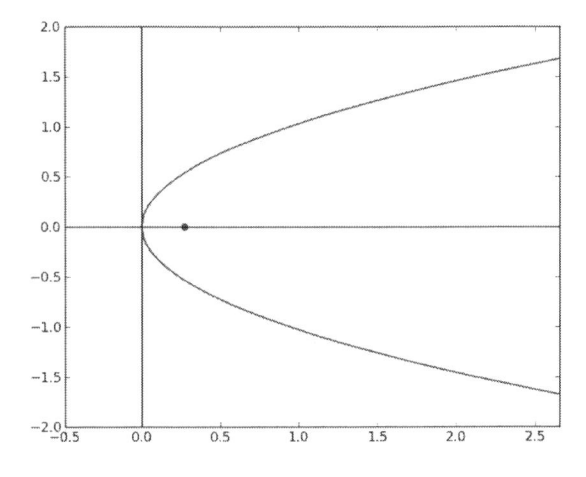

6.3. PARABOLAS

```
―――――――――――――――――――――――― output of the program ――――――
1 x^2 +   2 xy +   1 y^2 +   1 x +   -2 y +   1  = 0
[[ 1.   1.   0.5]
 [ 1.   1.  -1. ]
 [ 0.5 -1.   1. ]]
|A| =  -2.2500
rango(A) =  3
A33 =  [[ 1.  1.]
 [ 1.  1.]]
|A33| =  0.0
the equation represents a real parabola
y ** 2 = 2 *  0.530 x
focus: ( 0.265 , 0)
```

```python
# -*- coding: utf-8 -*-
"""
Mathematics and Python Programming    www.pysamples.com
p6f.py
Determines the parabola and plots it: discriminant != 0 and A33=0
"""

import numpy as np
import matplotlib.pyplot as plt

# example
c20 = 1
c11 = 2
c02 = 1
c10 = 1
c01 = -2
c00 = 1

print c20, 'x^2 + ', c11, 'xy + ', c02, 'y^2 + ', c10, 'x + ', c01, ' y + ', c00, ' = 0'
A = 0.5 * np.array([[2 * c20, c11, c10], [c11, 2 * c02, c01], [c10, c01, 2 * c00]])
print A
disc = np.linalg.det(A)
print '|A| = ', "%.4f" % disc
rankA = np.linalg.matrix_rank(A)
print 'rank(A) = ', rankA
A33 = 0.5 * np.array([[2 * c20, c11], [c11, 2 * c02]])
print 'A33 = ', A33
det33 = np.linalg.det(A33)
print '|A33| = ', det33
rank33 = np.linalg.matrix_rank(A33)

if rankA == 3:    # disc != 0
    if rank33 != 2:  # det33 = 0
        print 'the equation represents a real parabola'
        I = c20 + c02
        p = np.sqrt(-disc / I ** 3)
        print 'y ** 2 = 2 * ', "%.3f" % p, 'x'
        print 'focus: (', "%.3f" % (p / 2), ', 0)'
        y = np.zeros(200, float)
        x = np.linspace(0, 5 * p, 200)
        for i in range(0, 200):
            y[i] = np.sqrt(2 * p *x[i])
        plt.plot(x, y, 'r-', lw=1.5)
        plt.plot(x, -y, 'r-', lw=1.5)
        plt.plot(p / 2, 0, 'ro')
    #    plt.plot(-x, y, 'r-', lw=1.5)
```

```
#       plt.plot(-x, y, 'r-', lw=1.5)
        plt.axhline(color='black', lw=1)
        plt.axvline(color='black', lw=1)
        plt.xlim(-0.5, max(x))
        #plt.axis('equal')
        plt.show()
    else:
        print 'A33 is not null'
else:
    print 'Discriminant is not null'
```

Parabolic mirrors have the property to reflect in the focus of the parabola all light rays that are parallel to its axis. For example, if we focus a parabolic mirror toward the sun, the rays of light can be considered parallel, and they are reflected to a single point: the focus of the parabola. Conversely, if from the focal point of the parabola light is emitted, the light is reflected on the parabola and the reflected rays are parallel to the parabola axis.

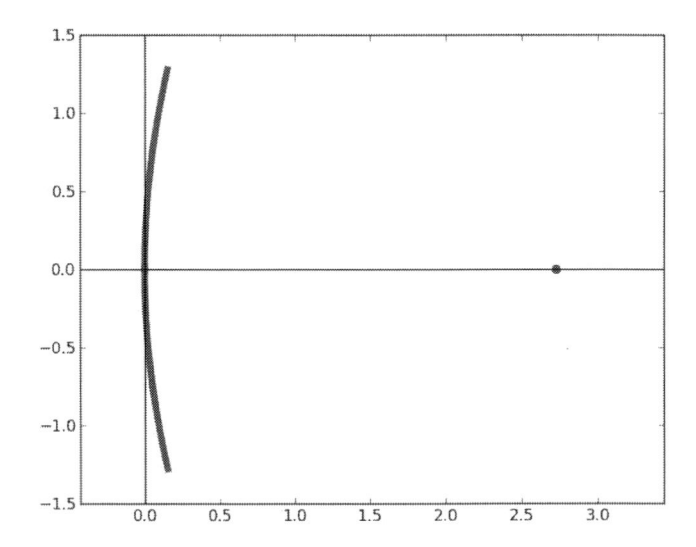

For example, with the previous program we can obtain the equation of a parabolic mirror about two meters in length, which focuses the light onto his focus, which is located to something less than three meters from mirror if we take the following parameter values:

```
c20 = 0.1
c02 = 0.5
c11 = np.sqrt(c20 * c02)
rank33 = 1
c10 = 5
c01 = 20
c00 = 30
```

In the figure the parabola has only been represented up to a width $x = 0.15m$.

```
——————————————————————— output of the program ———————————————————————
0.1 x^2 +  sqrt(0.05) xy +  0.5 y^2 +  5 x +  20  y +  30  = 0
[[ 0.1          0.1118034   2.5       ]
 [ 0.1118034    0.5        10.        ]
 [ 2.5         10.         30.        ]]
|A| =  -6.4098
rank(A) =  3
the equation represents a real parabola
y ** 2 = 2 *  5.447 x
focus: ( 2.724 , 0)
```

Bibliography for this chapter: [4], [21], [35], [39], [42], [53], [61]

7 | Exponential function

7.1 Power series

In the fourth chapter we saw the definition of a numerical series as the sum of an infinite number of terms of the sequence

$$x_1 + x_2 + x_3 + ... + x_k + ... = \sum_{i=1}^{\infty} x_i$$

The sum of an infinite number of terms of the sequence

$$u_1(x) + u_2(x) + u_3(x) + ... + u_n(x) + ...$$

is called a function series with respect to the variable x. If the values of the variable x, as well as the parameters of the functions $u_i(x)$ are real, the series is called real series. If the values and parameters can be complex numbers, the series is called complex series:

$$u_1(z) + u_2(z) + u_3(z) + ... + u_n(z) + ...$$

If we assign a value $x = x_0$ to the variable x, we obtain a numerical series:

$$u_1(x_0) + u_2(x_0) + u_3(x_0) + ... + u_n(x_0) + ...$$

The set of values of the variable x for which the series converges is called the field of convergence of the series. The value of the sum $S(x_0)$ of the series will depend on the value x_0 of the variable x taken.

The functions u_i can be of different forms, but at this moment we are interested in studying the series of functions whose form is the following:

$$a_0 + a_1 z + a_2 z^2 + a_3 z^3 + ... + a_n z^n + ... = \sum_{n=0}^{\infty} a_n z^n$$

Where the numbers a_j does not depend on the variable z, and are called coefficients. As with the functions series, you can have real power series, and complex power series. As in the case of the functions series, if we give a concrete value to the variable z, the power series becomes a numerical series:

$$a_0 + a_1 z_0 + a_2 z_0^2 + a_3 z_0^3 + ... + a_n z_0^n + ... = \sum_{n=0}^{\infty} a_n z_0^n$$

The series $\sum_{n=0}^{\infty} a_n z_0^n$ is absolutely convergent if the series

$$|a_0| + |a_1 z_0| + |a_2 z_0^2| + |a_3 z_0^3| + ... + |a_n z_0^n| + ... = \sum_{n=0}^{\infty} |a_n z_0^n|$$

is convergent.

Some examples of power series: $\sum z^n$, $\sum \frac{z^n}{n!}$, , $\sum \frac{z^n}{n}$, , $\sum n^n z^n$, etc. In the next chapter we will demonstrate that

$$\cos x = 1 - x^2 + \frac{x^4}{4!} - \frac{x^6}{6!} + ... + (-1)^n \frac{x^{2n}}{(2n)!} + ...$$

$$\sin x = x - \frac{x^3}{3!} + \frac{x^5}{5!} - ... + (-1)^n \frac{x^{2n+1}}{(2n+1)!} + ...$$

The following theorems of Abel describe the field of convergence of power series:

1. If the power series $\sum a_n z^n$ is convergent for some value $z = z_0$, it will also be absolutely convergent $\forall z$ such that $|z| < |z_0|$

2. If the power series $\sum a_n z^n$ is divergent for some value $z = z_0$, it will also be absolutely divergent $\forall z$ such that $|z| > |z_0|$

These theorems enable us to reach the next fundamental theorem: If $\sum a_n z^n$ is a power series which is not merely always convergent or always divergent for any value of z, then there is a number r such that $\sum a_n z^n$ converges for all value $|z| < r$, and is divergent for all $|z| > r$. The number r is called radius of convergence of the series.

The set of all complex numbers for which $|z| < r$, forms a circle of radius r in the plane of complex numbers, with its center in the point $z = 0$. This circle is called circle of convergence of the series. The series will be convergent for all points inside the circle; and will be divergent for all points outside the circle. The convergence or divergence for the z points located in the same circumference $|z| = r$ depends on the specific properties of each series. In the case of real variable x, we call it interval of convergence.

The previous theorem establishes the existence of the radius of convergence, but it does not say how to calculate it. The following theorem of Cauchy, published in 1821, and then rediscovered in 1892 by Hadamard, provides the magnitude of the radius of convergence: Let's take the power series $\sum a_n z^n$, and μ the upper limit $\mu = \overline{\lim} \sqrt[n]{|a_n|}$. Then:

1. If $\mu = 0$, the series is convergent for all the values of z.

2. If $\mu = +\infty$, the series is divergent for all the values of z.

3. If $0 < \mu < +\infty$, the power series converges absolutely for all $|z| < \frac{1}{\mu}$ and is divergent for all $|z| > \frac{1}{\mu}$.

So that we can take as radius of convergence r:

$$r = \frac{1}{\mu} = \frac{1}{\overline{\lim} \sqrt[n]{|a_n|}}$$

7.2 Exponential function $exp(z)$

Consider the function that we call $exp(z)$, and that is defined by the series:

$$exp(z) = \frac{1}{0!} + \frac{z}{1!} + \frac{z^2}{2!} + \frac{z^3}{3!} + ... + + \frac{z^n}{n!} + ... = \sum \frac{z^n}{n!}$$

We implement the Cauchy-Hadamard theorem that we have just seen:

$$\mu = \overline{\lim} \sqrt[n]{|a_n|} = \overline{\lim} \sqrt[n]{\left|\frac{z^n}{n!}\right|}$$

7.2. EXPONENTIAL FUNCTION $EXP(Z)$

In section 4.6 we proved:
$$\lim_{n \to \infty} \frac{a^n}{n!} = 0$$

and therefore we have now
$$\mu = \overline{\lim} \sqrt[n]{\left|\frac{z^n}{n!}\right|} = 0$$

and this series is convergent for all values of z.

```
# -*- coding: utf-8 -*-
"""
Mathematics and Python Programming    www.pysamples.com
p7a.py
"""

from numpy import e


def fact(x):
    if x == 0:
        return 1
    else:
        return x * fact(x - 1)


def u(i):
    termino = 1.0 / fact(i)
    return termino

x = 1.0
n = 12  # number of terms to calculate
s = 0
for i in range(0, n + 1):
    s += u(i)
    print 'S(' + str(i) + ') = ' + "%20.18f" % s

print 'real value of e: ' + "%20.18f" % e
print 'error: ' + "%10.8g" % (e - s)
```

```
────────────────────── output of the program ──────────────────────
S(0)  = 1.000000000000000000
S(1)  = 2.000000000000000000
S(2)  = 2.500000000000000000
S(3)  = 2.666666666666666519
S(4)  = 2.708333333333333037
S(5)  = 2.716666666666666341
S(6)  = 2.718055555555555447
S(7)  = 2.718253968253968367
S(8)  = 2.718278769841270037
S(9)  = 2.718281525573192248
S(10) = 2.718281801146384513
S(11) = 2.718281826198492901
S(12) = 2.718281828286168711
real value of e: 2.718281828459045091
error: 1.7287638e-10
0.141859054565  ms
```

We represent the functions e^x and e^{-x} through the Python program p7d.py:

```
# -*- coding: utf-8 -*-
"""
Mathematics and Python Programming    www.pysamples.com
```

```
p7d.py
represents y = e^x; y = e^-x; y=e^-(x^2)
"""

import numpy as np
import matplotlib.pyplot as plt

numpoints = 100
fig = plt.figure(facecolor='white')
x = np.linspace(-2.5, 2.5, numpoints)
y1 = np.zeros(numpoints, float)
y2 = np.zeros(numpoints, float)
y3 = np.zeros(numpoints, float)
for i in range(0, numpoints):
    y1[i] = np.exp(x[i])
    y2[i] = np.exp(-x[i])
    y3[i] = np.exp(-(x[i] ** 2))
ax = fig.add_subplot(1, 1, 1, aspect='equal')
p1, = plt.plot(x, y1, 'g', lw=2, label='$e^{x}$')
p2, = plt.plot(x, y2, 'r--', lw=2, label='$e^{-x}$')
#p3, = plt.plot(x, y3, 'b-', lw=2, label='$e^{-x^2}$')
plt.legend(('$e^{x}$', '$e^{-x}$',), loc='best')
#plt.legend(('$e^{-x^2}$',), loc='best')
ax.autoscale_view(tight=True)
plt.ylabel('y')
plt.xlabel('x')
ax.set_xlim(-2.1, 2.1)
ax.set_ylim(0, 12.5)
#ax.set_ylim(0, 1.1)
plt.axhline(color='black', lw=1)
plt.axvline(color='black', lw=1)
plt.show()
```

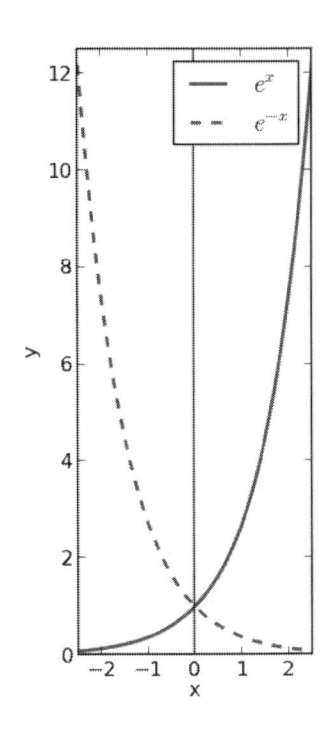

7.2. EXPONENTIAL FUNCTION $EXP(Z)$

If we modify slightly the same program p7d.py, we can obtain the graph of function $y = e^{-x^2}$:

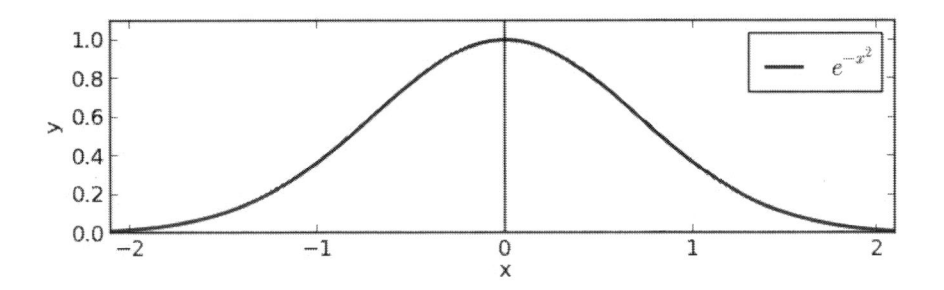

This function
$$exp(z) = \frac{1}{0!} + \frac{z}{1!} + \frac{z^2}{2!} + \frac{z^3}{3!} + ... + \frac{z^n}{n!} + ... = \sum \frac{z^n}{n!}$$
has some important properties:

1. $exp(0) = 1$.

2. $exp(z) \neq 0$, for all complex number z.

3. $exp(z + w) = exp(z) \cdot exp(w)$
   where z and w are any two complex number, because
   $$\sum_{k=0}^{\infty} \frac{z^k}{k!} \cdot \sum_{n=0}^{\infty} \frac{w^n}{n!} = \sum_{n=0}^{\infty} \frac{1}{n!} \sum_{k=0}^{n} \frac{n!}{k!(n-k)!} z^k w^{n-k} = \sum_{n=0}^{\infty} \frac{(z+w)^n}{n!}$$

4. $exp(-z) = \frac{1}{exp(z)}$
   since $1 = exp(0) = exp(z + (-z)) = exp(z) \cdot exp(-z)$.

5. $exp(z - w) = \frac{exp(z)}{exp(w)}$ as can be deduced from the properties 2 and 3.

6. $exp(k \cdot z) = (exp(z))^k$
   being $k \in \mathbf{Z}$, since if we repeatedly implement property 2 for $z = w$ we have
   $$exp(2z) = exp(z + z) = exp(z) \cdot exp(z) = (exp(z))^2$$

7. We define number $e = exp(1)$. If we implement the previous property, we have $exp(z \cdot 1) = e^z$.

The exponential function appears in many phenomena of populations growth. For example, the following program represents the historical data of the population of USA, and adjusts them to an exponential function. We have in fact adjust the data to two exponential functions, each function covering one hundred years:

```
# -*- coding: utf-8 -*-
"""
Mathematics and Python Programming    www.pysamples.com
p7b.py
"""

import numpy as np
import matplotlib.pyplot as plt
from scipy import stats

#sources: http://www.census.gov/population/www/censusdata/files/table-2.pdf
#http://www.census.gov/main/www/cen2000.html
#http://www.census.gov/2010census/popmap/
```

```python
pob1 = {1790: 3929214, 1800: 5308483, 1810: 7239881, 1820: 9638453, 1830: 12866020,
        1840: 17069453, 1850: 23191876, 1860: 31443321, 1870: 38558371, 1880: 50189209,
        1890: 62979766, 1900: 76212168}
pob2 = {1900: 76212168, 1910: 92228496, 1920: 106021537,
        1930: 123202624, 1940: 132164569, 1950: 151325798, 1960: 179323175,
        1970: 203302031, 1980: 226542199, 1990: 248709873, 2000: 281421906 , 2010: 308745538 }

xdata1 = np.sort(pob1.keys())
xdata2 = np.sort(pob2.keys())

ydata1 = np.sort(pob1.values())
ydata2 = np.sort(pob2.values())


fig = plt.figure(facecolor='white')
ax = fig.add_subplot(1, 1, 1)
ax.set_xlim(1780, 2020)
plt.ylabel('y = population')

p1, = plt.plot(xdata1, ydata1, 'ko')
p2, = plt.plot(xdata2, ydata2, 'ko')

#least squares fit:'
lny1 = np.zeros(12, float)
for i in range(0, 12):
    lny1 = np.log(ydata1)
lny2 = np.zeros(12, float)
for i in range(0, 12):
    lny2 = np.log(ydata2)

slope, intercept, r_value, p_value, std_err = stats.linregress(xdata1, lny1)
print 'r^2: ', "%8.6f" % (r_value ** 2)
print ('line: y =  ' + "%6.4f" % slope + 'x + ' + "%6.4f" % intercept)
print 'exponential y=Ae^bx:'
print ('y = ' + "%.4g" % np.exp(intercept) + ' e^(' + "%6.4f" % (slope) + 'x)')
ymc1 = np.zeros(12, float)
for i in range(0, 12):
    ymc1[i] = np.exp(intercept) * np.exp(slope * xdata1[i])
p3, = plt.plot(xdata1, ymc1, 'r', lw=2)

slope, intercept, r_value, p_value, std_err = stats.linregress(xdata2, lny2)
print 'r^2: ', "%8.6f" % (r_value ** 2)
print ('line: y =  ' + "%6.4f" % slope + 'x + ' + "%6.4f" % intercept)
print 'exponential y=Ae^bx:'
print ('y = ' + "%.4g" % np.exp(intercept) + ' e^(' + "%6.4f" % (slope) + 'x)')
ymc2 = np.zeros(12, float)
for i in range(0, 12):
    ymc2[i] = np.exp(intercept) * np.exp(slope * xdata2[i])
p4, = plt.plot(xdata2, ymc2, 'b', lw=2)
plt.show()
```

7.2. EXPONENTIAL FUNCTION $EXP(Z)$

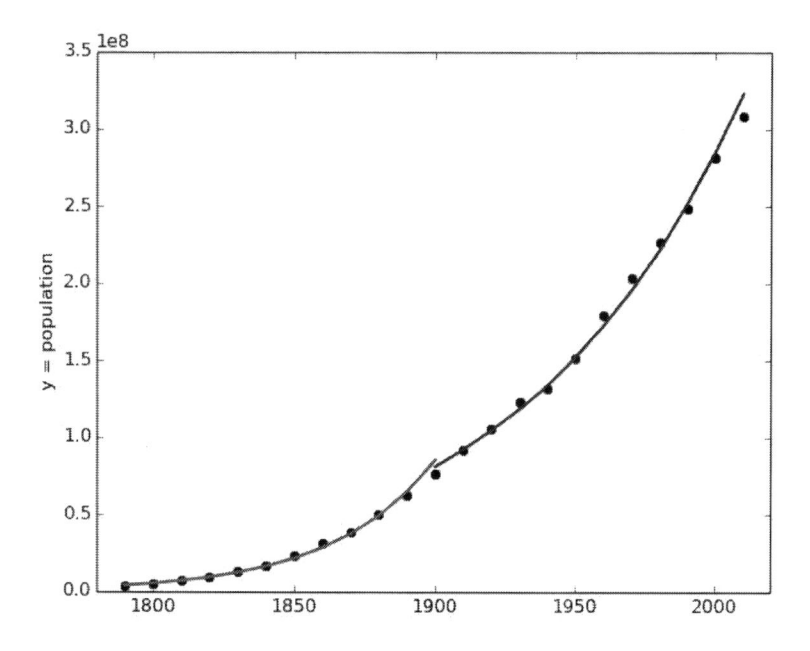

```
                                    output of the program
r^2:   0.996641
line: y =   0.0274x + -33.7860
exponential y=Ae^bx:
y = 2.123e-15 e^(0.0274x)
r^2:   0.994714
line: y =   0.0126x + -5.6400
exponential y=Ae^bx:
y = 0.003553 e^(0.0126x)
```

Now take $z = j\theta$, where θ is a real number:

$$exp(j\theta) = \frac{1}{0!} + \frac{(j\theta)}{1!} + \frac{(j\theta)^2}{2!} + \frac{(j\theta)^3}{3!} + ... + \frac{(j\theta)^n}{n!} + ... = \sum \frac{(j\theta)^n}{n!}$$

$$exp(j\theta) = 1 + j\theta - \theta^2 + \frac{(j\theta)^3}{3!} + \frac{\theta^4}{4!} + ... + \frac{(j\theta)^n}{n!} + ...$$

Since the series is absolutely convergent, we can group the real terms and the imaginary terms:

$$exp(j\theta) = e^{j\theta} = C(\theta) + jS(\theta)$$

being

$$C(\theta) = 1 - \theta^2 + \frac{\theta^4}{4!} - \frac{\theta^6}{6!} + ... = \cos\theta$$

$$S(\theta) = \theta - \frac{\theta^3}{3!} + \frac{\theta^5}{5!} - ... = \sin\theta$$

So we arrived at the Euler formula:

$$e^{j\theta} = \cos\theta + j\sin\theta$$

We now represent a complex number z:

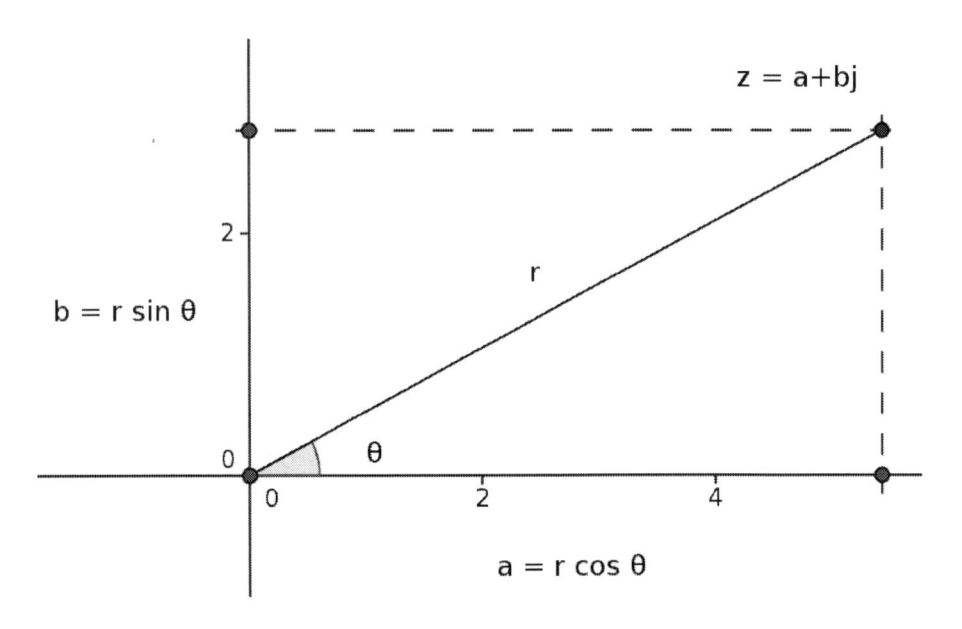

$$z = a + bj = r \cdot \cos\theta + j \cdot r \cdot \sin\theta$$

$$z = a + bj = r(\cos\theta + j\sin\theta) = r \cdot e^{j\theta}$$

And we have obtained another way to represent complex numbers as $z = re^{j\theta}$, which is called polar form of the complex number. Here $r = |z|$ is the modulus of z; and the angle θ is the argument: $arg(z) = \theta = arctan\frac{b}{a}$.

This way of representing complex numbers offers several advantages:

- The complex conjugate of $e^{j\theta}$ is $e^{-j\theta}$, since

$$e^{j\theta} \cdot e^{-j\theta} = e^0 = 1$$

- The exponential function is a periodical function whose period is equal to 2π.

$$e^{j\theta} = \cos\theta + j\sin\theta$$

$$e^{j2\pi} = \cos 2\pi + j\sin 2\pi = 1$$

$$e^{j2\pi n} = \cos 2\pi n + j\sin 2\pi n = 1 = e^{j2\pi}$$

This also means that:
$$e^{j\theta} \cdot e^{j2\pi} = e^{j(\theta + 2\pi)} = e^{j\theta}$$

- Multiplication of complex numbers:

$$z_1 \cdot z_2 = r_1 e^{j\theta_1} \cdot r_2 e^{j\theta_2} = r_1 r_2 e^{j(\theta_1 + \theta_2)}$$

When we multiply a complex number z by $e^{j\theta}$, the complex z rotates one angle θ.

7.2. EXPONENTIAL FUNCTION $EXP(Z)$

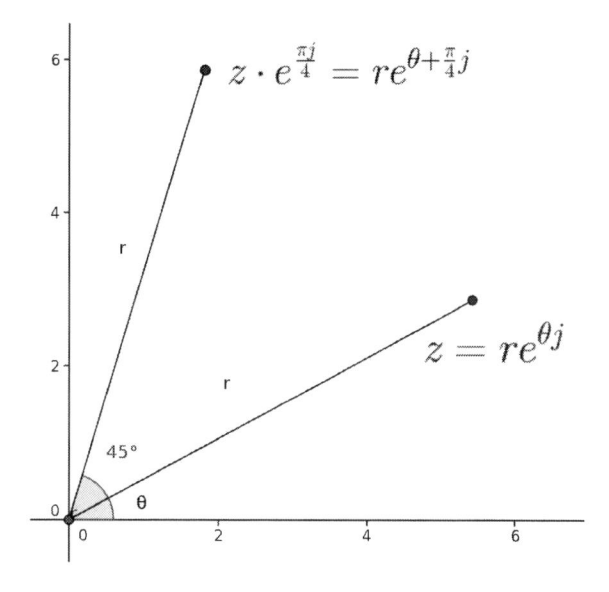

- Division of complex numbers:

$$\frac{z_1}{z_2} = r_1 e^{j\theta_1} \div r_2 e^{j\theta_2} = \frac{r_1}{r_2} e^{j(\theta_1 - \theta_2)}$$

- Exponentiation of a complex number z:

$$z^n = (re^{j\theta})^n = r^n e^{jn\theta}$$

Example 1: the figure represents[2] the powers of $z = 1 \cdot e^{j\frac{\pi}{6}}$. By multiplying each time by $e^{j\frac{\pi}{6}}$, the complex rotates an angle equal to $\frac{\pi}{6}$ in anticlockwise. When it comes to $z^{12} = e^{j\frac{12\pi}{6}} = e^{j2\pi} = 1$. In the figure we have joined all the points that represent the powers $z^1, z^2, ... z^{12} = 1$ and $z^{13} = 1 \cdot z = z$. If we calculate the powers of $e^{j\theta}$, where θ is a small angle, we'd get a circle whose radius is equal to one.

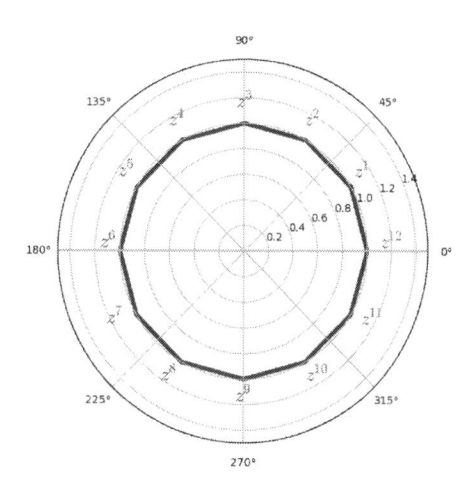

Example 2: the figure represents[3] the powers of $z = 1.02 \cdot e^{j\frac{\pi}{10}}$. In this case in which $r > 1$, the radius of each power is increasing while the complex rotates with each product, so that we get a spiral.

---

[2]Python program p7e.py at the end of the chapter
[3]Python program p7f.py at the end of the chapter

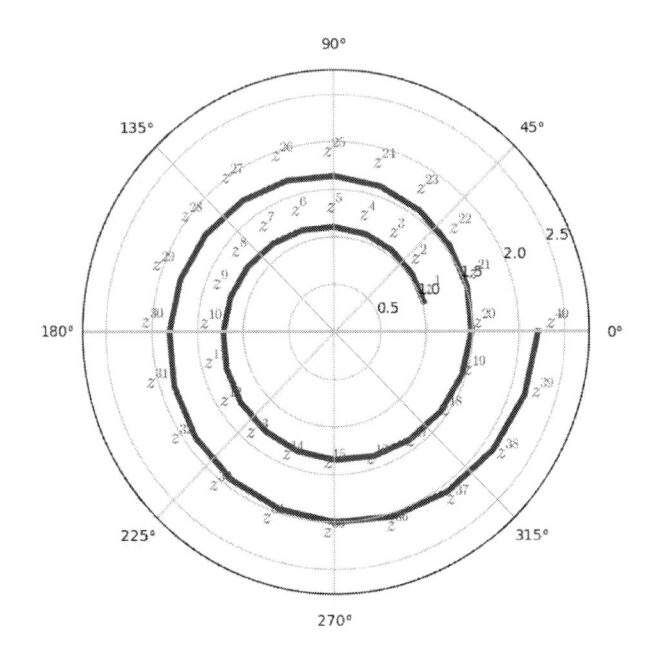

- The nth roots of z:
$$\sqrt[n]{z} = \sqrt[n]{re^{j\theta}} = r^{\frac{1}{n}} e^{j\frac{\theta}{n}}$$

Example 3: the figure represents the roots of $z = 1.05 \cdot e^{j\pi}$. With each root, the angle $\theta = \pi$ is divided by the index of the root, and successively to worth $\frac{\pi}{2}$, $\frac{\pi}{3}$, etc.

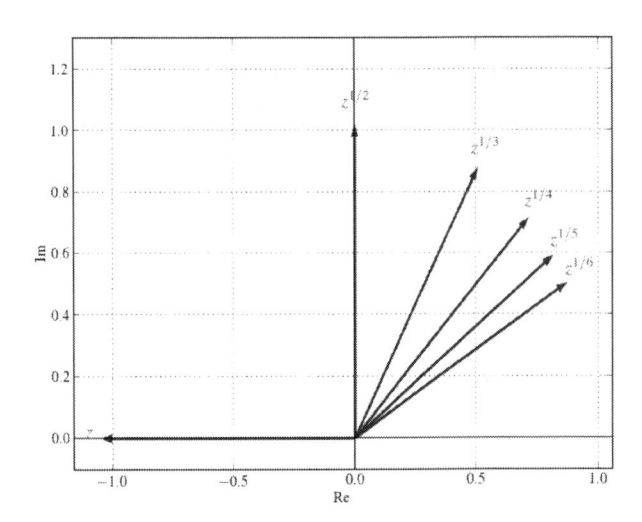

```
# -*- coding: utf-8 -*-
"""
Mathematics and Python Programming    www.pysamples.com
p7c.py
"""

from matplotlib import rc
import matplotlib.pyplot as plt
import numpy as np

j = complex(0, 1)
pi = np.pi
```

7.3. GRAPHICS PROGRAMS IN THIS CHAPTER

```python
z1 = [1.05, pi]
n = 5
rc('font', **{'family': 'serif', 'serif': ['Times']})
rc('text', usetex=True)
plt.figure()
plt.ylabel('Im')
plt.xlabel('Re')
plt.axhline(color='black', lw=1)
plt.axvline(color='black', lw=1)
plt.grid(b=None, which='major')
plt.ylim(-1, 2)
plt.xlim(-1.25, 1.25)


def flecha(z, text):
    dx = z.real
    dy = z.imag
    plt.arrow(0, 0, dx, dy, width=0.003, fc='b', ec='none', length_includes_head=True)
    if dx == 0:
        xtext = dx + 0.03
    else:
        xtext = np.sign(dx) * (abs(dx) * 1.05)
    if dy == 0:
        ytext = dy + 0.01
    else:
        ytext = np.sign(dy) * (abs(dy) * 1.05)
    plt.text(xtext, ytext, text, horizontalalignment='center',
             size='large', color='#EF0808', weight='bold')

for i in range(1, n + 2):
    r = np.power(z1[0], (1.0 / i))
    theta = z1[1] / i
    print (str(i) + ': ' + "%12.6f" % r + "%12.3f" % (theta / pi) + 'pi = ' +
           'pi/' + "%1.0f" % (pi / theta))
    a = r * np.cos(theta)
    b = r * np.sin(theta)
    z = complex(a, b)
    if i > 1:
        label = '$z^{1/' + str(i) + '}$'
    else:
        label = '$z$'
    flecha(z, label)
plt.show()
```

```
──────────────────────── output of the program ────────────────────────
1:     1.050000     1.000pi = pi/1
2:     1.024695     0.500pi = pi/2
3:     1.016396     0.333pi = pi/3
4:     1.012272     0.250pi = pi/4
5:     1.009806     0.200pi = pi/5
6:     1.008165     0.167pi = pi/6
```

7.3 Graphics programs in this chapter

potencias de $z = e^{\frac{\pi}{6}j}$

```python
# -*- coding: utf-8 -*-
"""
Mathematics and Python Programming    www.pysamples.com
p7e.py
"""

import matplotlib
import numpy as np
from matplotlib.pyplot import figure, show, rc, grid

pi = np.pi
z1 = [1.0, pi / 6]


pifraction = int(pi / z1[1])
r = []
theta = []
r.append(z1[0])
theta.append(z1[1])

# radar
rc('grid', color='#CACBD3', linewidth=1, linestyle='-')
rc('xtick', labelsize=10)
rc('ytick', labelsize=10)

# figure
width, height = matplotlib.rcParams['figure.figsize']
size = min(width, height)
#  square
fig = figure(figsize=(size, size))
ax = fig.add_axes([0.1, 0.1, 0.8, 0.8], polar=True, axisbg='#ffffff')

#powers of z
for i in range(2, 2 * pifraction + 2):
    radius = np.power(r[0], i)
    r.append(radius)
    angle = theta[0] * i
    theta.append(angle)
    if i < 2 * pifraction + 1:
        label = '$z^{' + str(i) + '}$'
    else:
        label = '$z^{' + str(1) + '}$'
    if i <= pifraction:
        radiuslabel = 1.1 * radius
        hlabel = 'center'
        vlabel = 'middle'
    elif i <= 2 * pifraction:
        radiuslabel = 1.2 * radius
        hlabel = 'center'
        vlabel = 'bottom'
    else:
        radiuslabel = 1.1 * radius
        hlabel = 'center'
        vlabel = 'middle'
    ax.annotate(label,
                xy=(angle, radiuslabel),   # theta, radius
                horizontalalignment=hlabel, verticalalignment=vlabel,
```

7.3. GRAPHICS PROGRAMS IN THIS CHAPTER

```
                    fontsize=18, color='#EE182E')

ax.plot(theta, r, color='#1821EE', lw=4)
ax.set_rmax(max(r) * 1.5)
grid(True)
show()
```

$$\text{powers of } z = 1.02 e^{\frac{\pi}{10} j}$$

```
# -*- coding: utf-8 -*-
"""
Mathematics and Python Programming    www.pysamples.com
p7f.py
"""

import matplotlib
import numpy as np
from matplotlib.pyplot import figure, show, rc, grid

pi = np.pi
z1 = [1.02, pi / 10]

pifraction = int(pi / z1[1])
r = []
theta = []
r.append(z1[0])
theta.append(z1[1])

# radar
rc('grid', color='#CACBD3', linewidth=1, linestyle='-')
rc('xtick', labelsize=10)
rc('ytick', labelsize=10)

width, height = matplotlib.rcParams['figure.figsize']
size = min(width, height)
#  hace un cuadrado
fig = figure(figsize=(size, size))
ax = fig.add_axes([0.1, 0.1, 0.8, 0.8], polar=True, axisbg='#ffffff')

#powers of z
for i in range(1, 4 * pifraction + 1):
    radius = np.power(r[0], i)
    r.append(radius)
    angle = theta[0] * i
    theta.append(angle)
    etiqueta = '$z^{' + str(i) + '}$'
    ax.annotate(etiqueta,
                xy=(angle, 1.1 * radius),  # theta, radius
                horizontalalignment='center', verticalalignment='middle',
                fontsize=14, color='#EE182E')

ax.plot(theta, r, color='#1821EE', lw=4)
ax.set_rmax(max(r) * 1.25)
grid(True)
show()
```

Bibliography for this chapter: [21], [29], [30], [36], [43], [45], [47], [51], [52], [54], [60], [61]

8 | Differentiation

8.1 Derivative of a function

Let $f(z)$ be a function of real or complex variable. We set a value of the variable $z = z_0$. To increase Δz of the independent variable corresponds an increment of the function:

$$\Delta f(z) = f(z_0 + \Delta z) - f(z_0)$$

We write the ratio between the increment of the function and the increment of its argument:

$$\frac{\Delta f(z)}{\Delta z} = \frac{f(z_0 + \Delta z) - f(z_0)}{\Delta z}$$

This quotient is defined for all $\Delta z \neq 0$. If this ratio approaches a fixed limit when $\Delta z \to 0$, then the limit is the derivative of the function $f(z)$, and we write it as $f'(z)$.

$$f'(z) = \lim_{\Delta z \to 0} \frac{\Delta f(z)}{\Delta z}$$

$$f'(z) = \lim_{\Delta z \to 0} \frac{f(z_0 + \Delta z) - f(z_0)}{\Delta z}$$

The following program calculates the value of the derivative of a polynomial function at a point x_0 we can choose, making Δx tends to zero. The program ends when $\Delta x < 10^{-6}$. In addition to calculate the value of the derivative of the function at a given point on the basis of the definition of the derivative, the program compares it with its exact value calculated with Numpy:

```
# -*- coding: utf-8 -*-
"""
Mathematics and Python Programming     www.pysamples.com
p8a.py
"""

import numpy as np

f = np.poly1d([3, -5, 1])   #f = 3 * x ** 2 - 5 * x + 1

x0 = -2.0
if abs(x0) > 0:
    deltax = abs(x0 / 2.0)
else:
    deltax = 0.1

print "        x            y / x "
while deltax >= 1e-6:
    deltay = np.polyval(f, (x0 + deltax)) - np.polyval(f, x0)
    c = deltay / deltax
    print "%10.6f" % deltax + "%12.6f" % c
```

8.1. DERIVATIVE OF A FUNCTION

```
        deltax = 0.1 * deltax
print
print 'x0 = ', x0
print 'f(x) = '
print f
Df = np.polyder(f, 1)
print 'Df(x): '
print Df
print 'Df(', x0, ') = ', np.polyval(Df, x0)
```

The function is entered in the program code, in this case

$$f(x) = 3x^2 - 5x + 1$$

and take $x_0 = -2.0$. The polynomial is written in Numpy as an array of the coefficients of x sorted in decreasing order of exponent x:

```
f = np.poly1d([3, -5, 1])
```

```
────────────────────────────── output of the program ──────────────────────────────
    Δx          Δy/Δx
 1.000000    -14.000000
 0.100000    -16.700000
 0.010000    -16.970000
 0.001000    -16.997000
 0.000100    -16.999700
 0.000010    -16.999970
 0.000001    -16.999997

x0 =  -2.0
f(x) =
   2
3 x - 5 x + 1
Df(x):

6 x - 5
Df( -2.0 ) =  -17.0
1.28197669983   ms
────────────────────────────────────────────────────────────────────────────────────
```

To define the derivative $f'(z)$ we will consider that the function $f(z)$ is defined for all values of its argument that belong to a certain neighbourhood of the point z_0, and in this way, to calculate the limit it is possible to take any value ζ of the argument that meets:

$$0 < |\zeta - z_0| < \varepsilon$$

That is: $\Delta z = \zeta - z_0$ and we can define the derivative of $f(z)$ as follows:

$$f'(z) = \lim_{\zeta \to z_0} \frac{f(\zeta) - f(z_0)}{\zeta - z_0}$$

In section 5.2 we saw that the function $\alpha(x) = f(x) - l$ is said to be infinitesimal when x tends to x_0 if the limit $lim_{x \to x_0} f(x) = l$. Now we take

$$\alpha(\zeta, z_0) = \frac{f(\zeta) - f(z_0)}{\zeta - z_0} - f'(z_0)$$

in this expression α and $(\zeta - z_0)$ are infinitesimal where $\zeta \to z_0$. If we call $\beta = \alpha \cdot (\zeta - z_0)$, we find $\beta = o(\zeta - z_0)$. We can now rewrite the above expression as:

$$f(\zeta) - f(z_0) = f'(z_0)(\zeta - z_0) + \beta$$

$$\Delta f(z) = f'(z_0)(\zeta - z_0) + \beta$$

And if $f'(z_0) \neq 0$, β tends to zero faster than the first term, by what the term $f'(z_0)(\zeta - z_0)$ is called the main part of the increment of the function $f(z)$. Now call differential of the independent variable z:

$$\Delta z = dz$$

$$\Delta f(z) = f'(z)dz + \beta$$

And we call differential of the function $f(z)$, denoted as $df(z)$ to the main part of the increment of the function:
$$df(z) = f'(z)dz$$

Or also, in the case that the argument is real, x:
$$dy = f'(x)dx$$
$$f'(z) = \frac{df(z)}{dz} = \frac{d}{dz}f(z)$$

Examples:

- We are going to calculate the derivative of the function $f(x) = 1 + x^2$ by using the definition of derivative:
$$f'(x) = \lim_{\Delta x \to 0} \frac{\Delta f(x)}{\Delta x} = \lim_{\Delta x \to 0} \frac{1 + (x + \Delta x)^2 - (1 + x^2)}{\Delta x}$$
$$f'(x) = \lim_{\Delta x \to 0} \frac{2x\Delta x + (\Delta x)^2}{\Delta x} = \lim_{\Delta x \to 0} 2x + \Delta x = 2x$$

- The definition of differential can also be used to calculate approximately a small change Δ. We have seen before:
$$\Delta f(z) = f'(z_0)(\zeta - z_0) + \beta$$

If we now take a function $y = f(x)$, its increment will be:
$$\Delta y = f'(x)dx + \beta = dy + \beta$$

And since β is a infinitesimal for small increments of the variable, we are going to take the approximation:
$$\Delta y \simeq dy = f'(x)dx$$

Let's look at a practical example: the volume of a sphere is given by $y = \frac{4}{3}\pi r^3$. How much will it increase, approximately, if the radius increases from 1 meter to 1.05 meters?. Here we have
$$y' = 4\pi r^2$$
$$dx = 1.05 - 1 = 0.05$$
$$\Delta y \simeq f'(x)dx = (4\pi \cdot 1) \cdot 0.05 = 0.2\pi \simeq 0.628$$

So that the fast approximate calculation, which is possible to perform without the help of a calculator, gives us a volume increase of some $0.63m^3$. The exact calculation made with Python would be:

```
>>> (4*pi/3) * (1.05 ** 3 - 1)
0.6602580560294553
```

That is, an increase of $0.66m^3$. The mistake made performing the approximation has been only $0.03m^3$. This procedure can be useful for making rapid estimates when there is no computer available.

Another example: What is the difference of height reached when launching a projectile vertically at a speed of $v = 20m/s$ instead of $19m/s$? (For the quick mental calculation we take $g = 10ms^{-2}$).
$$mgy = \frac{1}{2}mv^2$$
$$y = \frac{v^2}{2g}; \quad y' = \frac{v}{g}$$
$$\Delta y \simeq dy = f'(v)dv \simeq \frac{19}{10} \cdot (20 - 19) = 1.9$$

```
>>> (1 / (2 * 9.81)) * (20 ** 2 - 19 ** 2)
1.9877675840978593
```

The absolute error that was committed with the estimate is less than $9cm$: $1.9m$ compared to $1.99m$ real, and this for a final height of more than $20m$: the relative error that we have committed with the estimate is less than 0.05%.

- Suppose a certain magnitude y is determined by the equation $y = f(x)$, so that an error Δx to the measure x produces an error Δy in the magnitude y. For small values of Δx we can take $dy \simeq \Delta y$ and the relative error by measuring the magnitude y will be

$$\left| \frac{dy}{y} \right|$$

For example, suppose the magnitude Y is determined by measuring the angle that a needle of a indicator signals: $y = ctan\alpha$. The error $d\alpha$ in the determination of the angle that the needle marks will produce an error in the determination of the magnitude y equal to:

$$dy = \frac{d\alpha}{cos^2\alpha}$$

and the relative error will be:

$$\frac{dy}{y} = \frac{d\alpha}{cos^2\alpha \cdot ctan\alpha} = \frac{2}{sen2\alpha}d\alpha$$

Therefore, the relative error will depend on the angle, and will be smaller when $\alpha = \frac{\pi}{4}$, that is to say: $\frac{dy}{y} = 2d\alpha$

8.2 Geometric meaning

The following image represents a function, as well as the tangent line at a point D of the curve. The straight dashed lines that pass through the point D and another point on the curve increasingly close to him. Each one of these secant lines to the curve passes through a point on the curve each time closest to the point D, so that the secant line moves ever closer to the tangent line at the point D:

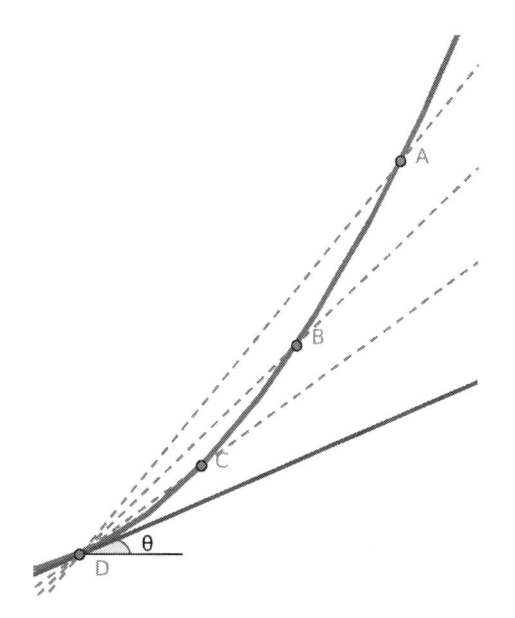

We'll call ϕ the angle that each one of these straight lines forms with the horizontal:

$$\Delta y = y_A - y_D$$

$$\Delta x = x_A - x_D$$

The tangent of the angle will be:

$$tan\phi = \frac{\Delta y}{\Delta x}$$

And as Δx decreases, the two points on the curve are closer, and the angle ϕ is approaching the angle θ which is a tangent line at that point D. So that in the limit:

$$tan\theta = \lim_{\Delta x \to 0} tan\phi = \lim_{\Delta x \to 0} \frac{\Delta y}{\Delta x}$$

By definition we have seen that this limit is precisely $f'(x)$:

$$f'(x) = \lim_{\Delta x \to 0} \frac{\Delta y}{\Delta x}$$

So that the value of the derivative at a point is equal to the slope of the tangent line at that point:

$$f'(x) = tan\theta$$

The following figure shows the geometric meaning of the differential dy. The function $y = 1 + x^2$, and its derivative; $y = 2x$ have been represented by the Python program p8k.py that is shown at the end of this section.

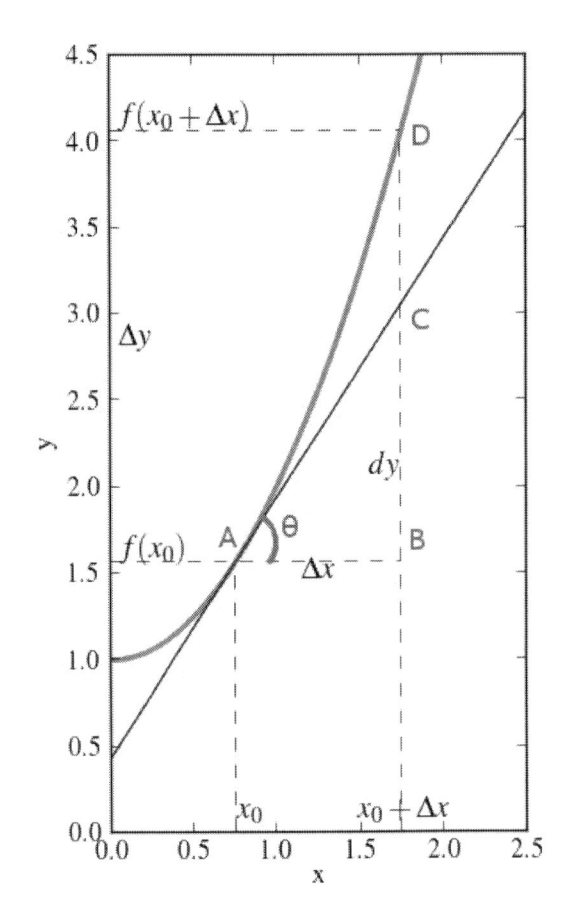

With the notation in the figure we have:

$$\Delta y = BD$$

$$\Delta x = dx = AB$$

8.2. GEOMETRIC MEANING

$$tan\theta = \frac{BC}{AB} = \frac{BC}{\Delta x} = f'(x)$$

And we have previously defined the differential as

$$dy = f'(x)dx$$

$$f'(x) = \frac{dy}{dx} = \frac{BC}{\Delta x}$$

And therefore, geometrically, the differential dy is equivalent to the length of the segment BC:

$$dy = BC$$

```python
# -*- coding: utf-8 -*-
'''
Mathematics and Python Programming    www.pysamples.com
p8k.py
y = 1+x^2; y' = 2x
'''

import numpy as np
import matplotlib.pyplot as plt
from matplotlib import rc

rc('font', **{'family': 'serif', 'serif': ['Times']})
rc('text', usetex=True)


def f(equis):
    fx = 1 + equis ** 2
    return fx


def df(equis):
    dfx = 2 * equis
    return dfx

x0 = 0.75
dx = 1
numpuntos = 50
x = np.linspace(0, 2.0, numpuntos)
y = np.zeros(numpuntos, float)
m = df(x0)
y0 = f(x0)
for i in range(0, numpuntos):
    y[i] = f(x[i])
fig = plt.figure(facecolor='white')
ax = fig.add_subplot(1, 1, 1, aspect='equal')
p1, = plt.plot(x, y, color='#F76429', lw=2.5,)
plt.ylabel('y')
plt.xlabel('x')
x = np.linspace(0.75, 1.25, 10)
y = np.zeros(10, float)
ceros = np.zeros(10, float)
plt.ylabel('y')
plt.xlabel('x')
ax.set_ylim(0, 4.5)
```

```
ax.set_xlim(0, 2.5)
#derivatives: y'(x0) = 1.75; y'(0.5)=1; y'(2)=4
#tangent line: y-y0 = m(x-x0) = 2x0*x - 2x0^2
#tangent line: y = y0 + mx - mx0
#vertical line
plt.plot([0, 2.5], [y0 - m * x0, y0 + m * (2.5 - x0)], 'b', lw=1)
plt.plot([x0, x0], [0, m], 'k--', lw=0.5)
plt.plot([x0 + dx, x0 + dx], [0, f(x0 + dx)], 'k--', lw=0.5)
plt.plot([0, x0 + dx], [y0, y0], 'k--', lw=0.5)
plt.plot([0, x0 + dx], [f(x0 + dx), f(x0 + dx)], 'k--', lw=0.5)
plt.text(0.05, y0, '$f(x_0)$', horizontalalignment='left',
        size='large', color='black', weight='bold')
plt.text(0.05, f(x0 + dx), '$f(x_0 + \Delta x)$', horizontalalignment='left',
        size='large', color='black', weight='bold')
plt.text(0.05, 0.5 * (y0 + f(x0 + dx)), '$\Delta y$', horizontalalignment='left',
        size='large', color='black', weight='bold')
plt.text(x0, 0.05, '$x_0$', horizontalalignment='left',
        size='large', color='black', weight='bold')
plt.text(x0 + dx, 0.05, '$x_0 + \Delta x$', horizontalalignment='center',
        size='large', color='black', weight='bold')
plt.text(x0 + 0.5 * dx, y0 - 0.1, '$\Delta x$', horizontalalignment='center',
        size='large', color='black', weight='bold')
plt.text(x0 + dx - 0.1, y0 + 0.5, '$dy$', horizontalalignment='center',
        size='large', color='black', weight='bold')
plt.show()
```

8.3 Obtaining derivatives

- We can see that the derivative and the slope of the tangent to the curve at a point are related. If the tangent takes an infinite value, the function does not have derivative at that point; if $f(x) = k$, where k is a constant, its graph is a straight horizontal line $y = k$ and the slope is 0, therefore, we have $f'(k) = 0$.

- The derivative has been defined as a limit. If the limit $\zeta \to z_0^+$ does not match the limit $\zeta \to z_0^-$, the curve does not have tangent at that point, and the function is said to have an angular point in x_0. For example, the function $f(x) = |x|$.

- Starting from the definition of the derivative we had the following expression:

$$f(\zeta) - f(z_0) = f'(z_0)(\zeta - z_0) + \beta$$

If we now take limits when $\zeta \to z_0$:

$$\lim_{\zeta \to z_0} (f(\zeta) - f(z_0)) = \lim_{\zeta \to z_0} (f'(z_0)(\zeta - z_0) + o(\zeta - z_0)) = 0$$

Therefore, if a function $f(z)$ is derivable in the point z_0, then it is also continuous at that point.

- Let $u(x)$ and $v(x)$ be two differentiable functions at a point x_0, and k a constant. Then the functions ku, $u + v$, $u - v$, uv and (if $v(x_0) \neq 0$), $\frac{u}{v}$ are derivable and its derivatives are given by:

$$(ku)' = ku'$$
$$(u \pm v)' = u' \pm v'$$
$$(uv)' = u'v + uv'$$
$$\left(\frac{u}{v}\right)' = \frac{u'v - uv'}{v^2}$$

8.3. OBTAINING DERIVATIVES

- If $f(x) = x^n$, then its derivative is $f'(x) = nx^{n-1}$, since if we apply the above rules:
  1. if $n = 1$: $f(x) = x$, and $f'(x) = 1 = nx^{n-1}$
  2. if $n = 2$: $f(x) = x^2 = x \cdot x$, and $f'(x) = 1x + x1 = 2x = nx^{n-1}$
  3. if $n = 3$: $f(x) = x^3 = x \cdot x^2$, and $f'(x) = f'(x^2)x + x^2 f'(x) = 2x^2 + x^2 = 3x^2 = nx^{n-1}$, etc.

- Derivative of the composite function: let's take the functions $x = x(t)$ and $y = y(x)$. If x is differentiable at the point t_0 and the function y is differentiable at the corresponding point $x_0 = x(t_0)$, then the composition of functions $y[x(t)]$ is differentiable at the point t_0, and the following formula is valid:
$$y[x(t)]' = y'(x_0)x'(t_0)$$
that is to say:
$$\frac{dy}{dt} = \frac{dy}{dx} \cdot \frac{dx}{dt}$$
This result is known as "chain rule". The proof is the following:
$$\frac{\Delta y}{\Delta t} = \frac{y[x] - y[x(t_0)]}{t - t_0} = \frac{y[x] - y[x(t_0)]}{x(t) - x(t_0)} \cdot \frac{x(t) - x(t_0)}{t - t_0}$$

- Derivative of a power series: Let's take the function $f(z)$ given by the following power series whose radius of convergence is $r > 0$
$$f(z) = a_0 + a_1 z + a_2 z^2 + a_3 z^3 + ... + a_n z^n + ...$$
Its derivative is:
$$f'(z) = a_1 + 2a_2 z + 3a_3 z^2 + ... + na_n z^{n-1} + ...$$

- Derivative of the exponential function:
$$e^z = \frac{1}{0!} + \frac{z}{1!} + \frac{z^2}{2!} + \frac{z^3}{3!} + ... + +\frac{z^n}{n!} + ... = \sum \frac{z^n}{n!}$$
if we now apply the above formula which we obtained for a power series, we get:
$$(e^z)' = 0 + 1 + z + \frac{z^2}{2!} + ... + \frac{z^{n-1}}{(n-1)!} + ... = \sum \frac{z^n}{n!} = e^z$$
$$(e^x)' = e^x$$
Let's have a look at the function $y = e^{x^2}$. We can consider it a composition of functions:
$$y = e^u, \quad u = x^2$$
The derivative will be:
$$\frac{dy}{dx} = \frac{dy}{du} \cdot \frac{du}{dx} = e^u \cdot 2x = 2xe^{x^2}$$
Therefore
$$(e^u)' = u' e^u$$

- Derivative of trigonometric functions: We are going to study now the function $y = e^{jx}$. As we have just seen, its derivative will be:
$$(e^{jx})' = (jx)' e^{jx} = je^{jx}$$
We now use the formula of Euler $e^{j\theta} = \cos\theta + j\sin\theta$ and we find:
$$(e^{j\theta})' = je^{j\theta} = j[\cos\theta + j\sin\theta] = j\cos\theta - \sin\theta$$
But if in the Euler formula we derive considering the derivative of a sum:
$$(e^{j\theta})' = (\cos\theta + j\sin\theta)' = (\cos\theta)' + j(\sin\theta)'$$
and matching the real and imaginary terms we arrive at the known result:
$$(\sin\theta)' = \cos\theta$$
$$(\cos\theta)' = -\sin\theta$$

8.4 Differential Calculus Theorems

Now we state several fundamental theorems for the differential calculus. The proofs of the theorems can be consulted in the literature cited at the end of this chapter. In this text we prefer just collecting the theorems one after the other to facilitate the student can establish relationships between the theorems and he don't miss in the details of the proofs.

1. Let's take $f(x)$ defined in a neighbourhood of the point c. If the function is differentiable at the point c, and $f'(c) > 0$ ($f'(c) < 0$), the function is increasing (decreasing) at the point c.

2. Let's take $f(x)$ defined in a neighbourhood of the point c. It is said that the function has a local maximum, (local minimum) at the point c, if there is a neighbourhood of the point c in which $f(c) > f(x)$ ($f(c) < f(x)$), $\forall x \in$ neighbourhood(c).

3. Fermat's Theorem: $f(x)$ defined on an open interval (a,b), and such that the function reaches a maximum or a minimum in a point $c \in (a,b)$. Then, if f is differentiable at the point c, its derivative is $f'(c) = 0$. We can see tat this theorem is a result of the previous two theorems: because the slope of the tangent to the graph match the derivative of the function, if the function is increasing at c, we will have that $f'(c) > 0$; if the function is decreasing at c, its derivative will be $f'(c) < 0$. If the function has a maximum or a minimum at the point c, its tangent at that point will be a horizontal line, which slope is $f'(c) = 0$.

4. Theorem of Rolle: $f(x)$ defined on a closed interval $[a,b]$, and such that it meets the following three conditions:

    (a) $f(x)$ is continuous in $[a,b]$.
    (b) $f(x)$ is derivable in (a,b).
    (c) $f(a) = f(b)$

    Then there exist a point $c \in (a,b)$ for which $f'(c) = 0$.
    This theorem is a consequence of the second Weierstrass theorem, which we saw in section 5.4. When f is continuous in $[a,b]$, it reaches in this interval their values maximum M and m, minimum. If both are equal, the function is constant in the interval and $f'(x) = 0$ for all the points in the interval; if they are different, $M \neq m$, and as $f(a) = f(b)$, at least one of the two values M, or m corresponds to a point of the open interval $c \in (a,b)$ and at that point c, according to the theorem of Fermat, the derivative is null: $f'(c) = 0$. Geometrically, the theorem of Rolle implies that in the given conditions, there is a point $c \in (a,b)$ for which the tangent line to the curve of the function is horizontal.

5. Lagrange Theorem: Let $f(x)$ be continuous on a closed interval $[a,b]$, and differentiable in (a,b). Then there is a point $c \in (a,b)$ for which the formula of Lagrange is valid:

$$f(b) - f(a) = f'(c)(b-a)$$

Note that the theorem of Rolle is a particular case of Lagrange's theorem for which $f(a) = f(b)$.

The geometrical interpretation is as follows: since $\frac{f(b)-f(a)}{b-a}$ is the slope of the straight line passing through the points A and B, and $f'(c)$ is the slope of the tangent to the curve at the point c, the theorem states that between the ends of the interval there is a point at which the tangent to the curve is parallel to the straight line that joins the secant ends A and B.

The following Python program provides a graphical representation for the theorems of Lagrange and Rolle. For the Lagrange theorem the function $f(x) = 6x^3 + x^2 + 5$ has been used.

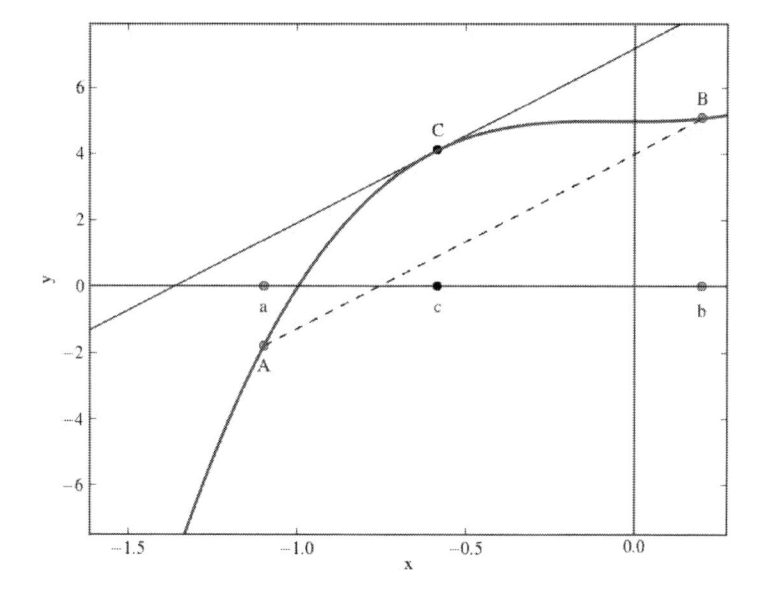

8.4. DIFFERENTIAL CALCULUS THEOREMS

---------- output of the program ----------

```
f(x)=
     3     2
  6 x + 1 x + 5
a = -1.1 ; A = f(a) = -1.776
b =  0.2 ; B = f(b) =  5.088
f ' (c) =  5.280  = slope of the secant line AB
P':      2
  18 x + 2 x
to find c we must find the roots of:
      2
  18 x + 2 x - 5
c = -0.59; C=f(c)=  4.14
the tangent line at f(c) = (-0.59,  4.14)
has the same slope that the secant AB
```

For the theorem of Rolle the function $f(x) = -x^2 + x + 1$ is represented.

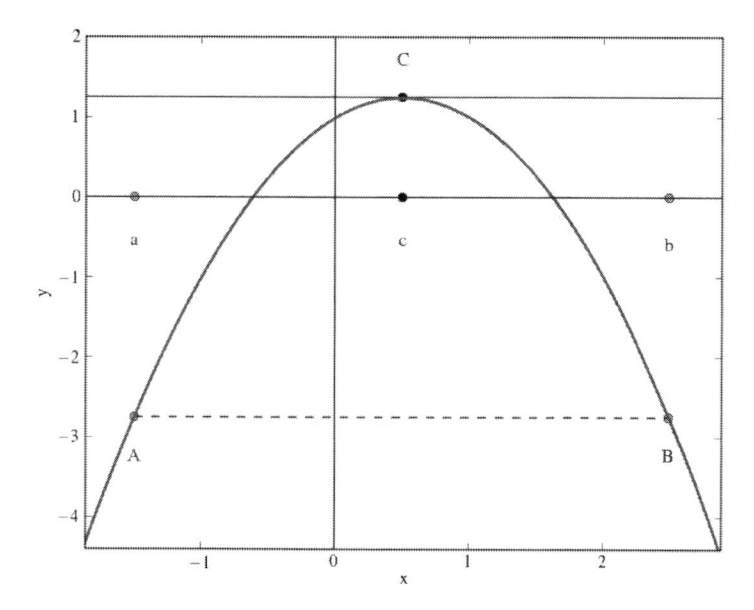

---------- output of the program ----------

```
posible b [ 2.5 -1.5]
f(x)=
      2
  -1 x + 1 x + 1
a = -1.5 ; A = f(a) = -2.75
```

```
b =  2.5 ;  B = f(b) =  -2.75
f ' (c) =  -0.000  = slope of the secant line AB
P':
-2 x + 1
to find c we must find the roots of:

-2 x + 1
c =  0.50; C=f(c)=  1.25
the tangent line at f(c) = ( 0.50,  1.25)
has the same slope that the secant AB
```

The code of the program is the following. For the example of Lagrange, uncomment the lines that appear between triple quotation marks, and comment the lines on which correspond to the example of Rolle.

```python
# -*- coding: utf-8 -*-
"""
Mathematics and Python Programming   www.pysamples.com
p8b.py
Theorem of Lagrange
"""

import numpy as np
import matplotlib.pyplot as plt
from matplotlib import rc

rc('font', **{'family': 'serif', 'serif': ['Times']})
rc('text', usetex=True)
pointsnum = 300


def datasample(poli, a, b, A, B):
    print 'f(x)='
    print poli
    print 'a = ', a, '; A = f(a) = ', A
    print 'b = ', b, '; B = f(b) = ', B


#--------------- Lagrange example ---------------
#coefficients an, an-1,... a0
#for the Lagrange example
fpoli = np.poly1d([6, 1, 0, 5])
a = -1.1
b = 0.2
A = np.polyval(fpoli, a)
B = np.polyval(fpoli, b)
#------------------------------------------------


'''
#--------------- Rolle exammple ---------------
#coefficients an, an-1,... a0
#for the example of Rolle
fpoli = np.poly1d([-1, 1, 1])
a = -1.5
A = np.polyval(fpoli, a)
fpoliA = 0 * fpoli
fpoliA[0] = A
ec = np.poly1d(fpoli - fpoliA)
listab = np.roots(ec)
```

8.4. DIFFERENTIAL CALCULUS THEOREMS

```python
    print 'posible b', listab
    b = listab[0]
    if b == a:
        b = listab[1]
    B = np.polyval(fpoli, b)
    #----------------------------------------------
    '''

    datasample(fpoli, a, b, A, B)
    characteristicx = [0]
    characteristicy = [0]
    characteristicx.append(a)
    characteristicx.append(b)


    def writex(m, texto):
        plt.text(m, -0.5, texto,
                 horizontalalignment='center', verticalalignment='top',
                 fontsize=13, color='blue', weight='bold')


    def writefx(m, M, texto):
        if M >= 0:
            posicion = 0.4 + M
            vertical = 'bottom'
        else:
            posicion = M - 0.4
            vertical = 'top'
        plt.text(m, posicion, texto,
                 horizontalalignment='center', verticalalignment=vertical,
                 fontsize=13, color='blue', weight='bold')

    writefx(a, A, 'A')
    writefx(b, B, 'B')
    plt.plot(a, 0, 'yo')
    plt.plot(b, 0, 'yo')
    writex(a, 'a')
    writex(b, 'b')
    plt.plot(a, A, 'yo')
    plt.plot(b, B, 'yo')
    plt.plot([a, b], [A, B], 'k—', lw=1)
    slope = (B - A) / (b - a)
    print 'f', "'", '(c) = ', "%5.3f" % slope, ' = slope of the secant line AB'

    #derivada
    d1 = np.polyder(fpoli)
    print 'P' + "'" + ': ' + str(d1)
    print 'to find c we must find the roots of:'
    d1[0] = d1[0] - slope
    print d1

    #raices de ec
    raices1 = np.roots(d1)
    for i in range(0, len(raices1)):
        if np.iscomplex(raices1[i]) is True:
            np.delete(raices1, i)
```

```
        else:
            if (raices1[i] < b) and (raices1[i] > a):
                c = raices1[i]
                C = np.polyval(fpoli, c)
                plt.plot(c, C, 'ko')
                plt.plot(c, 0, 'ko')
                writex(c, 'c')
                writefx(c, C, 'C')
                print ('c = ' + "%5.2f" % c + '; C=f(c)= ' + "%5.2f" % C)
                characteristicx.append(c)
                i = len(raices1) + 2

xmax = 1.5 * np.max(characteristicx)
xmin = 1.5 * np.min(characteristicx)
if xmin == 0:
    xmin = -1
if xmax == 0:
    xmax = 1
ymax = 2 * max([0.5, A, B, C])
ymin = 2 * min([-0.5, A, B, C])
#y-y0 = m(x-x0) = slope(x-x0)
#y = C + slope(x-c)
print 'the tangent line at f(c) = (' + "%5.2f" % c + ', ' + "%5.2f" % C + ')'
print 'has the same slope that the secant AB'
plt.plot([xmin, xmax], [C + slope * (xmin - c),
        C + slope * (xmax - c)], 'b-', lw=1)
x = np.linspace(xmin, xmax, pointsnum)
y = np.zeros(pointsnum, float)
for i in range(0, pointsnum):
    y[i] = np.polyval(fpoli, x[i])
plt.plot(x, y, 'r-', lw=2)
plt.axhline(color='black', lw=1)
plt.axvline(color='black', lw=1)
plt.ylim(ymin, ymax)
plt.ylabel('y')
plt.xlabel('x')
plt.show()
```

We can also use the theorem of Lagrange to prove some inequalities. For example, if we apply the theorem to the function $f(x) = \sin x$ in the interval $[a, b]$, we find

$$f(b) - f(a) = f'(c)(b - a)$$

$$\sin b - \sin a = f'(c)(b - a)$$

and since $|f'(x)| = |\cos x| \leq 1 \quad \forall x$ we obtain the following inequality:

$$|\sin b - \sin a| \leq |b - a|$$

6. **Cauchy's Theorem**: this theorem generalizes the previous theorem of Lagrange: Let $f(x)$ and $g(x)$ be two continuous functions in $[a, b]$ and differentiable in (a, b), and the function $g(x)$ in addition is such that its derivative is different from zero at all points in (a, b). Then in the interior of this interval there is a point c for which it is valid the following formula:

$$\frac{f(b) - f(a)}{g(b) - g(a)} = \frac{f'(c)}{g'(c)}$$

Lagrange's theorem is the particular case of the theorem of Cauchy if we take in this one $g(x) = x$.

8.4. DIFFERENTIAL CALCULUS THEOREMS

The geometrical interpretation of this theorem is as follows:
If we take the functions $f(t) = 2\sin^3 t$ and $g(t) = 2\cos^3 t$, and then we plot the points $A(x,y) = (g(t_1), f(t_1))$ and $B(x,y) = (g(t_2), f(t_2))$. Then

$$\frac{f(b) - f(a)}{g(b) - g(a)} = \frac{f(t_2) - f(t_1)}{g(t_2) - g(t_1)} = \frac{y_B - y_A}{x_B - x_A}$$

That is to say, the slope of the secant straight line AB.

The right member can be rewritten as

$$\frac{f'(c)}{g'(c)} = \frac{\frac{dy}{dt}}{\frac{dx}{dt}}$$

That is to say: the slope of the curve at a point $C(x,y) = (g(t_c), f(t_c))$. Therefore, the theorem states that there is a point C between A and B in which the tangent to the curve has the same slope as the secant straight line AB.

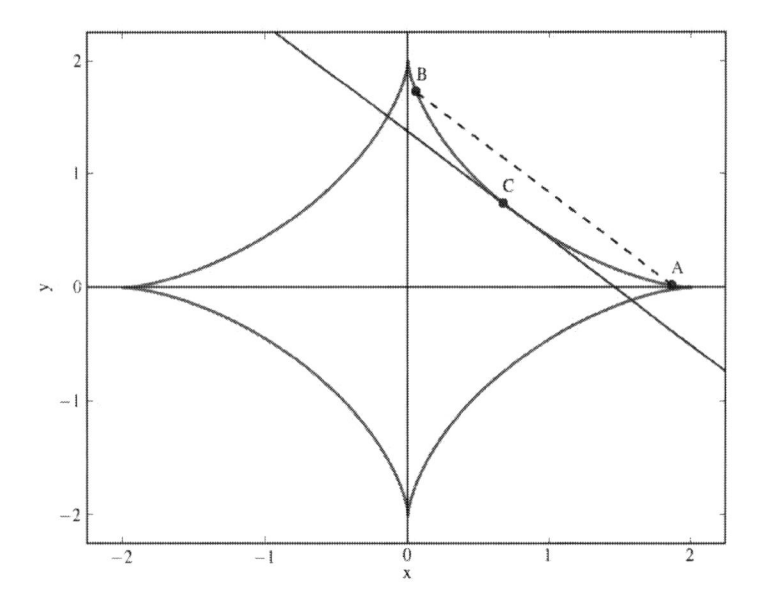

———————————— output of the program ————————————
```
t1 = pi / 15 = 0.209
t2 = pi / 2.5 = 1.257
A(1.87, 0.02)
B(0.06, 1.72)
slope = -0.94
tc = 0.801
C(0.67, 0.74)
```

```python
# -*- coding: utf-8 -*-
"""
Mathematics and Python Programming    www.pysamples.com
p8c.py
Theorem of Cauchy
"""

import numpy as np
import matplotlib.pyplot as plt
from matplotlib import rc

rc('font', **{'family': 'serif', 'serif': ['Times']})
```

```python
rc('text', usetex=True)

pi = np.pi
t1 = pi / 15
t2 = pi / 2.5
print 't1 = pi / 15 = ' + "%5.3f" % t1
print 't2 = pi / 2.5 = ' + "%5.3f" % t2


def f(t):
    fv = 2 * ((np.sin(t)) ** 3)
    return fv


def g(t):
    gv = 2 * ((np.cos(t)) ** 3)
    return gv

pointsnum = 360
tstart = 0
tend = pointsnum
x = np.zeros(pointsnum, float)
y = np.zeros(pointsnum, float)
t = tstart
while t < tend:
    radianes = np.deg2rad(t)
    x[t] = 2 * (np.cos(radianes) ** 3)
    y[t] = 2 * (np.sin(radianes) ** 3)
    t += 1
plt.plot(x, y, 'r-', lw=2)
plt.plot(g(t1), f(t1), 'bo')
plt.text(g(t1), f(t1) + 0.1, 'A', horizontalalignment='left',
        fontsize=12, color='black', weight='bold')
plt.plot(g(t2), f(t2), 'bo')
plt.text(g(t2), f(t2) + 0.1, 'B', horizontalalignment='left',
        fontsize=12, color='black', weight='bold')
plt.plot([g(t1), g(t2)], [f(t1), f(t2)], 'k-', lw=1.5)


def func_point(t, letra):
    print (letra + '(' + "%4.2f" % g(t) + ', ' + "%4.2f" % f(t) + ')')
func_point(t1, 'A')
func_point(t2, 'B')
slope = (f(t2) - f(t1)) / (g(t2) - g(t1))
print 'slope = ' + "%4.2f" % slope

tc = np.arctan(np.sqrt(-1 / slope))
print 'tc = ' + "%5.3f" % tc
func_point(tc, 'C')
plt.plot(g(tc), f(tc), 'bo')
plt.text(g(tc), f(tc) + 0.1, 'C', horizontalalignment='left',
        fontsize=12, color='black', weight='bold')

plt.plot([-1.5, 2.5], [f(tc) - slope * (1.5 + g(tc)),
        f(tc) + slope * (2.5 - g(tc))], 'b-', lw=1.5)
```

```
plt.xlim(-2.25, 2.25)
plt.ylim(-2.25, 2.25)
plt.axhline(color='black', lw=1)
plt.axvline(color='black', lw=1)
plt.ylabel('y')
plt.xlabel('x')
plt.show()
```

7. L'Hopital rule: let f and g be differentiable functions in the interval (a,b), in which g has non-zero derivative. Let's take $c \in (a,b)$. If

$$\lim_{x \to c} f(x) = 0 \quad \lim_{x \to c} g(x) = 0$$

or if

$$\lim_{x \to c} f(x) = \infty \quad \lim_{x \to c} g(x) = \infty$$

And also the limit of the quotient is

$$\lim_{x \to c} \frac{f'(x)}{g'(x)} = l$$

then

$$\lim_{x \to c} \frac{f(x)}{g(x)} = l$$

8.5 Taylor series

We consider now a function $f(x)$ defined on an interval, in which there are derivatives of the function up to the order n. Let A be a point of that interval. The polynomial

$$p_1(x) = f(a) + f'(a)(x - a)$$

has the same value and the same derivative that $f(x)$ at the point a. If we're looking for another polynomial, this time a second degree polynomial which meets the same conditions, i.e. $p_2(a) = f(a)$, and $p_2'(a) = f'(a)$, the polynomial that we seek will be:

$$p_2(x) = f(a) + f'(a)(x - a) + \frac{f''(a)}{2}(x - a)^2$$

In addition this polynomial satisfies $p_2''(a) = f''(a)$. So if we plot these polynomials, their graphic moves ever closer to the function, the greater the degree of the polynomial. If we use a polynomial of degree n, this would be $p_n(x)$:

$$f(a) + f'(a)(x - a) + \frac{f''(a)}{2!}(x - a)^2 + \frac{f^{(3)}(a)}{3!}(x - a)^3 + ... + \frac{f^{(n)}(a)}{n!}(x - a)^n$$

The following Python program represents the function $f(x) = 3\sin 2x$ and the Taylor polynomials $p1, p2, p3$ and $p4$ in a neighbourhood of the point $a = 60°$:

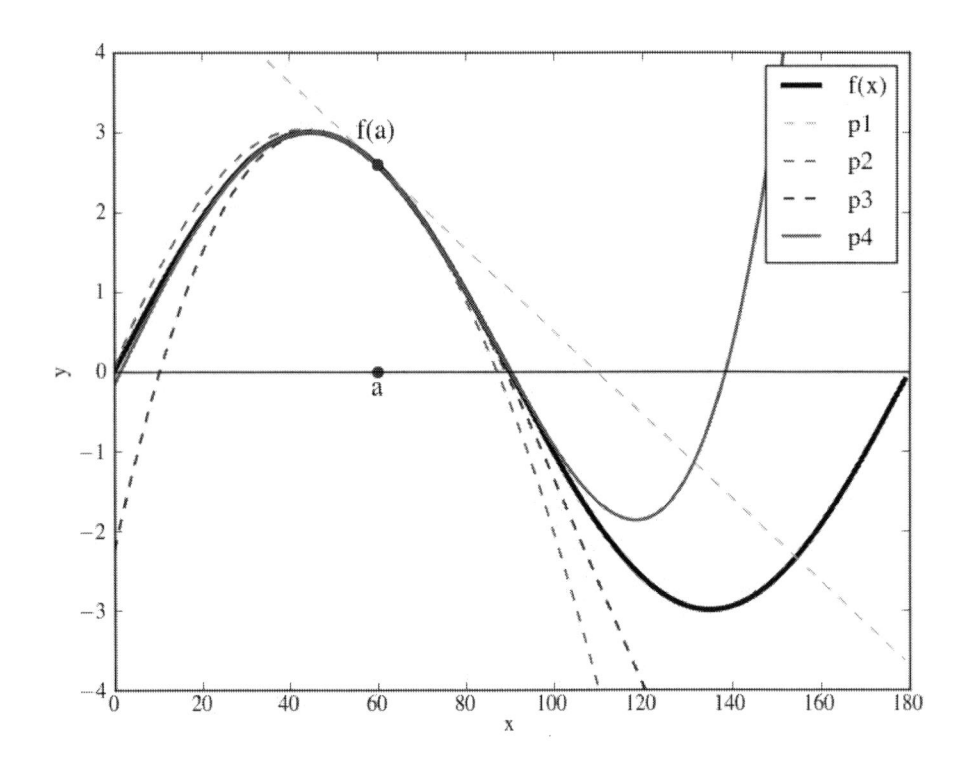

```
──────────────────────── output of the program ────────────────────────
f = 3*sin(2*x)
f(a) = 2.59808
D1 = 6*cos(2*x)
p1 =    2.60 + (-3.0/1.0)(x-a)^1
D2 = -12*sin(2*x)
p2 =    2.60 + (-3.0/1.0)(x-a)^1 + (-10./2.0)(x-a)^2
D3 = -24*cos(2*x)
p3 =    2.60 + (-3.0/1.0)(x-a)^1 + (-10./2.0)(x-a)^2 + (12./6.0)(x-a)^3
D4 = 48*sin(2*x)
p4 =    2.60 + (-3.0/1.0)(x-a)^1 + (-10./2.0)(x-a)^2 + (12./6.0)(x-a)^3 + (42./24.0)(x-a)^4
```

```python
# -*- coding: utf-8 -*-
"""
Mathematics and Python Programming    www.pysamples.com
p8d.py
Taylor series: first four derivatives
"""

import numpy as np
import matplotlib.pyplot as plt
from matplotlib import rc
import sympy as sy
from scipy.misc import factorial

x, y, z = sy.symbols('x y z')
sy.init_printing(use_unicode=True)

rc('font', **{'family': 'serif', 'serif': ['Times']})
rc('text', usetex=True)

a = 60.0    # degrees
aradians = np.deg2rad(a)
fx = 3 * sy.sin(2 * x)
print 'f = ' + str(fx)
fa = fx.subs(x, aradians).evalf(6)
```

8.5. TAYLOR SERIES

```python
print 'f(a) = ' + str(fa)

derivative = [0, 0, 0, 0, 0]
derivative[0] = fx
p = [0, 0, 0, 0, 0]
p[0] = "%6.2f" % fa
strp = p[0]
for d in range(1, 5):
    derivative[d] = sy.diff(fx, x, d)
    print 'D' + str(d) + ' = ' + str(derivative[d])
    p[d] = str(derivative[d].subs(x, aradians).evalf(2))
    strp = strp + ' + (' + p[d] + '/' + str(factorial(d)) + ')(x-a)^' + str(d)
    print 'p' + str(d) + ' = ' + strp


def f(x):
    fv = 3 * np.sin(2 * x)
    return fv


def D1(x):
    D1v = 6 * np.cos(2 * x)
    return D1v


def D2(x):
    D2v = -12 * np.sin(2 * x)
    return D2v


def D3(x):
    D3v = -24 * np.cos(2 * x)
    return D3v


def D4(x):
    D4v = 48 * np.sin(2 * x)
    return D4v

pointsnum = 180
x = np.zeros(pointsnum, float)
y = np.zeros(pointsnum, float)
p1 = np.zeros(pointsnum, float)
p2 = np.zeros(pointsnum, float)
p3 = np.zeros(pointsnum, float)
p4 = np.zeros(pointsnum, float)
i = 0
while i < pointsnum:
    x[i] = i
    radians = np.deg2rad(x[i])
    increment = np.deg2rad(i - a)
    y[i] = f(radians)
    p1[i] = f(aradians) + D1(aradians) * increment
    p2[i] = p1[i] + (D2(aradians) / 2) * (increment ** 2)
    p3[i] = p2[i] + (D3(aradians) / 6) * (increment ** 3)
    p4[i] = p3[i] + (D4(aradians) / 24) * (increment ** 4)
```

```
    #print x[i], y[i], p1[i]
    i += 1
plt.plot(x, y, 'k-', lw=3)
plt.plot(x, p1, 'y--', lw=1)
plt.plot(x, p2, 'g--', lw=1.25)
plt.plot(x, p3, 'b--', lw=1.5)
plt.plot(x, p4, 'r--', lw=1.75)
plt.ylim(-4, 4)
plt.plot(a, f(aradians), 'bo')
plt.text(a, f(aradians) + 0.3, 'f(a)', horizontalalignment='center',
        fontsize=15, color='black', weight='bold')
plt.plot(a, 0, 'bo')
plt.text(a, -0.3, 'a', horizontalalignment='center',
        fontsize=15, color='black', weight='bold')
plt.legend(('f(x)', 'p1', 'p2', 'p3', 'p4'), loc='best')
plt.axhline(color='black', lw=1)
plt.axvline(color='black', lw=1)
plt.ylabel('y')
plt.xlabel('x')
plt.show()
```

What is the error we commit to take $p_n(x)$ as an approximation of $f(x)$ in a neighbourhood of a point a? The error is called residual term $R_n(x)$ and is equal to (form of Lagrange):

$$R_n(x) = (x-a)^{n+1} \frac{f^{(n+1)}(\xi)}{(n+1)!}$$

ξ is a number between a and x.

If we take $a = 0$, the series is called MacLaurin formula:

$$f(0) + f'(0)x + \frac{f''(0)}{2!}x^2 + \frac{f^{(3)}(0)}{3!}x^3 + ... + \frac{f^{(n)}(0)}{n!}x^n$$

And you can often operate in a way such that the formula that we get is considerably simpler. In Python we can obtain directly the Maclaurin formula for a function. Let's look at an example for the function $f(x) = \sin x$:

```
# -*- coding: utf-8 -*-
"""
Mathematics and Python Programming    www.pysamples.com
p8e.py
Mc Laurin series
"""

import sympy as sy

x, y, z = sy.symbols('x y z')
sy.init_printing(use_unicode=True)
j = complex(0, 1)


#fx = sy.sin(x)
#fx = sy.cos(x)
#fx = sy.exp(x)
#fx = sy.cos(j * x)
#fx = sy.sin(j * x) / j
fx = sy.atan(x)
f0 = fx.subs(x, 0).evalf(3)
```

8.5. TAYLOR SERIES

```
maclaurin = str(f0) + ' + '
print 'f(x) = ' + str(fx)
for d in range(1, 11):
    #print maclaurin
    derivative = sy.diff(fx, x, d)
    derivative0 = derivative.subs(x, 0).evalf(3)
    sd = str(derivative0)
    print ('D' + str(d) + ' = ' + sd + '; D' + str(d) + '(0) = ' + sd)
    if sd != '0':
        maclaurin += '[' + sd + ' * (x**' + str(d) + ') / ' + str(d) + '!] + '
print
print maclaurin + '...'
serie = fx.series(x, 0, 11)   # series using sympy
print serie
```

────────────────── output of the program ──────────────────
```
f(x) = sin(x)
D1 = 1.00; D1(0) = 1.00
D2 = 0; D2(0) = 0
D3 = -1.00; D3(0) = -1.00
D4 = 0; D4(0) = 0
D5 = 1.00; D5(0) = 1.00
D6 = 0; D6(0) = 0
D7 = -1.00; D7(0) = -1.00

0 + [1.00 * (x**1) / 1!] + [-1.00 * (x**3) / 3!] + [1.00 * (x**5) / 5!] + [-1.00 * (x**7) / 7!] + ...
x - x**3/6 + x**5/120 - x**7/5040 + O(x**8)
```
──

We can obtain an upper bound of the error we make when we approximate

$$\sin x = x - \frac{x^3}{3!} + \frac{x^5}{5!} + \frac{x^7}{7!}$$

$$|R_9| = \left|\frac{x^9}{9!}\cos\xi\right| = \left|\frac{x^9}{362880}\cos\xi\right|$$

and since $|\cos\xi| \leq 1$ we have:

$$|R_9| \leq \left|\frac{x^9}{362880}\right|$$

and for example, for all values of x between 0 and $\frac{\pi}{3}$:

$$|R_9| \leq \frac{1}{362880}\left(\frac{\pi}{3}\right)^9 \simeq 4.1734e - 06 < \frac{1}{230000}$$

We take advantage of the previous program, to modify it and obtain the Taylor series of $f(x) = \cos x$:

────────────────── output of the program ──────────────────
```
f(x) = cos(x)
D1 = 0; D1(0) = 0
D2 = -1.00; D2(0) = -1.00
D3 = 0; D3(0) = 0
D4 = 1.00; D4(0) = 1.00
D5 = 0; D5(0) = 0
D6 = -1.00; D6(0) = -1.00
D7 = 0; D7(0) = 0

1.00 + [-1.00 * (x**2) / 2!] + [1.00 * (x**4) / 4!] + [-1.00 * (x**6) / 6!] + ...
1 - x**2/2 + x**4/24 - x**6/720 + O(x**8)
```
──

$$cos x = 1 - \frac{x^2}{2!} + \frac{x^4}{4!} - \frac{x^6}{6!} + ...$$

We calculate with the same program the series of e^x, Chx and Shx:

────────────────── output of the program ──────────────────
```
f(x) = exp(x)
D1 = 1.00; D1(0) = 1.00
D2 = 1.00; D2(0) = 1.00
D3 = 1.00; D3(0) = 1.00
```

```
D4 = 1.00; D4(0) = 1.00
D5 = 1.00; D5(0) = 1.00
D6 = 1.00; D6(0) = 1.00
D7 = 1.00; D7(0) = 1.00

1.00 + [1.00 * (x**1) / 1!] + [1.00 * (x**2) / 2!]
     + [1.00 * (x**3) / 3!] + [1.00 * (x**4) / 4!]
     + [1.00 * (x**5) / 5!] + [1.00 * (x**6) / 6!]
     + [1.00 * (x**7) / 7!] + ...
1 + x + x**2/2 + x**3/6 + x**4/24 + x**5/120 + x**6/720 + x**7/5040 + O(x**8)


f(x) = cosh(1.0*x)
D1 = 0; D1(0) = 0
D2 = 1.00; D2(0) = 1.00
D3 = 0; D3(0) = 0
D4 = 1.00; D4(0) = 1.00
D5 = 0; D5(0) = 0
D6 = 1.00; D6(0) = 1.00
D7 = 0; D7(0) = 0

1.00 + [1.00 * (x**2) / 2!] + [1.00 * (x**4) / 4!]
     + [1.00 * (x**6) / 6!] + ...
1 + 0.5*x**2 + 0.0416666666666667*x**4 + 0.00138888888888889*x**6 + O(x**8)


f(x) = 1.0*sinh(1.0*x)
D1 = 1.00; D1(0) = 1.00
D2 = 0; D2(0) = 0
D3 = 1.00; D3(0) = 1.00
D4 = 0; D4(0) = 0
D5 = 1.00; D5(0) = 1.00
D6 = 0; D6(0) = 0
D7 = 1.00; D7(0) = 1.00

0 + [1.00 * (x**1) / 1!] + [1.00 * (x**3) / 3!]
  + [1.00 * (x**5) / 5!] + [1.00 * (x**7) / 7!] + ...
1.0*x + 0.166666666666667*x**3 + 0.00833333*x**5 + 0.000198413*x**7 + O(x**8)
```

Note that

$$e^x = Chx + Shx \qquad e^{-x} = Chx - Shx$$

This allows us to find the expression of the functions hyperbolic sine, hyperbolic cosine and hyperbolic tangent, that we represent below with the following Python program:

```python
# -*- coding: utf-8 -*-
"""
Mathematics and Python Programming     www.pysamples.com
p81.py
Chx, Shx, Thx
"""

import numpy as np
import matplotlib.pyplot as plt

pointsnum = 100
fig = plt.figure(facecolor='white')
x = np.linspace(-3.5, 3.5, pointsnum)
y1 = np.zeros(pointsnum, float)
y2 = np.zeros(pointsnum, float)
y3 = np.zeros(pointsnum, float)
for i in range(0, pointsnum):
    y1[i] = (np.exp(x[i]) + np.exp(-x[i])) / 2
    y2[i] = (np.exp(x[i]) - np.exp(-x[i])) / 2
    y3[i] = y2[i] / y1[i]
ax = fig.add_subplot(1, 1, 1, aspect='equal')
p1, = plt.plot(x, y1, 'g', lw=2, label='$Chx$')
p2, = plt.plot(x, y2, 'r--', lw=2, label='$Shx$')
p3, = plt.plot(x, y3, 'b', lw=1.2, label='$Thx$')
plt.legend(('$Chx$', '$Shx$', '$Thx$'), loc='best')
```

8.5. TAYLOR SERIES

```
plt.ylabel('y')
plt.xlabel('x')
ax.set_xlim(-3.6, 3.6)
ax.set_ylim(-5.1, 5.1)
plt.axhline(color='black', lw=1)
plt.axvline(color='black', lw=1)
plt.show()
```

$$Chx = \frac{e^x + e^{-x}}{2} \qquad Shx = \frac{e^x - e^{-x}}{2} \qquad Thx = \frac{e^x + e^{-x}}{e^x - e^{-x}}$$

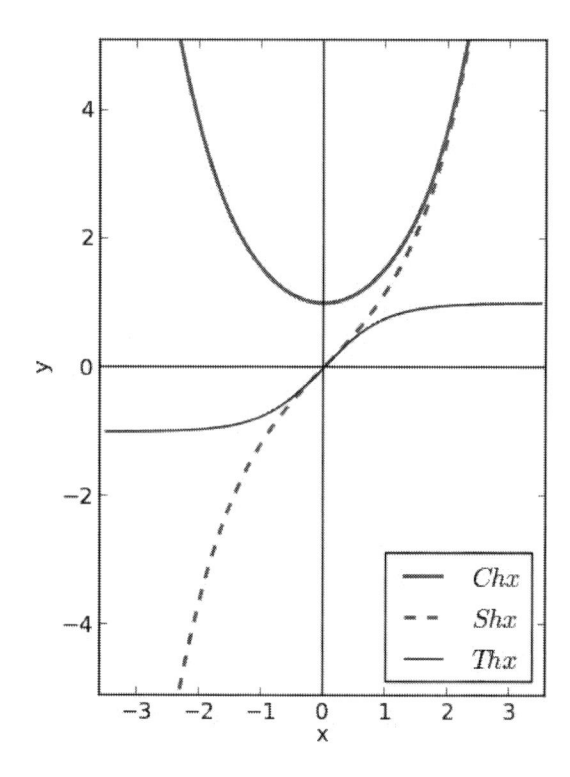

and their derivatives:

$$(Chx)' = \frac{e^x - e^{-x}}{2} = Shx \qquad (Shx)' = \frac{e^x + e^{-x}}{2} = Chx$$

$$(Thx)' = \left(\frac{Shx}{Chx}\right)' = \frac{Chx \cdot Chx - Shx \cdot Shx}{(Chx)^2} = \frac{1}{Ch^2 x}$$

since

$$e^x \cdot e^{-x} = 1 = (Chx + Shx) \cdot (Chx - Shx) = Ch^2 x - Sh^2 x$$

If we use a parameter t as a variable, and we represent through the Python program p8m.py the parametric function given by

$$x = Ch(t) \qquad y = Sh(t)$$

We get an equilateral hyperbola equilatera whose semi-axis is unit:

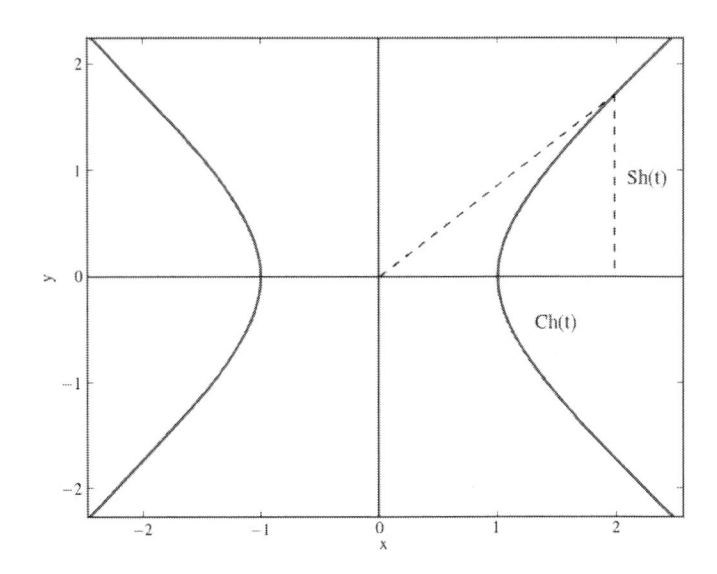

```python
# -*- coding: utf-8 -*-
"""
Mathematics and Python Programming    www.pysamples.com
p8m.py
"""

import numpy as np
import matplotlib.pyplot as plt
from matplotlib import rc

rc('font', **{'family': 'serif', 'serif': ['Times']})
rc('text', usetex=True)

t1 = np.deg2rad(75)
print 't1 = 75    = ' + "%5.3f" % t1 + ' rad'


def f(t):
    fv = np.sinh(t)
    return fv


def g(t):
    gv = np.cosh(t)
    return gv

pointsnum = 90
tstart = 0
tfinal = pointsnum
x = np.zeros(pointsnum, float)
y = np.zeros(pointsnum, float)
t = tstart
while t < tfinal:
    radians = np.deg2rad(t)
    x[t] = np.cosh(radians)
    y[t] = np.sinh(radians)
    t += 1
plt.plot(x, y, 'r-', lw=2)
```

8.5. TAYLOR SERIES

```
plt.plot(-x, y, 'r--', lw=2)
plt.plot(-x, -y, 'r--', lw=2)
plt.plot(x, -y, 'r--', lw=2)

plt.text(g(t1) + 0.1, f(t1) / 2, 'Sh(t)', horizontalalignment='left',
        fontsize=15, color='blue', weight='bold')
plt.text((1 + g(t1)) / 2, -0.5, 'Ch(t)', horizontalalignment='center',
        fontsize=15, color='blue', weight='bold')
plt.plot([g(t1), g(t1)], [f(t1), 0], 'k--', lw=1)
plt.plot([0, g(t1)], [0, f(t1)], 'k--', lw=1)
xmax = np.max(x)
ymax = np.max(y)
print xmax
plt.xlim(-xmax - 0.25, xmax + 0.25)
plt.ylim(-ymax - 0.25, ymax + 0.25)
plt.axhline(color='black', lw=1)
plt.axvline(color='black', lw=1)
plt.ylabel('y')
plt.xlabel('x')
plt.show()
```

An important series is that of *atanx*, which we can obtain with the same program, this time until the tenth derived. This series was discovered in 1670 by an English mathematician for $|x| \leq 1$:

```
―――――――――――――――――― output of the program ――――――――――――――――――
f(x) = atan(x)
D1 = 1.00; D1(0) = 1.00
D2 = 0; D2(0) = 0
D3 = -2.00; D3(0) = -2.00
D4 = 0; D4(0) = 0
D5 = 24.0; D5(0) = 24.0
D6 = 0; D6(0) = 0
D7 = -720.; D7(0) = -720.
D8 = 0; D8(0) = 0
D9 = 4.03e+4; D9(0) = 4.03e+4
D10 = 0; D10(0) = 0

0 + [1.00 * (x**1) / 1!] + [-2.00 * (x**3) / 3!] + [24.0 * (x**5) / 5!] +
  + [-720. * (x**7) / 7!] + [4.03e+4 * (x**9) / 9!] + ...
x - x**3/3 + x**5/5 - x**7/7 + x**9/9 + O(x**11)
```

$$atanx = x - \frac{x^3}{3} + \frac{x^5}{5} - \frac{x^7}{7} + \frac{x^9}{9} - ...$$

But this mathematician did not realized that if we take $x = 1$, the arctangent of 1 is precisely $\frac{\pi}{4}$, with what we reach the series discovered by Leibniz in 1673:

$$\frac{\pi}{4} = 1 - \frac{1}{3} + \frac{1}{5} - \frac{1}{7} + \frac{1}{9} - ...$$

This series converges, albeit slowly, to the value of $\frac{\pi}{4}$, and allows us to get an approximate value of π. The following Python program calculates the value of π by the 5000 first terms of the series and represents one out of every hundred terms. The exact value of π is represented as a straight red line:

```
―――――――――――――――――― output of the program ――――――――――――――――――
1      1.00000000    4.00000000         995   -0.00100503    3.13958462
3     -0.33333333    2.66666667         997    0.00100301    3.14359666
5      0.20000000    3.46666667         999   -0.00100100    3.13959266
7     -0.14285714    2.89523810        1001    0.00099900    3.14358866
9      0.11111111    3.33968254        1003   -0.00099701    3.13960062
11    -0.09090909    2.97604618        1005    0.00099502    3.14358072
13     0.07692308    3.28373848         ...
15    -0.06666667    3.01707182        4993    0.00020028    3.14199313
17     0.05882353    3.25236593        4995   -0.00020020    3.14119233
19    -0.05263158    3.04183962        4997    0.00020012    3.14199281
21     0.04761905    3.23231581        4999   -0.00020004    3.14119265
...
```

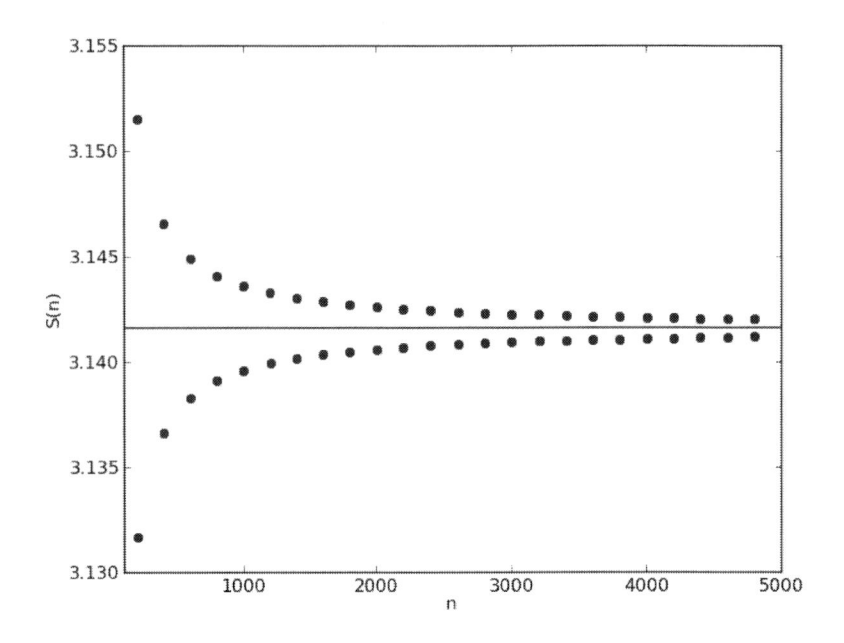

```python
# -*- coding: utf-8 -*-
"""
Mathematics and Python Programming    www.pysamples.com
p8j.py
Calculates number pi using Leibniz series
"""

import matplotlib.pyplot as plt
import numpy as np

sign = 1.0
piaprox = 0
piseries = []
x = []
step = 0
for i in range(1, 5000, 2):
    piaprox += 4 * sign / i
    if step == 100:
        piseries.append(piaprox)
        x.append(i)
        piseries.append(piaprox - 4.0 / (i + 1))
        x.append(i + 1)
        step = 0
    print str(i) + "%15.8f" % (sign / i) + "%15.8f" % piaprox
    sign *= -1
    step += 1
#print piseries
plt.plot(x, piseries, 'bo', lw=1.0)
plt.plot([99, 5000], [np.pi, np.pi], 'r-', lw=1.5)
plt.xlim(99, 5000)
plt.xlabel('n')
plt.ylabel('S(n)')
plt.show()
```

8.6 Applications of the derivative

8.6.1 Representation of polynomial functions

The Python program that is shown at the end of this section creates graphics of polynomial functions, also indicating the points where the function cuts the axes, and the maximum, minimum, and inflection points. The polynomial is introduced as a list at the line

a = [1, -1, -7, 1, 6]

That corresponds to the polynomial $1x^4 - x^3 - 7x^2 + x + 6$, for example.

The program also decides the maximum values of x and y to be graphed, and in the window in which the function appears represented, Matplotlib allows you to select with your mouse still over the area you want to show.

The function $y = 2x^5 - 4x^3 - 1$ has been represented in this example. The maximum is marked with the letter M; the minimum with m, and the inflection point with In:

```
─────────────────────────────── output of the program ───────────────────────────────
polynomial P: [-1, 0, 0, -4, 0, 2] =    2x^{5}     -4x^{3} -1
Cuts axis Y: (0, -1 )
cuts axis X: ( 1.469, 0)
cuts axis X: (-1.338, 0)
cuts axis X: (-0.690, 0)
P': [  0.   0. -12.   0.  10.   0.] =  10.0x^{4}    -12.0x^{2}
P'': [  0. -24.   0.  40.   0.   0.] =  40.0x^{3}   -24.0x^{1}
derivative is null when   x= -1.10
derivative is null when   x=  1.10
derivative is null when   x=  0.00
derivative is null when   x=  0.00
d2(-1.09544511501) = [-26.29068276]: Maximum: (-1.1,  1.1)
d2(1.09544511501)  = [ 26.29068276]: minimum: ( 1.1, -3.1)
d2(0.0) = [ 0.]: Inflection point: ( 0.0, -1.0)
d2(0.0) = [ 0.]: Inflection point: ( 0.0, -1.0)
```

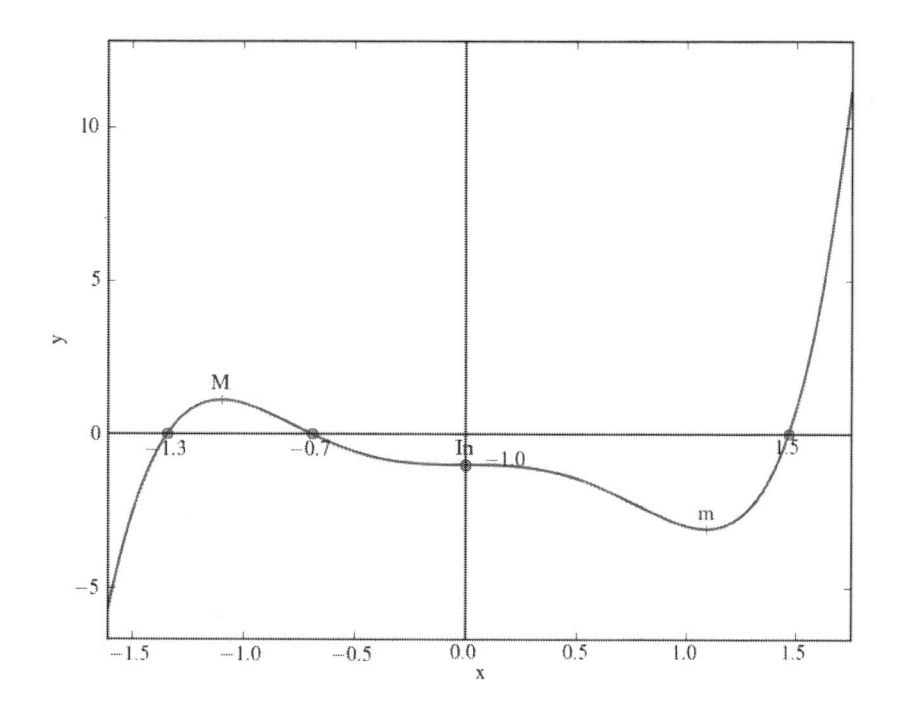

Representation of the function $y = x^4 - 2x^3 - x^2 + 2x - 3$:

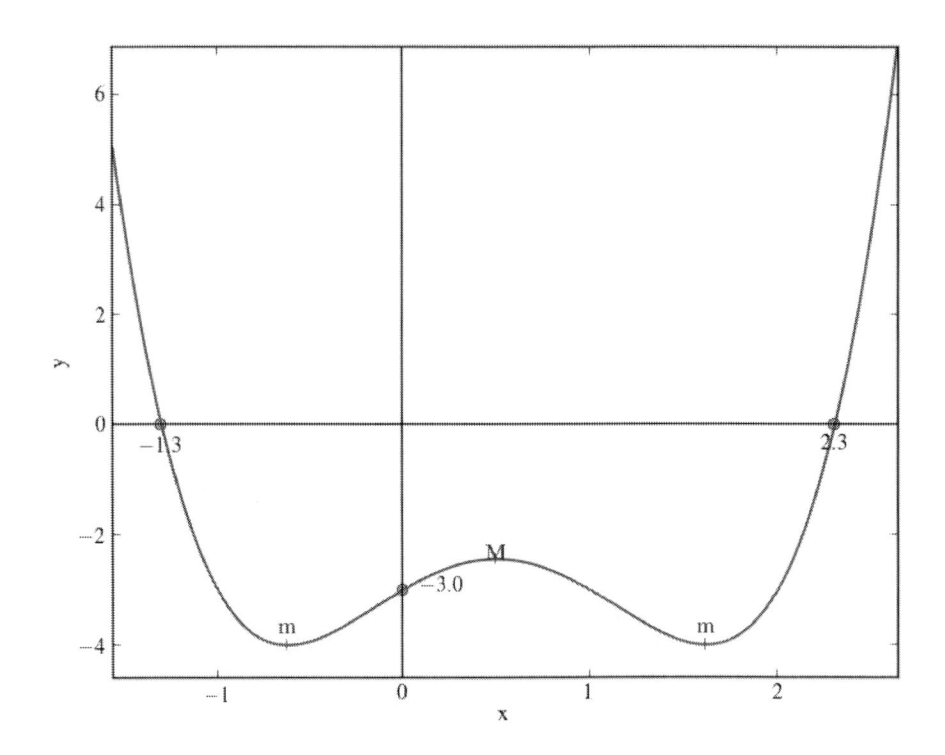

```
------------------------------ output of the program ------------------------------
polynomial P: [-3, 2, -1, -2, 1]  =   1x^{4}    -2x^{3}    -1x^{2} + 2x^{1} -3
Cuts axis Y: (0, -3 )
cuts axis X: ( 2.303, 0)
cuts axis X: (-1.303, 0)
P': [ 2. -2. -6.  4.  0.] =  4.0x^{3}   -6.0x^{2}   -2.0x^{1} +2.0
P'': [ -2. -12. 12.  0.  0.] = 12.0x^{2}   -12.0x^{1} -2.0
derivative is null when   x=  1.62
derivative is null when   x= -0.62
derivative is null when   x=  0.50
d2(1.61803398875) = [ 10.]: minimum: ( 1.6, -4.0)
d2(-0.61803398875) = [10.]: minimum: (-0.6, -4.0)
d2(0.5) = [-5.]: Maximum: ( 0.5, -2.4)
```

```python
# -*- coding: utf-8 -*-
"""
Mathematics and Python Programming     www.pysamples.com
p8f.py
represents polynomials, with M, m, axis cuts and inflection point
"""

import numpy as np
import matplotlib.pyplot as plt
from matplotlib import rc

rc('font', **{'family': 'serif', 'serif': ['Times']})
rc('text', usetex=True)

#coefficients a0, a1,... an

a = [-1, 0, 0, -4, 0, 2]

'''
fis_y0 = 0.0
```

8.6. APPLICATIONS OF THE DERIVATIVE

```
fis_v0 = 50.0
fis_g = -9.81
a = [fis_y0, fis_v0, 0.5 * fis_g]
'''

importantx = [0]
importanty = [0]


def latexpoly(p):
    if p[0] > 0:
        polynomial = '+' + str(p[0])
    elif p[0] < 0:
        polynomial = str(p[0])
    else:
        polynomial = ''
    n = 1
    while n <= (len(p) - 1):
        if p[n] == 0:
            n += 1
        else:
            sign = ''
            if p[n] < 0:
                sign = ' '
            elif p[n] > 0:
                sign = '+'
            if n > 0:
                polynomial = ' ' + str(p[n]) + 'x^{' + str(n) + '} ' + polynomial
            else:
                polynomial = ' ' + str(p[n]) + polynomial
            polynomial = sign + polynomial
            n += 1
    if (polynomial[0]) == '+':
        polynomial = polynomial[1:len(polynomial)]
    return polynomial


def derivative(p):
    d = np.zeros(len(p), float)
    i = len(p) - 1
    while i > 0:
        d[i - 1] = p[i] * i
        i -= 1
    return d


def f(x):
    y = a[0]
    i = 1
    while i < len(a):
        y = y + a[i] * x ** i
        i += 1
    return y


print 'polynomial P: ' + str(a) + ' = ' + latexpoly(a)
```

```python
print 'Cuts axis Y: (0,', f(0), ')'
plt.plot(0, f(0), 'yo')
phrase = '$' + "%3.1f" % f(0) + '$'
plt.text(0.1, f(0), phrase, horizontalalignment='left', fontsize=12,
         color='black', weight='bold')

#cuts axis X
cutX = np.roots(a[::-1])
cut_axis_X = False
cutXreal = []
for cut in cutX:
    if np.iscomplex(cut) == False:
        cut = float(np.real(cut))
        print 'cuts axis X: (' + "%6.3f" % cut + ', 0)'
        cutXreal.append(cut)
        cut_axis_X = True
        plt.plot(cut, f(cut), 'yo')
        phrase = '$' + "%3.1f" % cut + '$'
        plt.text(cut, -0.2, phrase, horizontalalignment='center',
                 verticalalignment='top', fontsize=12, color='black', weight='bold')
        importantx.append(cut)
        importanty.append(cut)
if cut_axis_X == False:
    print 'the function does not cut axis X'

#derivatives
d1 = np.zeros(len(a), float)
d2 = np.zeros(len(a), float)
d1 = derivative(a)
d2 = derivative(d1)
print 'P' + "'" + ': ' + str(d1) + ' = ' + latexpoly(d1)
print 'P' + "''" + ': ' + str(d2) + ' = ' + latexpoly(d2)

#roots of derivative d1
list1 = d1[::-1]
roots1 = np.roots(list1)
for i in range(0, len(roots1)):
    if np.iscomplex(roots1[i]) is True:
        np.delete(roots1, i)
    else:
        plt.plot(roots1[i], f(roots1[i]), 'k+')
        print 'derivative is null when  x= ' + "%5.2f" % roots1[i]
        importantx.append(roots1[i])
        importanty.append(f(roots1[i]))

for r in roots1:
    if np.iscomplex(r) == False:
        valor = d2[0]
        for i in range(1, len(d2)):
            valor = valor + d2[i] * r ** [i]
            position = '(' + "%4.1f" % r + ', ' + "%4.1f" % f(r) + ')'
            if valor < 0:
                point = 'Maximum: ' + position
                letters = 'M'
            elif valor > 0:
                point = 'minimum: ' + position
```

8.6. APPLICATIONS OF THE DERIVATIVE

```
                letters = 'm'
            else:
                point = 'Inflection point: ' + position
                letters = 'In'
        print 'd2(' + str(r) + ') = ' + str(valor) + ': ' + point
        plt.text(r, 0.3 + f(r), letters, horizontalalignment='center',
                verticalalignment='bottom', fontsize=13, color='blue', weight='bold')

pointsnum = 300
if len(a) < 3:
    xmax = 2 * np.max(importantx)
    xmin = 2 * np.min(importantx)
else:
    xmax = 1.2 * np.max(importantx)
    xmin = 1.2 * np.min(importantx)
if xmin == 0:
    xmin = -1
if xmax == 0:
    xmax = 1
importanty.append(f(xmax))
importanty.append(f(xmin))
ymax = 1.15 * np.max(importanty)
ymin = 1.15 * np.min(importanty)
if ymin == 0:
    ymin = -1
if ymax == 0:
    ymax = 1
x = np.linspace(xmin, xmax, pointsnum)
y = np.zeros(pointsnum, float)
for i in range(0, pointsnum):
    y[i] = f(x[i])
plt.xlim(xmin, xmax)
plt.ylim(ymin, ymax)
plt.plot(x, y, 'r-', lw=1.5)
plt.axhline(color='black', lw=1)
plt.axvline(color='black', lw=1)
plt.ylabel('y')
plt.xlabel('x')
plt.show()
```

8.6.2 Physics: parabolic trajectory

Using the known equations of kinematics, which we're not going to demonstrate here, we can calculate the range, the maximum height and the time it takes for a mobile to reach their goal after being launched with a speed v_0 and an angle α. The following program represents these equations for multiple angles, as well as the parabola of security, (shaded in the graphic), which is the region whose points can reach the mobile if it is launched with any angle at a fixed initial speed.

As is known, the maximum height is achieved with an angle 90 degrees, and is equal to $H = \frac{v_0^2}{2g}$; the maximum reach a is achieved with an angle 45 degrees, and is equal to $\frac{v_0^2}{g} = 2H$. The parabola of security is the parabola that passes through the following three points: $(-a, 0)$, $(0, H)$ and $(a, 0)$. Let's take, for example, the world record in javelin throw, which Jan Zelezný owns since the year 1996 with a distance of 98.48 meters. Suppose that he launched the javelin at the optimum angle of 45 degrees, and therefore its initial speed was

$$v_0 = \sqrt{98.48g} \sim 31.08 ms^{-1}$$

The following Python program calculates distances that would have been reached with other angles, and the parabola of security.

```
# -*- coding: utf-8 -*-
"""
Mathematics and Python Programming    www.pysamples.com
p8g.py
"""

import numpy as np
import matplotlib.pyplot as plt
from matplotlib import rc

rc('font', **{'family': 'serif', 'serif': ['Times']})
rc('text', usetex=True)

y0 = 0.0
v0 = 31.08
print 'v0 = ' + "%.2f" % v0 + ' m/s'
g = -9.81
H = (v0 ** 2) / (-2 * g)   # maximum attainable height
print 'maximum height = ', "%.1f" % H, ' m'
maximumrange = 2 * H
print 'maximum range = ', "%.1f" % maximumrange, ' m'
angles_list = [15.0, 30.0, 45.0, 60.0, 85.0]
#vy = v0y - g * t; time to reach the highest point: tfloor / 2
#vx = v0x; x = v0x * t; y = y0 + v0y*t + 0.5*g*t**2
#y = y0 + v0y*x/v0x + (0.5*g/v0x**2)*x**2
#ymax = y0 + v0y*tfloor/2 + 0.5*g*(tfloor/2)**2

print 'angle     ymax      range       time'
for j in range(0, 5):
    alfa = np.deg2rad(angles_list[j])
    v0y = v0 * np.sin(alfa)
    tfloor = - 2 * v0y / g
    v0x = v0 * np.cos(alfa)
    Range = - (v0 ** 2) * np.sin(2 * alfa) / g
    ymax = y0 + (v0y * tfloor / 2) + (0.5 * g * (tfloor / 2) ** 2)
    print (str(angles_list[j]) + "%11.1f" % ymax + "%11.1f" % Range + "%11.2f" % tfloor)


def f(x, beta):
    v0x = v0 * np.cos(np.deg2rad(beta))
    #coeficientes a0, a1,... an
    a = [y0, (np.tan(np.deg2rad(beta))), ((0.5 * g) / (v0x ** 2))]
    y = a[0]
    i = 1
    while i < len(a):
        y = y + a[i] * x ** i
        i += 1
    return y

pointsnum = 300
x = np.linspace(0, np.ceil(maximumrange), pointsnum)
y15 = np.zeros(pointsnum, float)
y30 = np.zeros(pointsnum, float)
y45 = np.zeros(pointsnum, float)
```

8.6. APPLICATIONS OF THE DERIVATIVE

```python
y60 = np.zeros(pointsnum, float)
y85 = np.zeros(pointsnum, float)
for i in range(0, pointsnum):
    y15[i] = f(x[i], 15.0)
    y30[i] = f(x[i], 30.0)
    y45[i] = f(x[i], 45.0)
    y60[i] = f(x[i], 60.0)
    y85[i] = f(x[i], 85.0)
plt.plot(x, y15, 'k:', lw=2, label='$15$')
plt.plot(x, y30, 'k-.', lw=2, label='$30$')
plt.plot(x, y45, 'r-', lw=2.5, label='$45$')
plt.plot(x, y60, 'k--', lw=2, label='$60$')
plt.plot(x, y85, 'k:', lw=2, label='$85$')

plt.plot(-x, y15, 'k:', lw=2, label='$15$')
plt.plot(-x, y30, 'k-.', lw=2, label='$30$')
plt.plot(-x, y45, 'r-', lw=2.5, label='$45$')
plt.plot(-x, y60, 'k--', lw=2, label='$60$')
plt.plot(-x, y85, 'k:', lw=2, label='$85$')
plt.axhline(color='grey', lw=1)
plt.axvline(color='grey', lw=1)
plt.legend(('$15\,^{\circ}$', '$30\,^{\circ}$', '$45\,^{\circ}$',
            '$60\,^{\circ}$', '$85\,^{\circ}$'), loc='best')
plt.ylabel('y')
plt.xlabel('x')

#security parabola
print 'security parabola: Ax^2 + Bx + C = 0'
A = -H / (maximumrange ** 2)
print 'A = -H / v0^2 = ', A
print 'B = 0'
print 'C = H = ', H
print ('y = ' + "%.6f" % A + 'x^2 + ' + "%.6f" % H)
#calculates 100 points of the parabola
xvalues = np.linspace(-maximumrange, maximumrange, 100)
yvalues = np.zeros(100, float)

for k in range(0, 100):
    yvalues[k] = A * xvalues[k] ** 2 + H
plt.plot(xvalues, yvalues, 'k-', lw=1.0)
plt.fill_between(xvalues, yvalues, 0, alpha=0.2, color='#BDD0D7')
plt.ylim(0, 1.1 * (v0 ** 2) / (-2 * g))
plt.show()
```

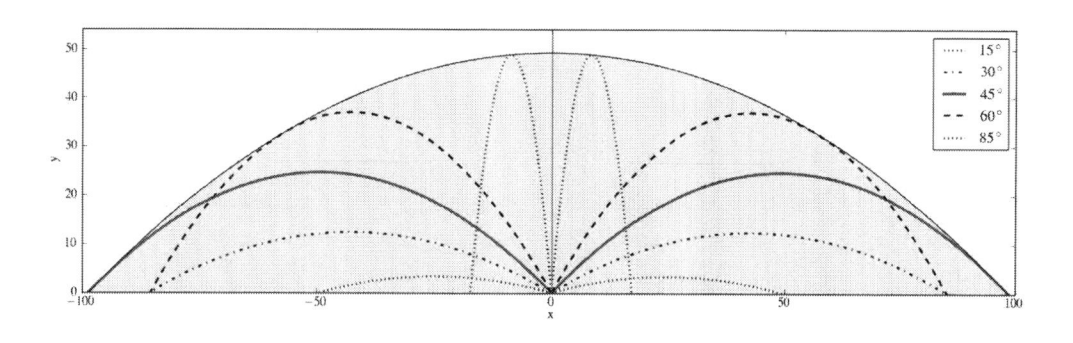

```
                                            ──────── output of the program ────────
v0 = 31.08 m/s
maximum height =   49.2  m
maximum range  =   98.5  m
angle       ymax        range       time
15.0         3.3         49.2        1.64
30.0        12.3         85.3        3.17
45.0        24.6         98.5        4.48
60.0        36.9         85.3        5.49
85.0        48.9         17.1        6.31
security parabola: Ax^2 + Bx + C = 0
A = -H / v0^2 =   -0.00507781637125
B = 0
C = H =   49.2337614679
y = -0.005078x^2 + 49.233761
```

8.6.3 Physics: Planck's law

According to the classical theory, the energy emitted by a radiant cavity must follow the formula of Raleigh-Jeans :

$$\rho_T(\nu)d\nu = \frac{8\pi\nu^2 kT}{c^3}d\nu$$

for our objectives, at this time we are only interested in the form of this function and not in their exact values. Call $y = \rho$, the energy density of J/m^3, $y > 0$; $x = \nu$, the frequency of the radiation, in s^{-1}, $x > 0$. If we group the constants of the formula in a constant that we call $C > 0$, we get:

$$y = CTx^2$$

that corresponds to a branch of a parabola for values of the temperature $T >= 0$.

According to this equation, the emitted energy at high frequencies by a radiant cavity should tend to infinity, however the experimental results show that the emitted energy is finite. This discrepancy is known as the "ultraviolet catastrophe", and points out the impossibility of the classical theory to explain it.

Max Planck tried to resolve the discrepancy between the classical theory and the experimental results and developed a theory according to which, the emitted radiation would be:

$$\rho_T(\nu)d\nu = \frac{8\pi\nu^2 kT}{c^3}\frac{h\nu}{e^{h\nu/kT}-1}d\nu$$

for a broad range of temperatures, we can take

$$e^{h\nu/kT} \gg 1$$

and grouping the constants as in the classic case, we get an equation of the form:

$$y = C_1 x^2 \frac{x}{e^{C_2 x/T}}$$

and for simplicity, we take the constants equal to the unit, as we are only interested in here the shape of the curve, and not its exact physical values:

$$y = x^3 e^{-x/T}$$

We can easily calculate their first two derivatives with Python:

```
# -*- coding: utf-8 -*-
"""
Mathematics and Python Programming    www.pysamples.com
p8h.py
"""

import sympy as sy
```

8.6. APPLICATIONS OF THE DERIVATIVE

```
x, T = sy.symbols('x T')
sy.init_printing(use_unicode=True)

fx = x ** 3 * sy.exp(-x / T)
print 'f = ' + str(fx)
for d in range(1, 3):
    derivative = sy.diff(fx, x, d)
    print ('D' + str(d) + ' = ' + str(derivative))

#sy.simplify(6*x*sy.exp(-x/T) - 6*x**2*sy.exp(-x/T)/T + x**3*sy.exp(-x/T)/T**2)
print 'D2(3T) = '
print sy.simplify(6 * 3 * T * sy.exp(-3 * T / T) -
                  6 * (3 * T) ** 2 * sy.exp(-3 * T / T) / T +
                  (3 * T) ** 3 * sy.exp(-3 * T / T) / T ** 2)
```

---------- output of the program ----------
```
f = x**3*exp(-x/T)
D1 = 3*x**2*exp(-x/T) - x**3*exp(-x/T)/T
D2 = 6*x*exp(-x/T) - 6*x**2*exp(-x/T)/T + x**3*exp(-x/T)/T**2
```

We can write the first derivative:

$$f'(x) = \left(3x^2 - \frac{x^3}{T}\right) e^{-x/T}$$

and for $f'(x) = 0$ there must be null the parenthess

$$3x^2 - \frac{x^3}{T} = x^2 \left(3 - \frac{x}{T}\right) = 0$$

$$x = 3T$$

We can quickly calculate the value of the second derivative for $x = 3T$ by a couple of lines of code of the program:

```
print 'D2(3T) = '
print sy.simplify(6 * 3 * T * sy.exp(-3 * T / T) -
                  6 * (3 * T) ** 2 * sy.exp(-3 * T / T) / T +
                  (3 * T) ** 3 * sy.exp(-3 * T / T) / T ** 2)
```

---------- output of the program ----------
```
D2(3T) =
-9*T*exp(-3)
```

Therefore, we have for $x = 3T$, $f'(x) = 0$; $f''(x) = -9Te^{-3} < 0$, which tells us that the function has a maximum for $x = 3T$.

We represent the function

$$y = x^3 e^{-x/T}$$

for several values of T:

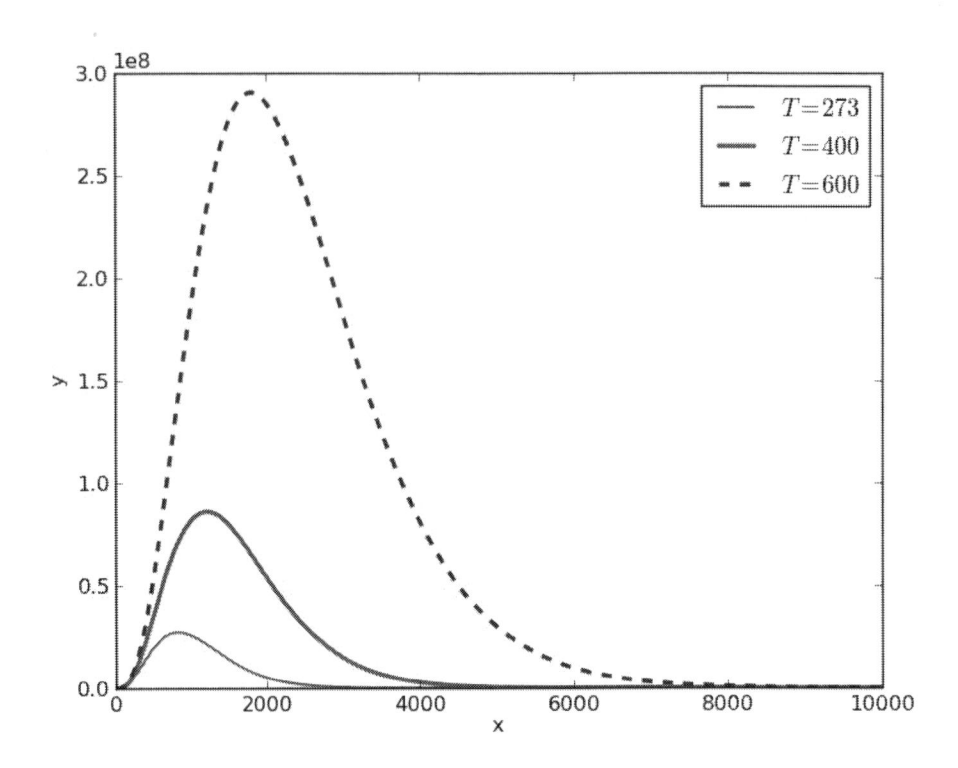

Therefore, according to the theory of Planck, the energy emitted at each temperature has a maximum, and the frequency at which this maximum appears increases with temperature.

```
# -*- coding: utf-8 -*-
"""
Mathematics and Python Programming    www.pysamples.com
p8i.py
represents approximately black body radiation, taking f = x**3*exp(-x/T)
"""

import numpy as np
import matplotlib.pyplot as plt

pointsnum = 500
fig = plt.figure(facecolor='white')
x = np.linspace(0, 1e4, pointsnum)
y1 = np.zeros(pointsnum, float)
y2 = np.zeros(pointsnum, float)
y3 = np.zeros(pointsnum, float)
for i in range(0, pointsnum):
    y1[i] = x[i] ** 3 * np.exp(-x[i] / 273.0)
    y2[i] = x[i] ** 3 * np.exp(-x[i] / 400.0)
    y3[i] = x[i] ** 3 * np.exp(-x[i] / 600.0)
p1, = plt.plot(x, y1, 'g-', lw=1.5, label='$T=273$')
p2, = plt.plot(x, y2, 'r-', lw=2.5, label='$T=400$')
p3, = plt.plot(x, y3, 'b--', lw=2.5, label='$T=600$')
plt.legend(('$T=273$', '$T=400$', '$T=600$'), loc='best')
plt.ylabel('y')
plt.xlabel('x')
plt.axhline(color='black', lw=1)
plt.axvline(color='black', lw=1)
plt.show()
```

8.6.4 Economics: production function

In a rice cultivation the amount of rice produced in a given area varies with the amount of fertilizer used. Suppose that in an agronomic study have been used fertilizer quantities ranging from the 50 and the 250 kg/ha, and the following relationship was found, where y is the rice production in kg/ha and x the amount of fertilizer used in kg:

$$y = 16.4x - 0.06x^2$$

It is a polynomial function which we can represent as we have seen in the previous section using the program p8f.py taking the values:

    a = [0, 16.4, -0.06]

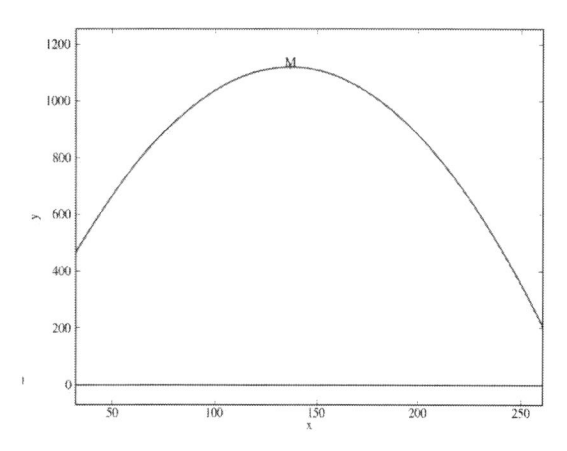

As it is shown below, the optimal quantity of fertilizer turns out to be $136.7 kg/ha$. If we use more, fertilizer is being wasted; if we use less, the production maximum of $1120 Kg/ha$ is not reached.

```
――――――――――――――――――――――――― output of the program ―――――――――――――――
polynomial P: [0, 16.4, -0.06] =   -0.06x^{2} + 16.4x^{1}
Cuts axis Y: (0, 0.0 )
cuts axis X: (273.333, 0)
cuts axis X: ( 0.000, 0)
P': [ 16.4  -0.12   0. ] =   -0.12x^{1} +16.4
P'': [-0.12  0.    0. ] = -0.12
derivative is null when   x= 136.67
d2(136.666666667) = [-0.12]: Maximum: (136.7, 1120.7)
```

8.6.5 Physics: Newton's law of cooling

The Newton's Law of cooling determines the speed at which a body cools:

$$\frac{dT}{dt} = -k(T - T_a)$$

being T the temperature of the body at the time t, T_0 the initial temperature of the body, and T_a the ambient temperature. This equation can be solved, and we can obtain the temperature T for each time instant t:

$$T = T_a + (T_0 - T_a) \cdot e^{-kt}$$

Now let's look at an example of its application: a pizza has been baked at 210 degrees Celsius. Five minutes after we take the pizza out of the oven and put it on the table, at an ambient temperature of 22 degrees Celsius, the temperature of the pizza is 60 degrees Celsius, so we can replace and calculate the constant k:

$$60 = 22 + (210 - 22) \cdot e^{-k*5}$$

$$-5k = \ln\left(\frac{60 - 22}{210 - 22}\right) = \ln\left(\frac{38}{188}\right)$$

$$k = 0.319771160621$$

What temperature will be the pizza a quarter of an hour after being pulled out of the oven?

$$T = 22 + (210 - 22) \cdot e^{-k*15} = 23.5$$

Bibliography for this chapter: [1], [2], [5], [7], [9], [13], [19], [24], [45], [47], [50], [55], [56], [60], [63]

9 | Integral

9.1 Riemann integral

Let $f(x)$ be a continuous function in an interval $[a, b]$. We arbitrarily divide the interval $[a, b]$ in n parts, taking as points of division:

$$a = x_0 < x_1 < x_2 < ... < x_{n-1} < x_n = b$$

We choose any point ξ_i in each interval $[x_{i-1}, x_i]$ and we perform the following sum:

$$I_n = \sum_{i=1}^{i=n} (x_i - x_{i-1}) f(\xi_i)$$

This sum is called a Riemann sum for $f(x)$ corresponding to the given partition of the interval. We carry this out for different partitions, with $n = 1, 2, 3, ...$, each time with an arbitrary division of the interval, but such that the range of all the intervals $[x_{i-1}, x_i]$ tends to zero as n increases. Then the limit

$$\lim_{n \to \infty} I_n = I$$

always exists and it is independent of the points of division and of the chosen points ξ_i. That is to say, for any number ε we can find a number $\delta(\varepsilon) > 0$ such that if

$$|x_i - x_{i-1}| < \delta$$

there is a number I such that

$$|I_n - I| < \varepsilon$$

and this number I is called definite integral of the function $f(x)$ between a and b:

$$I = \int_a^b f(x) dx$$

The following program computes the Riemann sum for a function. It sets a maximum interval length δ, and the length of each interval is a random number less than δ. Within each interval it also chooses a number ξ at random. It also displays the exact value of the integral, by using the fundamental theorem of integral calculus that we will prove in section 9.4. We show now this program just to show how much the Riemann sum approaches the exact value of the area under the function, as we take a smaller interval δ.

```
# -*- coding: utf-8 -*-
"""
Mathematics and Python Programming      www.pysamples.com
p9a.py
"""

import numpy as np
import sympy as sy
import matplotlib.pyplot as plt
```

9.1. RIEMANN INTEGRAL

```python
x = sy.symbols('x')
sy.init_printing(use_unicode=True)

a = 2.0
b = 10.0
#delta = 2.5
delta = 1.0
#delta = 0.25
z = [a]

print 'f(x) = 3 * x ** 2 + 1'
print 'Integral: '
integral = sy.integrate(3 * x ** 2 + 1)
Fb = float(integral.subs(x, b).evalf(6))
Fa = float(integral.subs(x, a).evalf(6))
print integral
print ('I = ' + "%6.4f" % Fb + ' - ' +
       "%6.4f" % Fa + ' = ' + str(Fb - Fa))
print

j = 1
i = a
width = []
while i < b:
    randomlength = np.random.rand() * delta
    z.append(z[j - 1] + randomlength)
    if z[j] > b:
        z[j] = b
    width.append(z[j] - z[j - 1])
    i = z[j]
    j += 1
z = np.around(z, 6)
n = len(z) - 1
print '[a, b] = [', a, ', ', b, ']'
print 'delta = ', delta
print 'number of intervals: ', n
width.append(0)
width = np.around(width, 6)
norma = np.max(width)
print 'maximum length of an interval = ', norma


def f(t):
    fx = (3 * t ** 2 + 1)
    return fx


print 'Riemann Sum:'
xi = np.zeros(n + 1, float)
y = np.zeros(n + 1, float)
suma = 0
for i in range(1, n + 1):
    xi[i] = z[i - 1] + np.random.rand() * (z[i] - z[i - 1])
    #print i, z[i - 1], z[i], 'anchura:' , z[i] - z[i-1], xi[i]
    y[i] = f(xi[i])
```

```
        suma += (z[i] - z[i - 1]) * y[i]
print 'endpoints of the intervals:'
print z
print 'length of the intervals:'
print width[0: -1]
print 'total length: ', width.sum()
print 'values of xi inside every interval:'
print xi[1: len(xi)]
#print y
print 'I = ', suma
#graph
heights = np.zeros(n, float)
heights = np.delete(y, 0)
heights = np.append(heights, 0)
#print heights
fig = plt.figure()
ax = plt.subplot(111)
ax.bar(z, heights, width=width, alpha=0.4, color='#3E9ECB')
for k in range(1, len(xi)):
    ax.plot(xi[k], 0, 'yo')
#plot the function
graphpoints = 200
funcion = np.zeros(graphpoints + 1, float)
equis = np.zeros(graphpoints + 1, float)
increment = (b - a) / graphpoints
for i in range(0, graphpoints + 1):
    equis[i] = a + i * increment
    funcion[i] = f(equis[i])
plt.plot(equis, funcion, 'r--', lw=3)
plt.ylabel('y')
plt.xlabel('x')
plt.show()
```

The following graphs represent the function and the Riemann sum for the values of $\delta = 2.5, 1.0$ and 0.25. The values of $x = \xi_i$ are represented as points in yellow on the axis X:

———————————————— output of the program ————————————————
```
f(x) = 3 * x ** 2 + 1
Integral:
x**3 + x
I = 1010.0000 - 10.0000 = 1000.0
[a, b] = [ 2.0 , 10.0 ]
delta =  2.5
number of intervals:  7
maximum length of an interval =  2.439395
Riemann Sum:
endpoints of the intervals:
[  2.        3.601538  4.432247  4.735722  6.389128  7.333767  9.773162  10.]
length of the intervals:
[ 1.601538  0.830709  0.303475  1.653405  0.944639  2.439395  0.226838]
total length:  7.999999
values of xi inside every interval:
[ 2.00034735  3.85167607  4.7075039   5.44779322  7.11435243  9.54236249  9.78302267]
I =  1106.51986245
```

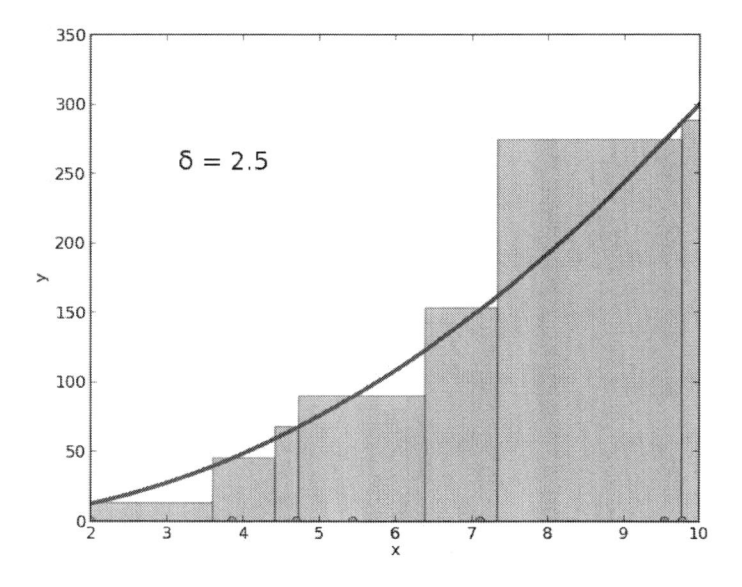

9.1. RIEMANN INTEGRAL

δ = 2.5

---------- output of the program ----------
```
f(x) = 3 * x ** 2 + 1
Integral:
x**3 + x
I = 1010.0000 - 10.0000 = 1000.0
[a, b] = [ 2.0 , 10.0 ]
delta = 1.0
number of intervals: 15
maximum length of an interval =  0.912919
Riemann Sum:
endpoints of the intervals:
[ 2.       2.102812  2.916805  3.611286  4.324224  4.770668  5.121727  6.034647
  6.246615  6.983055  7.553192  8.256003  8.564931  9.029536  9.688369 10.]
length of the intervals:
[ 0.102812  0.813993  0.694481  0.712938  0.446444  0.351059  0.912919  0.211969
  0.736439  0.570138  0.70281   0.308928  0.464605  0.658833  0.311631]
total length: 7.999999
values of xi inside every interval:
[ 2.09493555  2.37200072  3.20069441  3.81526692  4.46616876  5.05451924
  5.94030017  6.08136339  6.58219484  7.28602608  7.67271085  8.34500137
  8.69308617  9.19668583  9.93856481]
I =   989.380858446
```

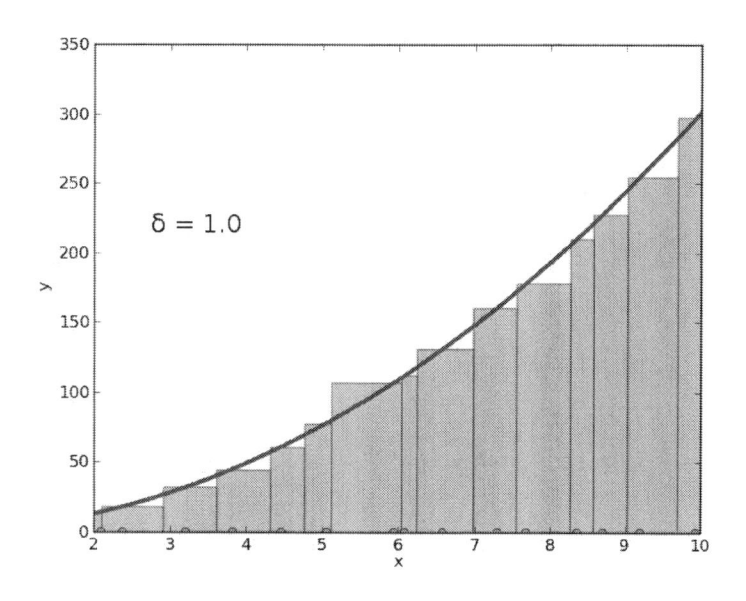

δ = 1.0

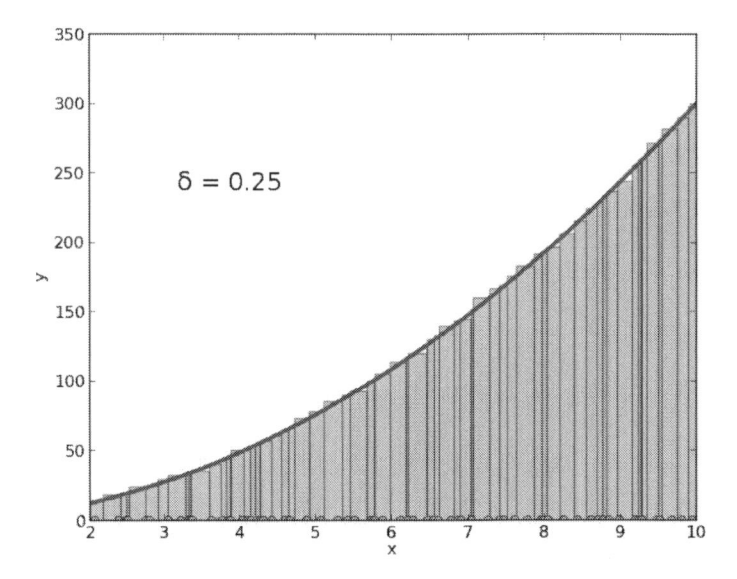

```
─────────────── output of the program ───────────────
 f(x) = 3 * x ** 2 + 1
Integral:
x**3 + x
I = 1010.0000 - 10.0000 = 1000.0
[a, b] = [ 2.0 , 10.0 ]
delta = 0.25
number of intervals: 67
maximum length of an interval = 0.24928
Riemann Sum:
endpoints of the intervals:
[ 2.         2.174407   2.423687   ...   9.758952   9.896344  10. ]
length of the intervals:
[ 0.174407  0.24928   0.08318  ...  0.214383  0.137392  0.103656]
total length: 8.000004
values of xi inside every interval:
[ 2.05893415  2.38708876  2.49601634  ...  9.81097521  9.95009989]
I = 1000.49729438
```

9.2 Indefinite integral

In the program of the previous chapter we have made use of a feature of Python that enables us to obtain a function that is called antiderivative, primitive integral or indefinite integral. We have already seen that the derivative of a function $f(x)$ is

$$f'(x) = \frac{df(x)}{dx}$$

The inverse operation of the derivative is to search for a function $F(x)$ such that $F'(x) = f(x)$ and is expressed as:

$$F(x) = \int f(x)dx$$

Every function $F(x)$ which satisfies the condition that its derivative is $F'(x) = f(x)$, is called primitive function of the function $f(x)$. So the integration of a function $f(x)$ is the search of a primitive function.

If $F(x)$ is a primitive function of $f(x)$, then the function $F(x) + C$ also will be, where C a constant, because $(F(x) + C)' = f(x)$. The constant C of integration must be added if we want to get the expression of all the primitives, which would differ from one another in the value of a constant. If we take $C = 0$ we obtain the expression of only one of the primitives of $f(x)$.

Although all continuous function $f(x)$ in an interval has a primitive function in that interval, it is often impossible to get the expression of the primitive function $F(x)$, and in such cases we have

9.2. INDEFINITE INTEGRAL

to resort to approximate methods. For another series of cases, there exist a method for obtaining the expression of the primitive $F(x)$.

This is not the place, nor is the purpose of this book detailing all of the possible methods of integration. We are interested now in noting the tools that Python offers to get the expression of the primitive of a function in certain cases, and see some examples:

```
# -*- coding: utf-8 -*-
"""
Mathematics and Python Programming     www.pysamples.com
p9b.py
"""

import sympy as sy

x = sy.symbols('x')
sy.init_printing(use_unicode=True)

print 'f(x) = x ** 4 - 3 * x ** 2 + 0.5 *x + 10'
integral = sy.integrate(x ** 4 - 3 * x ** 2 + 0.5 * x + 10)
print 'F(x) = ', integral
```

──────────────────────────── program outputs ────────────────────────────
```
f(x) = x ** 4 - 3 * x ** 2 + 0.5 *x + 10
F(x) =   0.2*x**5 - 1.0*x**3 + 0.25*x**2 + 10.0*x

f(x) = cos x
F(x) =   sin(x)

f(x) = sen 5x cos 5x
F(x) =   sin(5*x)**2/10

f(x) = x exp(x ** 2)
F(x) =   exp(x**2)/2

f(x) = sqrt(x**3)
F(x) =   2*x*(x**3)**(1/2)/5

f(x) = (sen x) ** 2
F(x) =   x/2 - sin(x)*cos(x)/2

f(x) = cos x (senx)3
F(x) =   sin(x)**4/4

f(x) = 1 / (3x+1) ** 4
F(x) =   -1/(9*(3*x + 1)**3)

f(x) = cos3x exp(sen3x)
F(x) =   exp(sin(3*x))/3

f(x) = x cos x
F(x) =   x*sin(x) + cos(x)

f(x) = (x + 2) * sin(x ** 2 + 4 * x + 6)
F(x) =   -cos(x**2 + 4*x + 6)/2

f(x) = 1 / (x * (log(x)) ** 3)
F(x) =   -1/(2*log(x)**2)

f(x) = (6 - x) / ((x - 3) * (2 * x + 5))
F(x) =   3*log(x - 3)/11 - 17*log(x + 5/2)/22
```

9.3 Definite Integral

Let $f(x)$ be a function defined in the closed interval $[a, b]$; the following expression is called definite integral of the function f between a and b:

$$\int_a^b f(x)dx = \lim_{n \to \infty} \sum_{i=1}^n f(\xi_i)\Delta x_i$$

The number a is called lower limit of integration, b is the upper limit. The symbol $\int$ comes from the letter S indicating that the integral is the limit of a sum. We will now enumerate some properties of the definite integral:

- If a function $f(x)$ is continuous on an interval $[a, b]$, then it is integrable in that interval.
- $\int_a^b f(x)dx = -\int_b^a f(x)dx$
- $\int_a^a f(x)dx = 0$
- If $f(x) \geq 0 \quad \forall x \in [a, b]$, the definite integral can be geometrically interpreted as the area A between the curve of the function $f(x)$ and axis X. The following Python program represents the circle of radius equal to 1 centered at the origin $f(x) = \sqrt{1 - x^2}$ in the interval $x \in [0, 1]$ and the area between the curve and the axis X:

$$A = \int_0^1 \sqrt{1 - x^2}$$

output of the program

```
f(z) = (1.0 - z ** 2) ** (1 / 2)
F(z) = pi/4
```

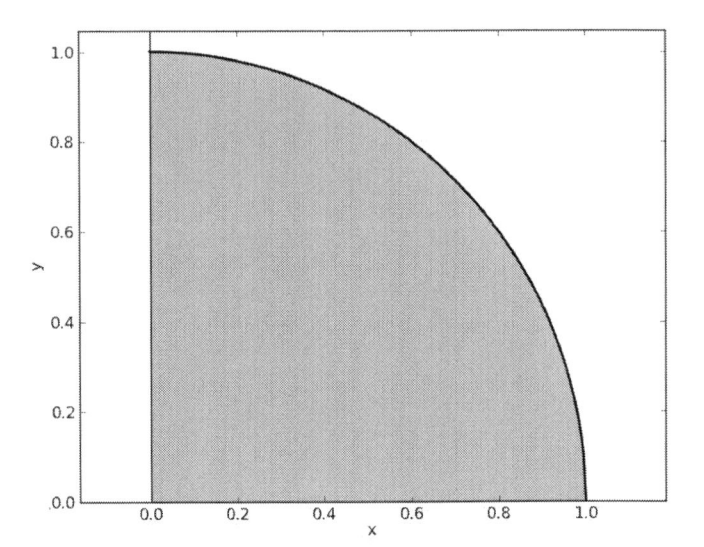

```
# -*- coding: utf-8 -*-
"""
Mathematics and Python Programming    www.pysamples.com
p9c.py
"""

import numpy as np
import matplotlib.pyplot as plt
import sympy as sy
```

9.3. DEFINITE INTEGRAL

```
z = sy.symbols('z')
sy.init_printing(use_unicode=True)

print 'f(z) = (1.0 - z ** 2) ** (1 / 2)'
integral = sy.integrate(sy.sqrt(1 - z ** 2), (z, 0, 1))
print 'F(z) = ', integral

pointsnum = 200
x = np.linspace(0.0, 1.0, pointsnum)
y = np.sqrt(1 - x ** 2)
plt.plot(x, y, color='k', lw=2.0)
plt.fill_between(x, y, 0, alpha=1.0, color='#BDD0D7')
plt.ylabel('y')
plt.xlabel('x')
plt.axis('equal')
plt.axhline(color='black', lw=1)
plt.axvline(color='black', lw=1)
plt.xlim(0, 1.05)
plt.ylim(0, 1.05)
plt.show()
```

If $f(x) < 0$ in the interval $[c, d]$ we find that the area between the curve of the function and the axis X is $A = -\int_c^d f(x)$.

For example the function

$$f(x) = (x+4)(x-1)(x-3) = x^3 - 13x + 12$$

has the following representation and area, which has been obtained with the program of Python that is displayed after the results.

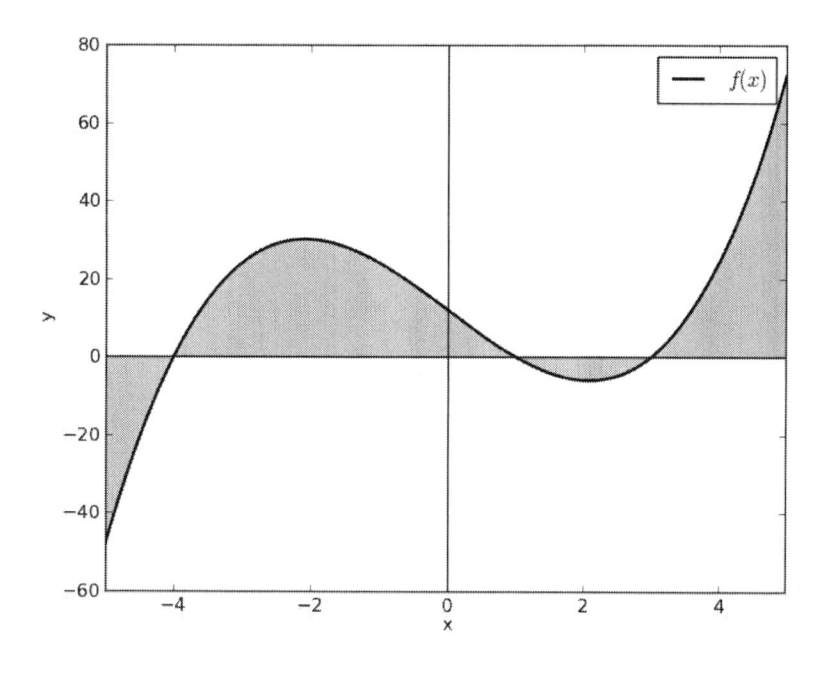

The area between $f(x)$ and the axis X is:

$$A = -\int_{-5}^{-4} f(x) + \int_{-4}^{-1} f(x) - \int_{1}^{3} f(x) + \int_{3}^{5} f(x)$$

```
_____ output of the program _____
f(x) =
x**3 - 13*x + 12
F(x) =
x**4/4 - 13*x**2/2 + 12*x

axis X cuts:  [-4.  1.  3.]
ends of the integration intervals:  [-5. -4.  1.  3.  5.]
surfaces of the intervals:
[-21.75  93.75  -8.   56. ]
absolute values of the surfaces of the intervals:
[ 21.75  93.75   8.   56. ]
Total Area =  179.5
```

Another example: $f(x) = x^5 + 2x^4 + \frac{x^3}{4} - \frac{3x^2}{4}$

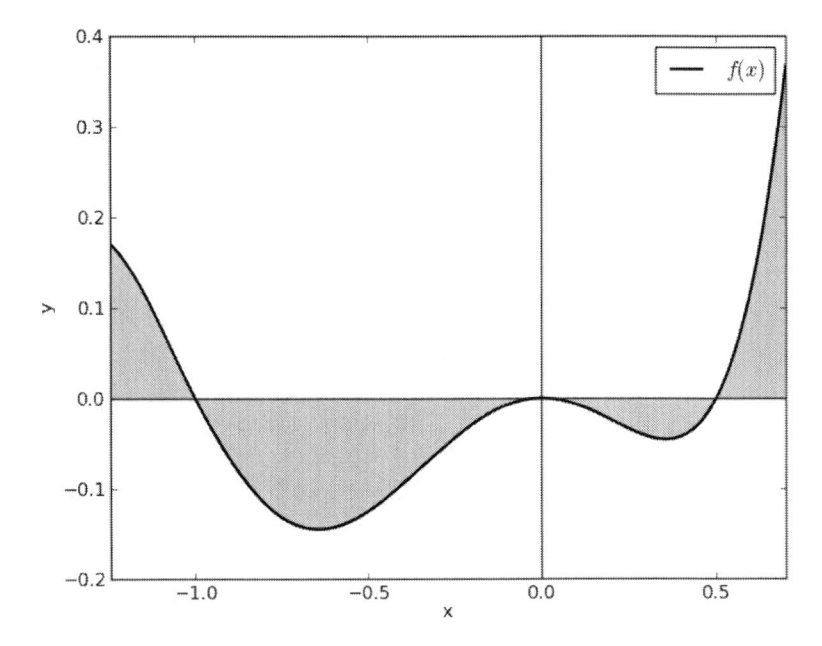

```
_____ output of the program _____
f(x) =
x**2*(4*x**3 + 8*x**2 + x - 3)/4
F(x) =
x**6/6 + 2*x**5/5 + x**4/16 - x**3/4
axis X cuts:  [-1.5 -1.   0.   0.   0.5]
ends of the integration intervals:  [-1.25 -1.    0.   0.5  0.7 ]
surfaces of the intervals:
[ 0.02321777 -0.07916667 -0.01223958  0.028332  ]
absolute values of the surfaces of the intervals:
[ 0.02321777  0.07916667  0.01223958  0.028332  ]
Total Area =   0.142956023437
```

```python
# -*- coding: utf-8 -*-
"""
Mathematics and Python Programming    www.pysamples.com
p9d.py
"""

import numpy as np
import matplotlib.pyplot as plt
import sympy as sy

x = sy.symbols('x')
sy.init_printing(use_unicode=True)
```

9.3. DEFINITE INTEGRAL

```
a = -5.0
b = 5.0
function = sy.simplify(x ** 3 - 13 * x + 12)  # polynomial
npfunction = [1, 0, -13, 12]  # polynomial coefficients



a = -1.25
b = 0.7
function = sy.simplify(x ** 5 + 2 * x ** 4 + (x ** 3) / 4 - (3 * x ** 2) / 4)
npfunction = [1, 2, 0.25, -0.75, 0, 0]


#sympy
print 'f(x) = '
print function
I = sy.integrate(function)
print 'F(x) = '
print I
print

#numpy
roots = np.round(np.sort(np.roots(npfunction)), 2)
print 'axis X cuts: ', roots
ends = [a]
i = 0
while i < len(roots):
    #print roots[i], ends[i]
    if (roots[i] > ends[i] and roots[i] < b):
        ends.append(roots[i])
    else:
        ends.append(ends[i])
    #print ends
    i += 1
ends.append(b)
ends = np.unique(ends)
print 'ends of the integration intervals: ', ends
areas = np.zeros(len(ends) - 1, float)
for i in range(0, len(ends) - 1):
    areas[i] = sy.integrate(function, (x, ends[i], ends[i + 1]))
print 'surfaces of the intervals: '
print areas
areas = np.absolute(areas)
print "absolute values of the surfaces of the intervals: "
print areas
print 'Total Area = ', np.sum(areas)

pointsnum = 200
x = np.linspace(a, b, pointsnum)
f = np.zeros(pointsnum, float)
for i in range(0, pointsnum):
    f[i] = np.polyval(npfunction, x[i])
plt.plot(x, f, 'k-', lw=2)
plt.fill_between(x, f, 0, alpha=0.8, color='#BDD0D7')
plt.ylabel('y')
plt.xlabel('x')
```

```python
plt.axhline(color='black', lw=1)
plt.axvline(color='black', lw=1)
plt.xlim(a, b)
plt.legend(('$f(x)$',), loc='best')
plt.show()
```

- For any partition of the interval $[a,b]$, the following condition is met:

$$\lim_{n\to\infty}\sum_{i=1}^{n}\Delta_i x = b - a$$

- $\int_a^b k \cdot f(x)dx = k\int_a^b f(x)dx$ because

$$\lim_{n\to\infty}\sum_{i=1}^{n} kf(\xi_i)\Delta x_i = k \cdot \lim_{n\to\infty}\sum_{i=1}^{n} f(\xi_i)\Delta x_i$$

- If f and g are two integrable functions in $[a,b]$ then:

$$\int_a^b [f(x)+g(x)]dx = \int_a^b f(x)dx + \int_a^b g(x)dx$$

$$\int_a^b [f(x)-g(x)]dx = \int_a^b f(x)dx - \int_a^b g(x)dx$$

The area bounded by the two functions is $\int_a^b [f(x)-g(x)]dx$.

For example, if $f(x) = 2senx$ and $g(x) = cosx$, the intersection of their graphs corresponds to:

$$2tanx = 1$$
$$x = \arctan(\frac{1}{2})$$

Then, we take as bounds of the integration interval: $a = 26.56; b = 206.56$.

$$\int_a^b [f(x)-g(x)]dx = \int_a^b [2senx - cosx]dx = [-2cosx - senx]_a^b$$

The following Python program calculates the area bounded by the two functions, in the interval $[a,b]$:

```
# -*- coding: utf-8 -*-
"""
Mathematics and Python Programming     www.pysamples.com
p9e.py
"""

import numpy as np
import matplotlib.pyplot as plt
import sympy as sy
from matplotlib import rc

rc('font', **{'family': 'serif', 'serif': ['Times']})
rc('text', usetex=True)

z = sy.symbols('z')
sy.init_printing(use_unicode=True)
a = sy.atan(0.5)
```

9.3. DEFINITE INTEGRAL

```python
b = sy.pi + a
#sympy
function = 2 * sy.sin(z) - sy.cos(z)
print 'f(x) = '
print function
I = sy.integrate(function)
print 'F(x) = '
print I
Area = sy.integrate(function, (z, a, b))
print 'Area bounded by the functions:'
print 'A = ', "%6.3f" % Area
#graph
pointsnum = 500
a = (180 * np.arctan(0.5) / np.pi)
b = a + 180
fa = 2 * np.sin(np.arctan(0.5))
fb = 2 * np.sin(np.pi + np.arctan(0.5))
print 'a = arctan(1/2) = ', "%6.3f" % np.arctan(0.5), ' rad = ', "%6.2f" % a
print 'b = a + 180 = ', "%6.2f" % b
Fb = -np.sin(np.deg2rad(b)) - 2 * np.cos(np.deg2rad(b))
Fa = -np.sin(np.deg2rad(a)) - 2 * np.cos(np.deg2rad(a))
print 'F(b) = ', "%6.3f" % Fb
print 'F(a) = ', "%6.3f" % Fa
print 'F(b) - F(a) = ', "%6.3f" % (Fb - Fa)
x = np.linspace(0, 270, pointsnum)
f = np.zeros(pointsnum, float)
g = np.zeros(pointsnum, float)
for i in range(0, pointsnum):
    f[i] = 2 * np.sin(np.deg2rad(x[i]))
    g[i] = np.cos(np.deg2rad(x[i]))
#graph 1
plt.fill_between(x, f, g, where=(x >= a), color='#BDD0D7', alpha=0.5)
plt.fill_between(x, f, g, where=(x > b), color='#FFFFFF')
plt.plot(x, f, 'b-', lw=2)
plt.plot(x, g, 'r-', lw=2)
'''
#graph 2
plt.fill_between(x, f, 0, where=(x >= a), color='blue', alpha=1.0)
plt.fill_between(x, f, 0, where=(x > b), color='#FFFFFF')
plt.fill_between(x, g, 0, where=(x >= a), color='#F8EF01', alpha=0.6)
plt.fill_between(x, g, 0, where=(x > b), color='#FFFFFF')
plt.plot(x, f, 'k-', lw=2)
plt.plot(x, g, 'k--', lw=2)
'''
plt.plot([a, a], [0, fa], 'k--', lw=1)
plt.plot([b, b], [0, fb], 'k--', lw=1)
plt.text(a, -0.2, 'a', horizontalalignment='center', fontsize=15, color='black', weight='bold')
plt.text(b, 0.1, 'b', horizontalalignment='center', fontsize=15, color='black', weight='bold')
plt.legend(('f(x)=2sinx', 'g(x)=cosx',), loc='best')
plt.ylabel('y')
plt.xlabel('x')
plt.axhline(color='black', lw=1)
plt.axvline(color='black', lw=1)
plt.xlim(0, 270)
plt.ylim(-2, 2.5)
plt.show()
```

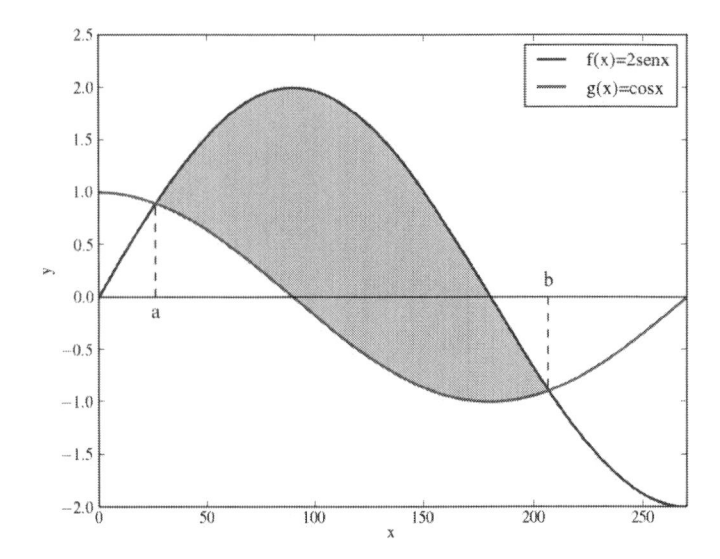

```
―――――――――――――――――――――――――― output of the program ――――――――――――――――
f(x) =
2*sin(z) - cos(z)
F(x) =
-sin(z) - 2*cos(z)
Area bounded by the functions:
A =    4.472
a = arctan(1/2) =   0.464  rad =   26.57
b = a + 180 =   206.57
F(b) =    2.236
F(a) =   -2.236
F(b) - F(a) =    4.472
―――――――――――――――――――――――――――――――――――――――――――――――――――――――――――――
```

The area bounded by the two curves is shown in the following image, which shows in blue the area under the curve of the function f, in yellow the area under the curve of the function g, and in brown the areas that are cancelled by subtracting both. This second graph is obtained with the commented lines with triple quotation marks in the previous program of Python.

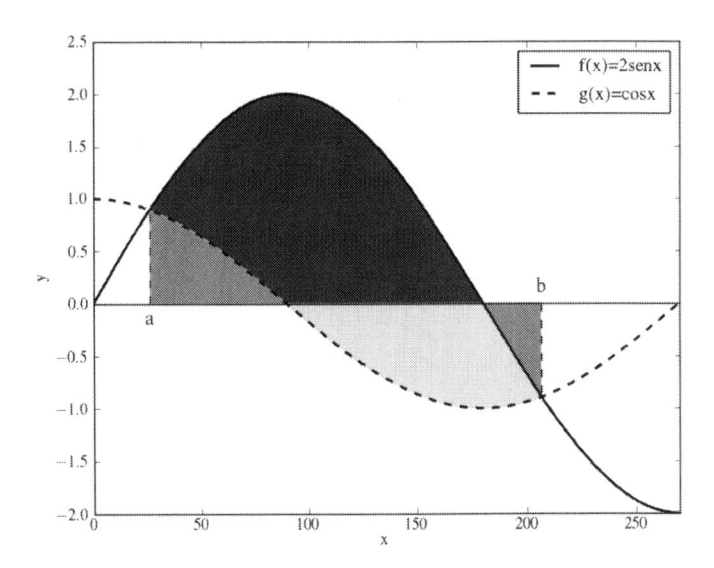

- If the function f is integrable in several intervals $[a, c], [c, b], [a, b]$, being $a < c < b$, then

$$\int_a^b f(x)dx = \int_a^c f(x)dx + \int_c^b f(x)dx$$

9.3. DEFINITE INTEGRAL

This formula is also valid if f is integrable in a closed interval which contains the three numbers a, b, c in any order.

- Si $f(x) = k \quad \forall x \in [a, b]$, we have

$$\int_a^b f(x)dx = \int_a^b k\,dx = k \cdot \int_a^b dx = k \cdot (b-a)$$

since we saw before the property

$$\lim_{n \to \infty} \sum_{i=1}^n \Delta_i x = b - a$$

- If the functions $f(x)$ and $g(x)$ are integrable in the interval $[a, b]$ and if $f(x) \geq g(x) \quad \forall x \in [a, b]$ then:

$$\int_a^b f(x)dx \geq \int_a^b g(x)dx$$

For example, we represent now the functions $f(x) = 1$ and $g(x) = \sin x$ in the interval $[0, \pi]$:

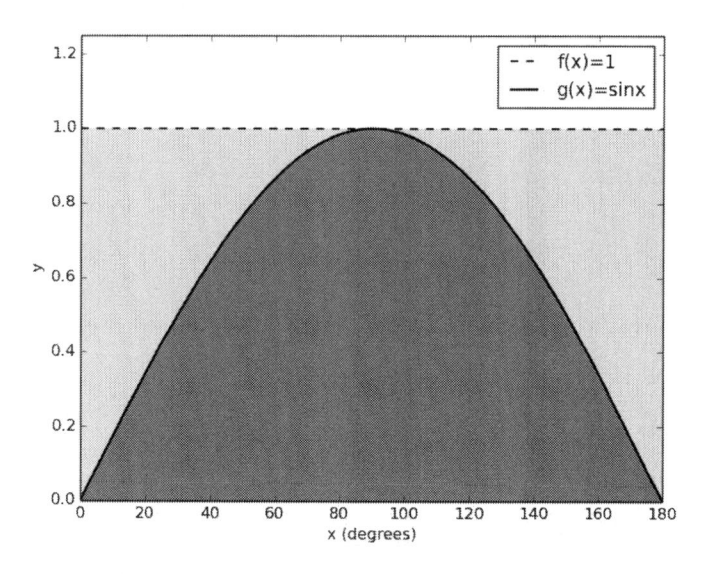

```
# -*- coding: utf-8 -*-
"""
Mathematics and Python Programming     www.pysamples.com
p9f.py
"""

import numpy as np
import matplotlib.pyplot as plt

pointsnum = 180
x = np.linspace(0, 180, pointsnum)
f = np.zeros(pointsnum, float)
g = np.zeros(pointsnum, float)
for i in range(0, pointsnum):
    f[i] = 1.0
    g[i] = np.sin(np.deg2rad(x[i]))
plt.plot(x, f, 'k--', lw=1.5)
plt.fill_between(x, f, 0, alpha=0.5, color='#BDD0D7')
plt.plot(x, g, 'k-', lw=2)
```

```
plt.fill_between(x, g, 0, alpha=1, color='#F5591E')
plt.ylabel('y')
plt.xlabel('x (degrees)')
plt.axhline(color='black', lw=1)
plt.axvline(color='black', lw=1)
plt.ylim(0, 1.25)
plt.legend(('f(x)=1', 'g(x)=sinx'), loc='best')
plt.show()
```

- Let $f(x)$ be a continuous function in an interval $[a,b]$, in which the function reaches its absolute maximum M and its absolute minimum m:

$$m \leq f(x) \leq M, \quad \forall x \in [a,b]$$

then it meets

$$m(b-a) \leq \int_a^b f(x)dx \leq M(b-a)$$

- Mean value theorem for integrals: Let $f(x)$ be a continuous function in an interval $[a,b]$. Then there is a number $c \in [a,b]$ such that

$$\int_a^b f(x)dx = f(c)(b-a)$$

The value $f(c)$ is called average value of the function $f(x)$ in the interval $[a,b]$.

The following Python program shows an example for $f(x) = x^2$, and $[a,b] = [-3,5]$. The rectangle of height $f(c)$ has an area equal to A. The points c and $f(c)$ are shown in yellow.

```
# -*- coding: utf-8 -*-
"""
Mathematics and Python Programming    www.pysamples.com
p9g.py
"""

import numpy as np
import matplotlib.pyplot as plt
import sympy as sy

z = sy.symbols('z')
sy.init_printing(use_unicode=True)
a = -3
b = 5
#sympy
function = z ** 2
print 'f(x) = '
print function
I = sy.integrate(z ** 2)
print 'F(x) = '
print I
Area = sy.integrate(z ** 2, (z, a, b))
print 'Total area calculated using Sympy:'
print 'A = ', Area, ' = ', "%6.3f" % Area
c = np.sqrt(float(Area) / (b - a))
print 'c = A/(b-a) = ', "%6.3f" % c
fc = float(function.subs(z, c).evalf(9))
print 'f(c) = ', "%6.3f" % fc
print 'f(c)(b-a)= ', "%6.3f" % (fc * (b - a))
```

9.3. DEFINITE INTEGRAL

```
#graph
pointsnum = 200
x = np.linspace(a, b, pointsnum)
f = np.zeros(pointsnum, float)
g = np.zeros(pointsnum, float)
for i in range(0, pointsnum):
    f[i] = x[i] ** 2
    g[i] = fc
plt.plot(x, f, 'k-', lw=2)
plt.fill_between(x, f, 0, alpha=1, color='#F5591E')
plt.plot(x, g, 'k--', lw=1.5)
plt.fill_between(x, g, 0, alpha=0.6, color='#5A7FAE')
plt.plot(c, 0, 'yo')
plt.plot(c, fc, 'yo')
plt.ylabel('y')
plt.xlabel('x')
plt.axhline(color='black', lw=1)
plt.axvline(color='black', lw=1)
plt.ylim(0, b ** 2)
plt.show()
```

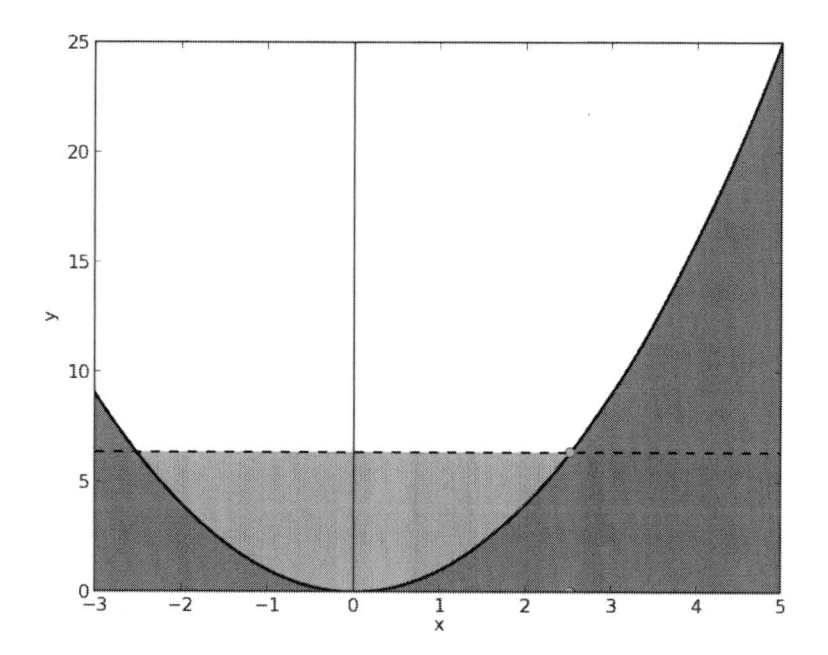

──────────── output of the program ────────────
```
f(x) =
z**2
F(x) =
z**3/3
Total area calculated using Sympy:
A =  152/3  =   50.667
c = A/(b-a) =    2.517
f(c) =    6.333
f(c)(b-a)=   50.667
```

9.4 Fundamental theorem of integral calculus

Before we will prove this theorem we are going to study the concept of definite integral with a variable upper bound. Let's take a function $f(x)$ defined in the interval $[a,b]$. Let us call $\Phi(\zeta)$ the integral calculated not between a and b but between a and a variable number ζ such that $a \leq \zeta \leq b$:

$$\Phi(\zeta) = \int_a^\zeta f(x)dx$$

Geometrically, this function $\Phi(\zeta)$ represents the area under the function $f(x)$ from a to ζ. However, the function $\Phi(\zeta)$ meets

$$\Phi'(\zeta) = \left[\int_a^\zeta f(x)dx\right]' = f(\zeta)$$

because

$$\Phi'(\zeta) = \lim_{\Delta\zeta \to 0} \frac{\Phi(\zeta + \Delta\zeta) - \Phi(\zeta)}{\Delta\zeta}$$

$$\Phi'(\zeta) = \lim_{\Delta\zeta \to 0} \frac{\int_\zeta^{\zeta+\Delta\zeta} f(\zeta)d\zeta}{\Delta\zeta}$$

if we now apply the mean value theorem, there exists a number c between ζ and $\zeta + \Delta\zeta$ such that

$$f(c)\Delta\zeta = \int_\zeta^{\zeta+\Delta\zeta} f(\zeta)d\zeta$$

therefore:

$$\Phi'(\zeta) = \lim_{\Delta\zeta \to 0} f(c)$$

and when $\Delta\zeta \to 0$, $c \to \zeta$ and we reach the point we wanted to prove:

$$\Phi'(\zeta) = f(\zeta)$$

Therefore, $\Phi(\zeta)$ is a primitive for the function $f(\zeta)$. If $F(\zeta)$ is another primitive, both of them will differ only in a constant C:

$$\Phi(\zeta) = F(\zeta) + C, \qquad a \leq \zeta \leq b$$

If in this equality we take $\zeta = a$:

$$\Phi(a) = \int_a^a f(x)dx = F(a) + C = 0$$

and therefore:

$$C = -F(a)$$

That is to say, that for any $\zeta \in [a,b]$, $\Phi(\zeta) = F(\zeta) - F(a)$. If we now take $\zeta = b$ we obtain the fundamental theorem of calculus:

$$\int_a^b f(x)dx = F(b) - F(a)$$

and it is usually written:

$$\int_a^b f(x)dx = F(x)]_a^b = F(b) - F(a)$$

Let's take for example the ellipse:

$$\frac{x^2}{a^2} + \frac{y^2}{b^2} = 1$$

$$y = \frac{b}{a}\sqrt{1 - \frac{x^2}{a^2}} = \frac{b}{a^2}\sqrt{a^2 - x^2}$$

9.4. FUNDAMENTAL THEOREM OF INTEGRAL CALCULUS

if we put x as a function of the angle: $x = a \cdot cost$, $dx = -a \cdot sentdt$:

$$y = \frac{b}{a^2}\sqrt{a^2 - a^2 cos^2 t} = \frac{b}{a}\sqrt{1 - cos^2 t} = \frac{b}{a} sent$$

By simplicity we are going to take $a = 1$, and $b = \frac{1}{2}$, and we are going to calculate the area of the ellipse in the first quadrant.

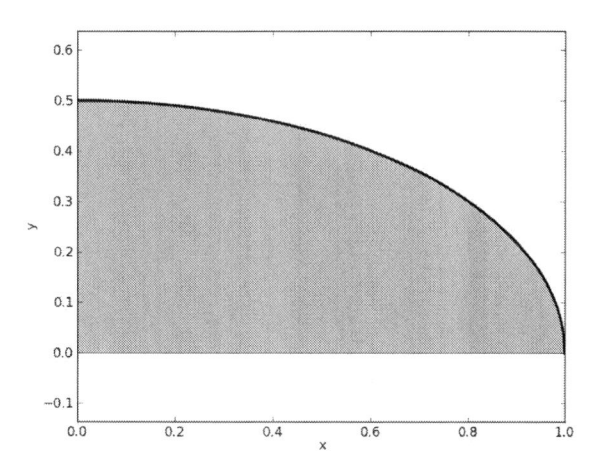

Now we must integrate the function:

$$y = \frac{1}{2} sent$$

$$\int_a^b f(x)dx = \frac{-1}{2}\int_0^{\pi/2} sen^2 t \, dt$$

and we can do it using the program we saw in section 9.2:

```
────────────────────────── output of the program ──────────────────────────
f(x)  =  sin(x) ** 2
F(x)  =  x/2 - sin(x)*cos(x)/2
```

and therefore:

$$\int_a^b f(x)dx = \frac{-1}{2}\left[\frac{t}{2} - \frac{sen(2t)}{4}\right]$$

in the interval $t \in [\frac{\pi}{2}, 0]$, That corresponds to the interval $x \in [0, a]$:

$$\int_a^b f(x)dx = \frac{-1}{2}\left[\frac{0}{2} - \frac{sen(0)}{4}\right] + \frac{1}{2}\left[\frac{\pi}{4} - \frac{sen(\pi)}{4}\right]$$

$$\int_a^b f(x)dx = \frac{1}{2}\left[\frac{\pi}{4}\right] = \frac{\pi}{8}$$

The area of an ellipse is known by geometric reasoning that is equal to πab, in our case the area of the whole ellipse would be $\frac{\pi}{2}$ and the area of the first quadrant $A = \frac{\pi}{8}$, so now we can see that the area calculated by the integral coincides with the actual area. Let's look at another example with Python:

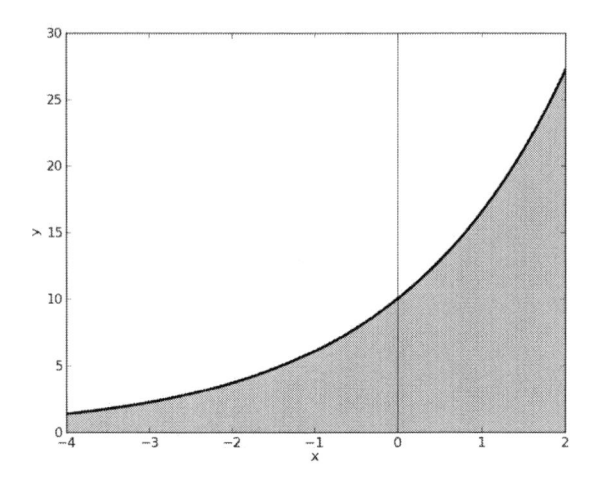

```
                                            output of the program
f(x) =
10*exp(x/2)
F(x) =
20*exp(x/2)
F(b) = F( 2.0 ) =  54.3671875
F(a) = F( -4.0 ) =  2.70654296875
F(b) - F(a) =  51.6606445312
```

```python
# -*- coding: utf-8 -*-
"""
Mathematics and Python Programming    www.pysamples.com
p9h.py
"""

import numpy as np
import matplotlib.pyplot as plt
import sympy as sy

x = sy.symbols('x')
sy.init_printing(use_unicode=True)
a = -4.0
b = 2.0
#sympy
function = 10 * sy.exp(x / 2)
print 'f(x) = '
print function
I = sy.integrate(function)
print 'F(x) = '
print I
Fb = float(I.subs(x, b).evalf(3))
Fa = float(I.subs(x, a).evalf(3))
print 'F(b) = F(', b, ') = ', Fb
print 'F(a) = F(', a, ') = ', Fa
print 'F(b) - F(a) = ', Fb - Fa

#graph
pointsnum = 100
x = np.linspace(a, b, pointsnum)
f = np.zeros(pointsnum, float)
for i in range(0, pointsnum):
    f[i] = 10 * np.exp(x[i] / 2)
plt.plot(x, f, 'k-', lw=2.5)
```

9.5. LENGTHS, AREAS AND VOLUMES

```
plt.fill_between(x, f, 0, alpha=1, color='#BDD0D7')
plt.ylabel('y')
plt.xlabel('x')
plt.axhline(color='black', lw=0.6)
plt.axvline(color='black', lw=0.6)
plt.xlim(a, b)
plt.show()
```

If we use the Python program of section 9.1 to calculate the sum of Riemann, we get with a value of $\delta = 0.01$ and dividing the interval $[a, b]$ into 600 intervals, a value for the area equal to $I = 51.6666$.

9.5 Lengths, areas and volumes

9.5.1 Area of a curvilinear sector

We call curvilinear sector to the flat shape bounded by a curve expressed in polar coordinates, and two polar radii. Let's take a curve given in polar coordinates: $\rho = \rho(\varphi)$. The area bounded by two radii of polar angles α and β is:

$$A = \frac{1}{2} \int_\alpha^\beta \rho^2(\varphi) d\varphi$$

For example, the area between the first and second turn of Archimedean spiral $\rho = k\varphi$ is:

---------- output of the program ----------
```
f(x) =
k*z
F(x) =
k**2*z**3/6
Area[ 2*pi ,  4*pi ] =  28*pi**3*k**2/3
```

$$A = \frac{28}{3} \pi^3 k^2$$

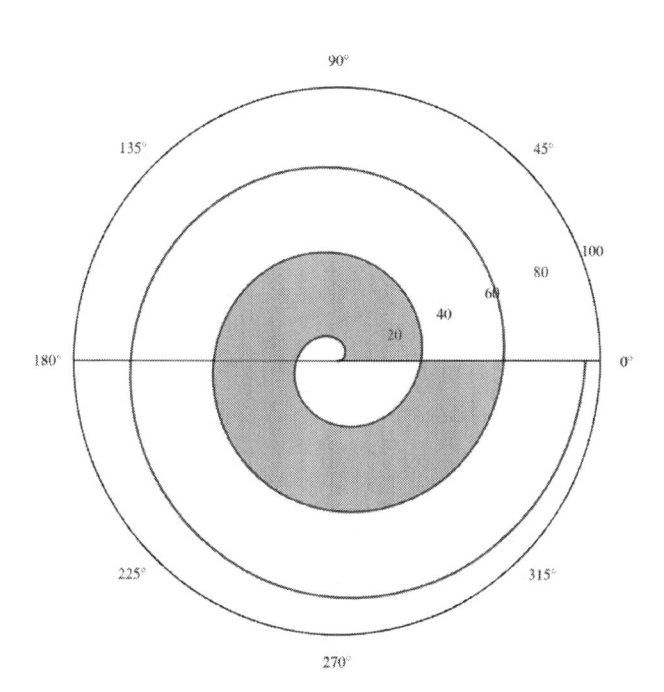

The limits of integration that has used the program are $\alpha = 2\pi$ and $\beta = 4\pi$. The program has taken $k = 5$ to plot the curve. The desired area has then been shaded with an image manipulation program.

```python
# -*- coding: utf-8 -*-
"""
Mathematics and Python Programming    www.pysamples.com
p9i.py
"""

import matplotlib
import numpy as np
import sympy as sy
from matplotlib.pyplot import figure, show, rc, grid

rc('font', **{'family': 'serif', 'serif': ['Times']})
rc('text', usetex=True)

z, k = sy.symbols('z, k')
sy.init_printing(use_unicode=True)
a = 2 * sy.pi
b = 4 * sy.pi
#sympy
function = k * z
print 'f(x) = '
print function
f_to_integrate = sy.Pow(function, 2)
print 'F(x) = '
print sy.simplify((sy.integrate(f_to_integrate, (z))) / 2)
#print f_to_integrate
print
Area = (sy.integrate(f_to_integrate, (z, a, b))) / 2
print 'Area[', a, ',', b, '] = ', Area


#graph
r = []
theta = []
x = []
y = []
r.append(0.0)
theta.append(0.0)

# radar
rc('grid', color='#CACBD3', linewidth=1, linestyle='-')
rc('xtick', labelsize=10)
rc('ytick', labelsize=10)

width, height = matplotlib.rcParams['figure.figsize']
size = min(width, height)
# makes a square
fig = figure(figsize=(size, size))
ax = fig.add_axes([0.1, 0.1, 0.8, 0.8], polar=True, axisbg='#ffffff')
angulo = 0
grid(False)
while angulo <= 6 * np.pi:
    angulo += np.pi / 36
    theta.append(angulo)
    radio = 5.0 * angulo
    r.append(radio)
```

9.5.2 Arc length

Let $f(x)$ be a continuous function in $[a, b]$. The length of the arc of curve between a and b is:

$$L = \int_a^b \sqrt{1 + [f'(x)]^2} \, dx$$

if the curve is given in parametric form as $x = x(t)$, $y = y(t)$, and $a = x(\alpha)$, $b = x(\beta)$ and $t \in [\alpha, \beta]$ is necessary to carry out the following change of variable:

$$x = x(t) \qquad dx = x'(t) dt$$

$$f'(x) = \frac{dy}{dx} = \frac{dy}{dt}\frac{dt}{dx} = \frac{dy/dt}{dx/dt} = \frac{y'(t)}{x'(t)}$$

$$L = \int_a^b \sqrt{1 + [f'(x)]^2}\, dx = \int_\alpha^\beta \sqrt{1 + \left[\frac{y'(t)}{x'(t)}\right]^2}\, x'(t) dt$$

$$L = \int_\alpha^\beta \sqrt{[x'(t)]^2 + [y'(t)]^2}\, dt$$

For example, the following program plots and calculates the length of the arc of the cycloid $x = k(t - sent)$, $y = k(1 - cost)$ for $t \in [0, 2\pi]$:

```
                         output of the program
t1 = 0.000
t2 = 6.283 = 2 pi
k =   1.5
r = c^2 (1-cosz)^2 + (senz)^2 =
  =   c**2*(-2*cos(z) + 2)
  = 2 * c**2 * (1-cos(z) = 4 * c**2 * [sin(z/2)] ** 2
L =   8*c
If c =   1.5   then L =   12.0
a =   0.000
b =   9.425
```

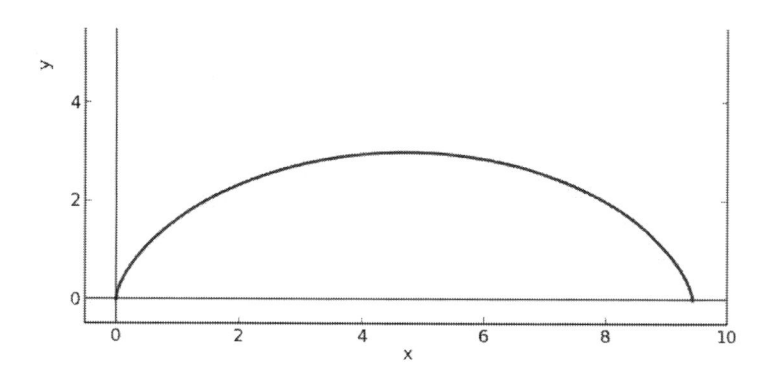

```
# -*- coding: utf-8 -*-
"""
Mathematics and Python Programming    www.pysamples.com
p9j.py
"""

import numpy as np
```

```python
import matplotlib.pyplot as plt
import sympy as sy

t1 = 0.0
t2 = 2 * np.pi
print 't1 = ' + "%5.3f" % t1
print 't2 = ' + "%5.3f" % t2 + ' = 2 pi'

k = 1.5  # for numpy, the constant is k, and for sympy it is c
print 'k = ', k

z, c = sy.symbols('z, c')
sy.init_printing(use_unicode=True)

print 'r = c^2 (1-cosz)^2 + (senz)^2 = '
r = sy.simplify((c ** 2) * ((1 - sy.cos(z)) ** 2 + (sy.sin(z) ** 2)))
print ' = ', r
print ' = 2 * c**2 * (1-cos(z) = 4 * c**2 * [sin(z/2)] ** 2'
integral = sy.integrate(2 * c * sy.sin(z / 2), (z, 0, 2 * sy.pi))
print 'L = ', integral
print 'If c = ', k, ' then L = ', 8 * k


def f(t):
    fv = k * (t - np.sin(t))
    return fv


def g(t):
    gv = k * (1 - np.cos(t))
    return gv

print 'a = ' + "%5.3f" % f(t1)
print 'b = ' + "%5.3f" % f(t2)
pointsnum = 360
tinicial = 0
tfinal = pointsnum
x = np.zeros(pointsnum, float)
y = np.zeros(pointsnum, float)
t = tinicial
while t < tfinal:
    radians = np.deg2rad(t)
    x[t] = f(radians)
    y[t] = g(radians)
    t += 1
plt.plot(x, y, 'b-', lw=2)
plt.axhline(color='black', lw=1)
plt.axvline(color='black', lw=1)
plt.axis([-0.5, 10, -0.5, 10])
#plt.axis('equal')
plt.ylabel('y')
plt.xlabel('x')
plt.show()
```

9.5.3 Area of a surface of revolution

If we rotate the arc of the previous section around the axis X, we get a surface of revolution whose area is given by

$$S = 2\pi \int_a^b f(x)\sqrt{1 + [f'(x)]^2}dx$$

and if the curve is given in its parametric form $x = x(t); y = y(t) \geq 0 \quad \forall t \in [\alpha, \beta]$:

$$S = 2\pi \int_a^b y(t)\sqrt{[x'(t)]^2 + [y'(t)]^2}dt$$

If the curve is given in polar coordinates $\rho = \rho(\varphi)$:

$$S = 2\pi \int_a^b \rho sen\varphi \sqrt{\rho^2 + [\rho']^2}d\varphi$$

In the previous cycloid example: $x = k(t - \sin t), y = k(1 - \cos t)$ para $t \in [0, 2\pi]$

$$S = 2\pi \int_\alpha^\beta k(1 - \cos t)\sqrt{[k(t - \sin t)']^2 + [k(1 - \cos t)']^2}dt$$

$$S = 2\pi \int_0^{2\pi} k(1 - \cos t)\sqrt{[k(1 - \cos t)]^2 + [k \sin t]^2}dt$$

$$S = 2\pi k^2 \int_0^{2\pi} (1 - \cos t)\sqrt{(1 + \cos^2 t - 2\cos t) + (\sin^2 t)}dt$$

$$S = 2\pi k^2 \sqrt{2} \int_0^{2\pi} (1 - \cos t)\sqrt{1 - \cos t}dx$$

$$S = 2\pi k^2 \sqrt{2} \int_0^{2\pi} (1 - \cos t)^{3/2}dt$$

and since $1 - \cos t = 2\sin^2 \frac{t}{2}$, we obtain the following integral:

$$S = 2\pi k^2 \sqrt{2} \int_0^{2\pi} 2\sqrt{2} \left(\sin \frac{t}{2}\right)^3 dt$$

$$S = 8\pi k^2 \int_0^{2\pi} \left(\sin \frac{t}{2}\right)^3 dt$$

We can calculate easily this integral through a few Python code lines:

```
import sympy as sy
t = sy.symbols('t')
print sy.integrate((sy.sin(t/2))**3, (t, 0, 2*sy.pi))
```

──────────────── output of the program ────────────────

8/3

And therefore the surface of revolution created by the rotation of the arc is:

$$S = \frac{64}{3}\pi k^2$$

If we now take $k = 1.5 = \frac{3}{2}$ the area is $S = 48\pi$. The following two 3D views have been made by a Python program which uses Maya (http://mayavi.sourceforge.net/install.html):

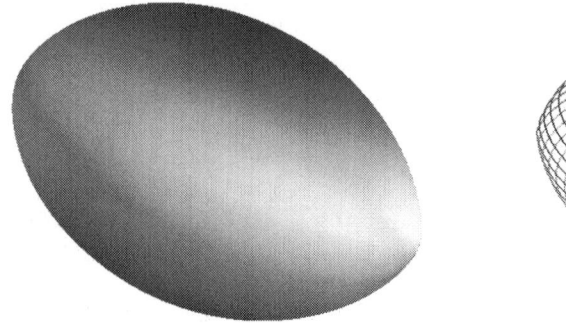

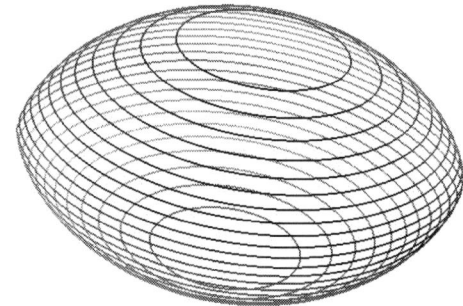

```
# -*- coding: utf-8 -*-
"""
Mathematics and Python Programming     www.pysamples.com
p9k.py
"""

import numpy as np
# create the data
dt, dv = np.pi / 180.0, np.pi / 180.0
k = 1.5
[t, v] = np.mgrid[0:2 * np.pi + dt * 1.5:dt, 0:2 * np.pi + dv * 1.5:dv]
x = k * (t - np.sin(t))
y = k * (1 - np.cos(t)) * np.cos(v)
z = k * (1 - np.cos(t)) * np.sin(v)

# represents using mayavi
from mayavi import mlab
s = mlab.mesh(x, y, z)
mlab.show()
```

9.5.4 Volume of a solid of revolution

Be $f(x)$ a continuous function on the interval $[a, b]$ such that $f(x) \geq 0 \quad \forall x \in [a, b]$. The function $y = f(x)$ defines a surface between its curve, the axis X and the straight lines $x = a$ and $x = b$. If we rotate this surface around the axis X, we get a solid of revolution whose volume is:

$$V = \pi \int_a^b [f(x)]^2 dx$$

If we are interested in the surface between the curves of both functions $f(x)$ and $g(x)$, the volume of revolution is:

$$V = \pi \int_a^b [f^2(x) - g^2(x)] dx$$

The following two Python programs represent the functions $f(x) = x^2$ and $g(x) = x^5$ in the interval $x \in [0, 1]$ and calculate the area between the two curves; they also also represent in 3D the volume of revolution obtained by rotating the surface around the axis X. The vertical axis is called z in the 2D plot to match its name in 3D.

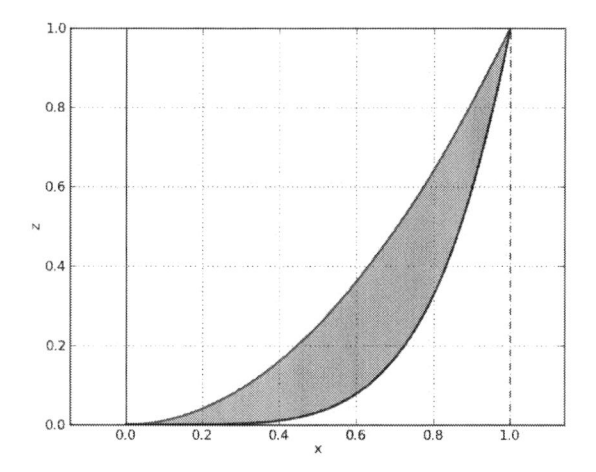

9.5. LENGTHS, AREAS AND VOLUMES

─────────── output of the program ───────────
```
f(x) = x ** 2
g(x) = x ** 5
A = Area(f) - Area(g) =  1/3  -  1/6  =  1/6
V = [ -pi*x**11/11 + pi*x**5/5 ] from a to b
V = 6*pi/55 - 0
V =  6*pi/55
```

```python
# -*- coding: utf-8 -*-
"""
Mathematics and Python Programming    www.pysamples.com
p91.py
"""

import numpy as np
import matplotlib.pyplot as plt
import sympy as sy
x = sy.symbols('x')
sy.init_printing(use_unicode=True)

a = 0
b = 1
f = x ** 2
#f = sy.sqrt(x)
g = x ** 5
print 'f(x) = x ** 2'
area1 = sy.integrate(f, (x, a, b))
print 'g(x) = x ** 5'
area2 = sy.integrate(g, (x, a, b))

print 'A = Area(f) - Area(g) = ', area1, ' - ', area2, ' = ', area1 - area2

indefinitevolume = sy.integrate(sy.pi * (f ** 2 - g ** 2))
print 'V = [', indefinitevolume, '] from a to b'
print ('V = ' + str(sy.integrate(sy.pi * (f ** 2 - g ** 2), (x, b))) +
       ' - ' + str(sy.integrate(sy.pi * (f ** 2), (x, a))))
volume = sy.integrate(sy.pi * (f ** 2 - g ** 2), (x, a, b))
print 'V = ', volume
pointsnum = 200
x = np.linspace(a, b, pointsnum)
yf = x ** 2
yg = x ** 5
plt.plot(x, yf, color='r', lw=2.0)
```

```python
plt.fill_between(x, yf, 0, alpha=1.0, color='#F8C31C')
plt.plot(x, yg, color='b', lw=2.0)
plt.fill_between(x, yg, 0, alpha=1.0, color='white')
plt.grid(True)
plt.axis('equal')
plt.ylabel('z')
plt.xlabel('x')
plt.axhline(color='black', lw=1)
plt.axvline(color='black', lw=1)
plt.plot([b, b], [0, b], 'k—', lw=1.0)
plt.show()
```

These two surfaces of revolution are rendered in 3D below using the application Maya from Python. The volume is bounded by the two surfaces. The surface generated by $f(x) = x^2$ is represented in dark tones, and the surface generated by $g(x) = x^5$ in light tones. The graphic generated by the Python program can be edited and run in Maya.

```
# -*- coding: utf-8 -*-
"""
Mathematics and Python Programming    www.pysamples.com
p9m.py
"""

import numpy as np

# create data
dr, dtheta = 0.01, np.pi / 360
#[r,theta] = np.mgrid[1:2+dr:dr, 0:2*np.pi+dtheta*1.5:dtheta]
[r, theta] = np.mgrid[0:1 + dr:dr, 0:2 * np.pi + dtheta * 1.5:dtheta]
x = r * np.cos(theta)
y = r * np.sin(theta)
z2 = np.sqrt(r)
z5 = np.power(r, 0.2)
# represents using mayavi
from mayavi import mlab
s2 = mlab.mesh(x, y, z2)
s5 = mlab.mesh(x, y, z5)
mlab.show()
```

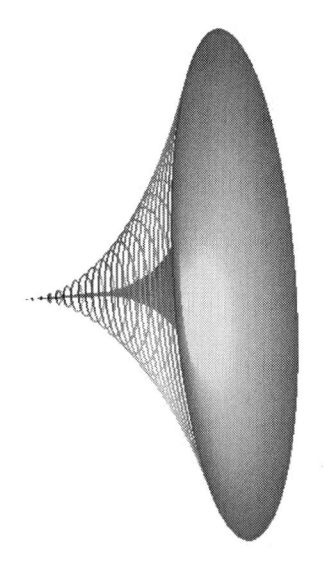

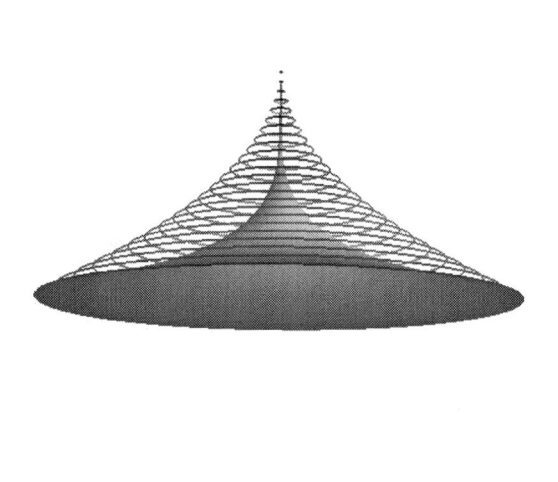

9.5. LENGTHS, AREAS AND VOLUMES

If on the other hand we would rotate the curve around the axis Y the volume of the solid of revolution would be:

$$V = 2\pi \int_a^b x f(x) dx$$

For the solids whose cross sectional area $A(h)$ is known for each height h we can obtain the volume of the solid using the integral:

$$V = \int_a^b A(h) dh$$

For example: a cylinder of radius R and H has a section $A(h) = \pi R^2$ which does not vary with the height, and its volume will be:

$$V = \int_0^H \pi R^2 dh = \pi R^2 H$$

But a cone of the same radius R and height H has a circular cross-section πr^2 where the radius r varies with height. Consider the inverted cone represented below. The slope of an edge of the cone is equal to $\frac{H}{R}$ and therefore we can relate the height with the radio using the equation of the straight line $h = \frac{H}{R} r$:

We can get the following expression for the area of a section as a function of height: $A(h) = \pi \frac{R^2}{H^2} h^2$. We are now ready to calculate the volume of the cone:

$$V = \int_0^H \pi \frac{R^2}{H^2} h^2 dh = \frac{\pi R^2 H}{3}$$

The two previous formulas, which we have obtained through integration, are already known in geometry from primary education. Now let's look at a more complicated example: the clepsydra.

The clepsydra is a clock of water that was used in antiquity. It consists of a container filled with water, and whose shape is such that the water level drops to constant speed $a = \frac{dy}{dt}$ by a small hole cut in its bottom. By means of physical reasoning we will not discuss here, it is known that the clepsydra has the form generated by rotating the curve $y = cx^4$ around the vertical axis Y. The constant c is

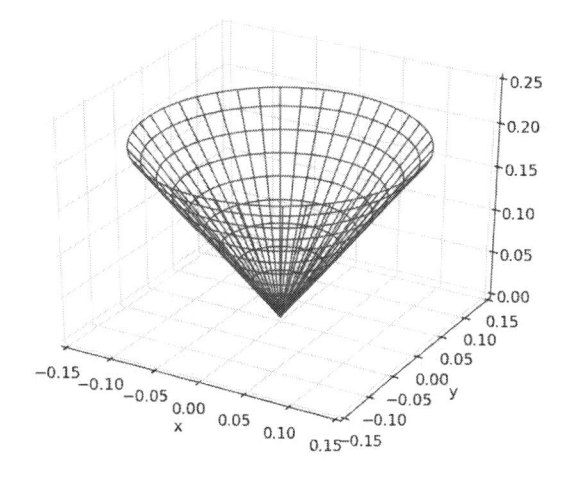

$$c = \frac{\pi^2 a^2}{2gks^2}$$

Here a is the speed of drop in the level of water, g is the acceleration due to gravity, $k = 0.6$ for water, and s is the surface of the hole through which the water goes out.

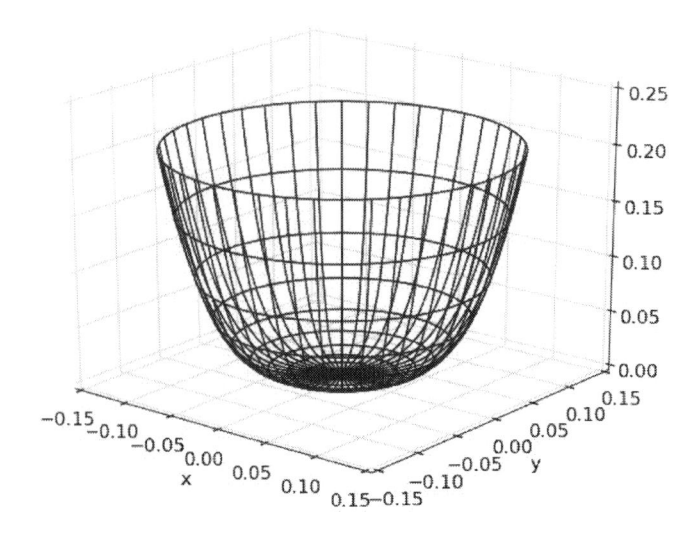

In the first place we are going to calculate the volume of the clepsydra through the integration of the area of its section as we have done in the case of the cone and the cylinder. The clepsydra also has a circular cross-section and its radius and height are related by the equation

$$h = cr^4$$

$$r = \left(\frac{h}{c}\right)^{1/4}$$

Hence we can relate the area to the height:

$$A(h) = \pi r^2 = \pi \left(\frac{h}{c}\right)^{2/4} = \frac{\pi}{\sqrt{c}}\sqrt{h}$$

$$V = \int_a^b A(h)dh = \frac{\pi}{\sqrt{c}} \int_0^H h^{1/2} dh$$

$$V = \frac{2\pi}{3\sqrt{c}} H^{3/2}$$

$$V = \frac{2\pi}{3\sqrt{c}} c^{3/2} R^{12/2} = \frac{2}{3}\pi c R^6 = \frac{2}{3}\pi c R^4 R^2$$

$$V = \frac{2}{3}\pi R^2 H$$

Therefore the volume of the clepsydra is twice the volume of a cone with the same radius and height and two-thirds of the volume of a cylinder which also has the same radius and height.

9.5. LENGTHS, AREAS AND VOLUMES

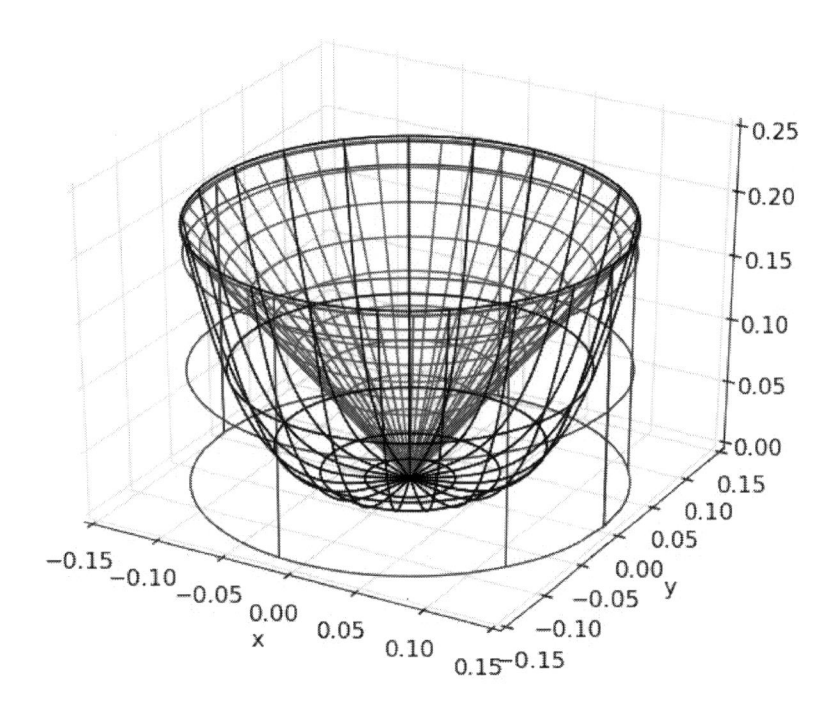

We can also calculate easily with Python the volume of the clepsydra when we rotate the curve $y = cx^4$ around the axis Y using the formula that we have already mentioned: $V = 2\pi \int_a^b x f(x) dx$. The picture shows the surface that generates the same clepsydra:

```
# -*- coding: utf-8 -*-
"""
Mathematics and Python Programming   www.pysamples.com
p9n.py
"""

import numpy as np
import matplotlib.pyplot as plt
import sympy as sy

#z = sy.symbols('z')
sy.init_printing(use_unicode=True)

c, x, H = sy.symbols('c, x, H')
#y = c * x ** 4

integ1 = 2 * sy.pi * ( x * H)
V1 = sy.integrate(integ1, x)
integ = 2 * sy.pi * ( x * c * x ** 4)
V2 = sy.integrate(integ, x)
V = V1 - V2
print 'V = ', V

S = sy.integrate(H - c * x ** 4, x)
print 'Area bounded by the curve and axis Y = ', S

pointsnum = 200
a = 0
b = 14.6266250136  # maximum radius R of the clepsydra (cm)
cte = 436.971875791 * (1.0 / 1e6)  # (cm-3)
```

```
x = np.linspace(a, b, pointsnum)
y = cte * x ** 4
plt.plot(x, y, color='k', lw=2)
ymax = np.max(y)
plt.fill_between(x, ymax, 0, alpha=1.0, color='#F8C31C')
plt.fill_between(x, y, 0, alpha=1.0, color='white')
plt.axis('equal')
plt.ylabel('y')
plt.xlabel('x')
plt.grid(True)
plt.axhline(color='black', lw=1)
plt.axvline(color='black', lw=1)
plt.plot([a, b], [ymax, ymax], 'k—', lw=1.0)
plt.ylim(0, 1.1 * ymax)
plt.show()
```

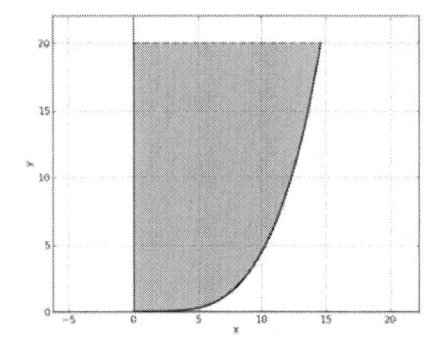

```
─────────────────────────────────── output of the program ───
V =  pi*H*x**2 - pi*c*x**6/3
Area bounded by the curve and axis Y =   H*x - c*x**5/5
─────────────────────────────────────────────────────────────
```

The volume that we get is the following:

$$V = \pi H R^2 - \frac{\pi c R^6}{3} = \pi H R^2 - \frac{\pi H R^2}{3} = \frac{2}{3}\pi R^2 H$$

Which coincides with the volume calculated before through the integration of the cross-cutting areas.

The following program performs the 3D renderings that we have seen as well as a 3D representation of clepsydra using Mayavi (shown above). The program also computes the time it should take to empty the recipient, the radius of the hole and dimensions of the clepsydra, using the height H we desire. We have use this program to calculate a clepsydra 20 cm in height that takes one hour to drain through a hole whose radius is 1mm:

```
─────────────────────────────────── output of the program ───
time to empty:  3600   seconds
clepsydra height:  200.0   mm
diameter of the hole:  2.0 mm
the water level goes down at a constant speed = 3.0 mm / minute
c =  436.971875791
clepsydra radius:  146.266250136   mm
clepsydra volume:    0.008961 m3 =   8.961 litres
Volume calculate by sections:   0.008961 m3 =   8.961 litres

Volume of a cone of same radius and height:   4.481 litres
Volume of a cylinder of same radius and height: 13.442 litres
Area bounded by the curve and axis Y =    234.0 cm2
Relations among cone, clepsydra and cylinder:
V clepsydra=  2.0 Vcone
V cylinder =  3.0 Vcono
─────────────────────────────────────────────────────────────
```

9.5. LENGTHS, AREAS AND VOLUMES

```
# -*- coding: utf-8 -*-
"""
Mathematics and Python Programming    www.pysamples.com
p9o.py
"""

import numpy as np

# create data
#constants:
k = 0.6   # fluid constant, for water k=0.6
g = 9.81  # gravity

#data:
seconds = 3600  # time to empty the clepsydra, in seconds
print 'time to empty: ', seconds, ' seconds'
H = 0.2  # height in meters
print 'clepsydra height: ', H * 1000, ' mm'
rs = 1.0  # radius of the hole, in mm

#calculations:
s = np.pi * (rs * 1e-3) ** 2   # area
print 'diameter of the hole: ', 2 * rs, 'mm'
a = H / seconds  # speed of the water level, in m/s
print ('the water level goes down at a constant speed = ' +
       str(np.round(a * 60 * 1000, 0)) + ' mm / minute')
c = (np.pi ** 2 * a ** 2) / (2 * g * k ** 2 * s ** 2)
#c = a ** 2 / ( 2 * g * (rs * 1e-3) ** 4)
print 'c = ', c
#h = c r ** 4
R = np.power(H / c, 0.25)
print 'clepsydra radius: ', 1000 * R, ' mm'
V = np.pi * H * R ** 2 - (np.pi * c * R ** 6) / 3
print ('clepsydra volume: ' + "%10.6f" % V + ' m3 = ' +
       "%6.3f" % (V * 1000) + ' litres')

Vsec = (np.pi / c ** 0.5) * (2.0 / 3) * H ** (1.5)
print ('Volume calculate by sections: ' +
       "%10.6f" % Vsec + ' m3 = ' +
       "%6.3f" % (Vsec * 1000) + ' litres')
print
print ('Volume of a cone of same radius and height: ' +
       "%6.3f" % ((1000 * H * np.pi / 3.0) * R ** 2) + ' litres')
print ('Volume of a cylinder of same radius and height: ' +
       "%6.3f" % (1000 * H * np.pi * R ** 2) + ' litres')
print ('Area bounded by the curve and axis Y = ' +
       "%7.1f" % ((H * R - ((c * R ** 5) / 5)) * 1e4) + ' cm2')
print 'Relations among cone, clepsydra and cylinder:'
print 'V clepsydra= ', Vsec / ((H * np.pi / 3.0) * R ** 2), 'Vcone'
print 'V cylinder = ', 3.0, 'Vcono'

# matplotlib plotting
from mpl_toolkits.mplot3d import Axes3D
import matplotlib.pyplot as plt
import numpy as np
fig = plt.figure()
```

```python
ax = fig.add_subplot(111, projection='3d')
ax.w_xaxis.set_pane_color((1.0, 1.0, 1.0, 1.0))
ax.w_yaxis.set_pane_color((1.0, 1.0, 1.0, 1.0))
ax.w_zaxis.set_pane_color((1.0, 1.0, 1.0, 1.0))
dr = 1e-3
dtheta = 0.01
[r, theta] = np.mgrid[0:R + dr:dr, 0:2 * np.pi + dtheta * 1.5:dtheta]
x = r * np.cos(theta)
y = r * np.sin(theta)
#clepsydra
z = c * ((x ** 2 + y ** 2)) ** 2
ax.plot_wireframe(x, y, z, rstride=15, cstride=15, color='blue')
ax.plot_wireframe(x, y, z, rstride=15, cstride=15, color='blue')
#cone
zcono = (H / R) * np.sqrt(x ** 2 + y ** 2)
ax.plot_wireframe(x, y, zcono, rstride=15, cstride=15, color='red')
#cylinder x**2 + y**2 = R ** 2
xcil = np.linspace(-R, R, 200)
zcil = np.linspace(0, H, 200)
X, Z = np.meshgrid(xcil, zcil)
Y = np.sqrt(R ** 2 - X ** 2)
ax.plot_wireframe(X, Y, Z, rstride=40, cstride=40, color='green')
ax.plot_wireframe(X, -Y, Z, rstride=40, cstride=40, color='green')

plt.xlabel('x')
plt.ylabel('y')
plt.show()

# clepsydra plotting using mayavi
z = c * r ** 4
from mayavi import mlab
s = mlab.mesh(x, y, z)
mlab.show()
```

9.6 Applications of the integral to Physics

The work done by a force $f(x)$ along a displacement $\Delta x = x_2 - x_1$ is equal to the area under the curve $f(x)$ in that interval x_1, x_2:

$$W = \int_{x_1}^{x_2} f(x)dx$$

This Python program has been used to make the plots for the following sections:

```python
# -*- coding: utf-8 -*-
"""
Mathematics and Python Programming    www.pysamples.com
p9p.py
"""

import numpy as np
import matplotlib.pyplot as plt
import sympy as sy

x, k = sy.symbols('x, k')
sy.init_printing(use_unicode=True)
```

9.6. APPLICATIONS OF THE INTEGRAL TO PHYSICS

```python
# unidades en SI
'''
# constant force f=k
a = 1
b = 2
k = 10
f = k
# spring F=-kx
a = 15e-2   # 10 cm
b = 2e-2    # 2 cm
k = 400
f = -k * x
# isothermal expansion
a = 1.0e-3  # 1 liter
b = 5.0e-3  # 5 liters
n = 0.01    # 0.01 mol
R = 8.314   # J/K.mol
T = 298.0
f = n * R * T / x   # aqui x= V
'''
# adiabatic expansion
a = 1.0e-3  # 1 liter = V1
b = 5.0e-3  # 5 liters
P1 = 0.25 * 1.013 * 1e5  # 0.25 atmosferas
gamma = 1.66
K = P1 * (a ** gamma)
f = K * (x ** (-1 * gamma))

print 'f(x) =', f
integral = sy.integrate(f, x)
print 'W = ', integral, ' from ', a, ' to ', b
#print 'W = (', sy.N(integral, b), ') - (', sy.N(integral, a), ')'
area = sy.integrate(f, (x, a, b))
print 'W = ', "%7.3f" % area

#graph
pointsnum = 200
equis = np.linspace(a, b, pointsnum)
y = np.zeros(pointsnum, float)


def f(z):
    # constant force f=k
    #fx = k
    # spring F=-kx
    #fx = -k * z
    # isothermal expansion
    #fx = n * R * T / z
    #adiabatic expansion
    fx = K * (z ** (-1 * gamma))
    return fx
for i in range(0, pointsnum):
    y[i] = f(equis[i])

plt.plot(equis, y, color='k', lw=2)
plt.fill_between(equis, y, 0, alpha=1.0, color='#A1BCC3')
```

```
#plt.axis('equal')
plt.ylabel('$P \quad (Nm^-2)$')
plt.xlabel('$V \quad (m^3)$')
plt.axhline(color='black', lw=1)
plt.axvline(color='black', lw=1)
plt.plot([a, a], [0, f(a)], 'k--', lw=1)
plt.plot([b, b], [0, f(b)], 'k--', lw=1)
#plt.xlim(0.9 * np.min(equis), 1.1 * np.max(equis))
plt.xlim(0, 1.1 * np.max(equis))
plt.ylim(0, np.max(y))
plt.show()
```

9.6.1 W done by a constant force

$F = 10N$, $x_1 = 1m$, $x_2 = 2m$

$$W = \int_{x_1}^{x_2} 10 dx$$

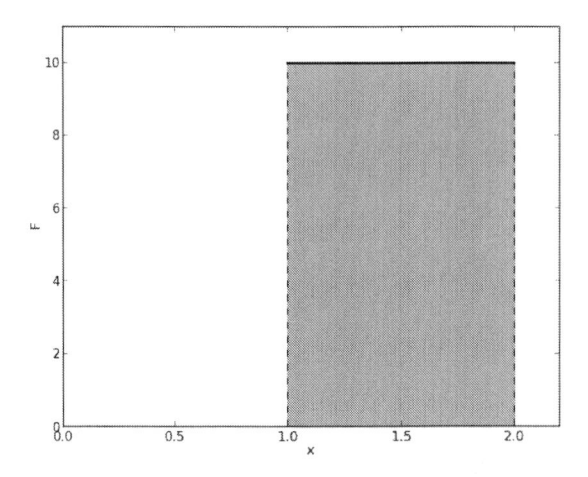

─────────────────────── W constant force ───────────────────────
```
f(x) = 10
W   =  10*x   from  1  to  2
W   =    10.000
```
──

9.6.2 W done to compress a spring

$F = -kx$, $x_1 = 0.15m$, $x_2 = 0.02m$, $k = 400Nm^{-1}$

$$W = \int_{x_1}^{x_2} -kx\, dx$$

─────────────────────── W to compress a spring ───────────────────────
```
f(x) = -400*x
W   =  -200*x**2   from  0.15  to  0.02
W   =     4.420
```
──

9.6. APPLICATIONS OF THE INTEGRAL TO PHYSICS 197

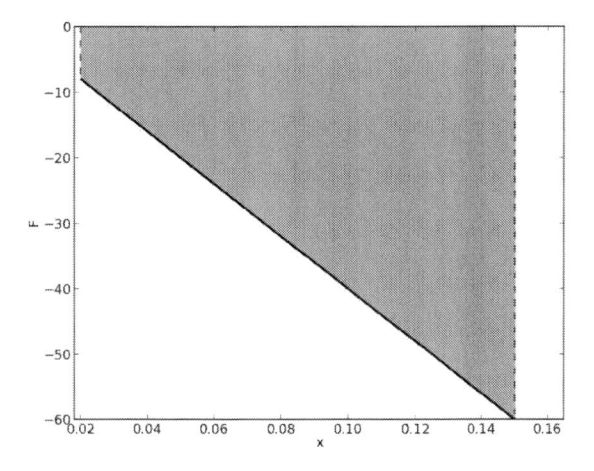

9.6.3 W during an isothermal expansion

We calculate in this example the work done during the expansion of 0.01 moles of an ideal gas at a temperature $T = 298K$; $R = 8.314 JK^{-1}mol^{-1}$:

$$W = \int_{V_1}^{V_2} PdV = nRT \int_{V_1}^{V_2} \frac{dV}{V}$$

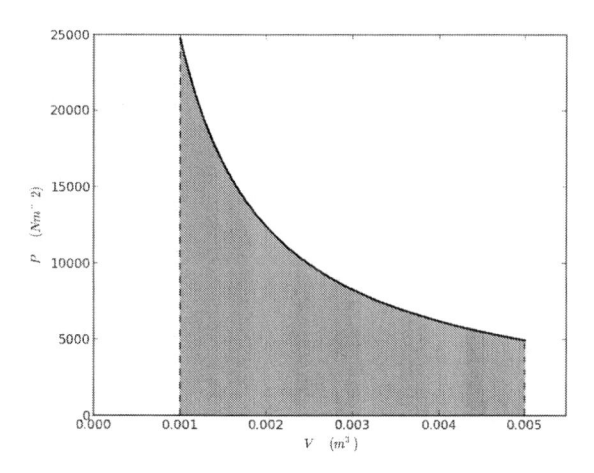

──────────────── W expansión isoterma ────────────────
```
f(x) = 24.77572/x
W  =  24.77572*log(x)   from  0.001  to  0.005
W  =     39.875
```
──

9.6.4 W during an adiabatic expansion

In an adiabatic expansion, $PV^\gamma = K =$ constant. Therefore, $P = KV^{-\gamma}$. We calculate in the next example, the work done during the expansion of 0.001 moles of Helium at a temperature $T = 298K$; $\gamma = 1.66$:

$$W = \int_{V_1}^{V_2} PdV = \int_{V_1}^{V_2} KV^{-\gamma}dV$$

──────────────── W adiabatic expansion ────────────────
```
f(x) = 0.26518530479389*x**-1.66
W  =  -0.40179591635438*x**-0.66   from  0.001  to  0.005
W  =     25.107
```
───

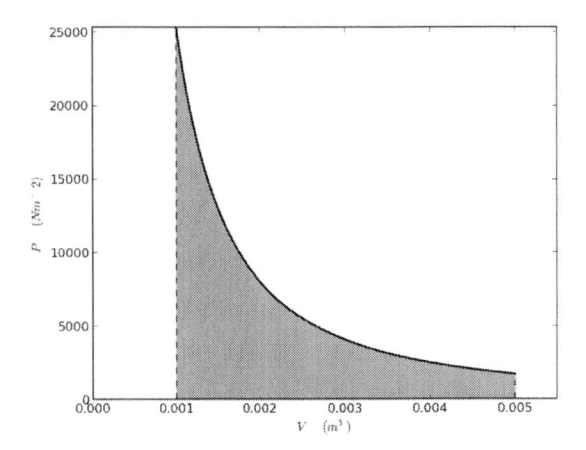

9.6.5 Carnot Cycle

The next program represents the Carnot cycle of 0.5 moles of Helium gas when it works between $273K$ and $373K$, under an initial pressure of 1 atm (101325 Pa). It is possible to choose different conditions in the program.

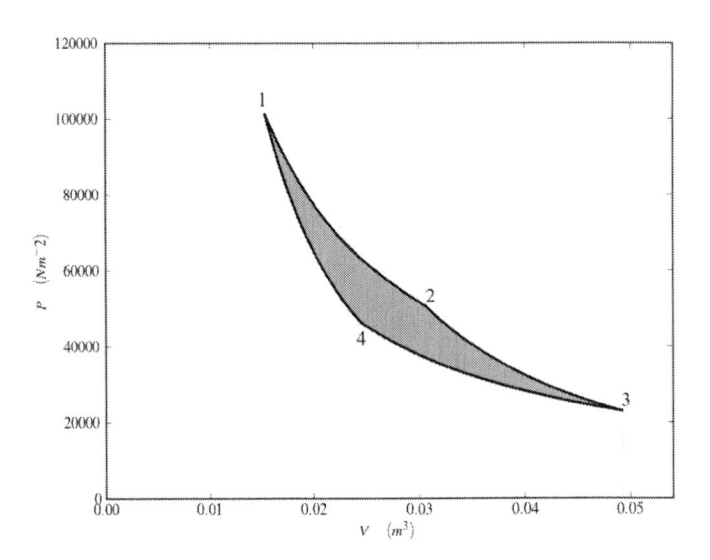

```
_____ Carnot cycle _____
Th, Tc, efficiency : 373.0  273.0   0.27
P1, V1, TH = 101325.0; 15.30liters; 373.0
f1(x) = 1550.647163/x
W1 =   1550.647163*log(x)   entre    0.0   y   0.0306
W1 =   1074.827
P2, V2, Th = 50662.5; 30.61liters; 373.0
---------------------------------------
gamma =   1.66
K =   155.295329505
f2(x) = 155.295329504909*x**(-1.66)
W2 =   -235.295953795316*x**(-0.66)   entre   0.0306   y   0.0491
W2 =   629.883
P3, V3, Tc = 23108.3; 49.11liters; 273.0
---------------------------------------
f3(x) = 1134.924063/x
W3 =   1134.924063*log(x)   entre   0.0491   y   0.0246
W3 =   -786.669
P4, V4, Tc = 46216.5; 24.56liters; 273.0
---------------------------------------
gamma =   1.66
K =   98.2830436668
f4(x) = 98.2830436668127*x**(-1.66)
W4 =   -148.913702525474*x**(-0.66)   entre   0.0246   y   0.0153
W4 =   -629.883
---------------------------------------
```

9.6. APPLICATIONS OF THE INTEGRAL TO PHYSICS

```
W = W1 + W2 + W3 + W4 =  288.157
checkings:
W = efficiency * Q12 = 288.157
Q12 absorbed by the hot reservoir during isothermal expansion = 1074.827 = 1074.827
Q34 taken by the cold reservoir during isothermal expansion = -786.669 = -786.669
|Q34| / Q12 = Tc / Th :
0.73190  =  0.73190
```

```python
# -*- coding: utf-8 -*-
"""
Mathematics and Python Programming    www.pysamples.com
p9q.py
"""

import numpy as np
import matplotlib.pyplot as plt
import sympy as sy
from matplotlib import rc

rc('font', **{'family': 'serif', 'serif': ['Times']})
rc('text', usetex=True)

x, k = sy.symbols('x, k')
sy.init_printing(use_unicode=True)

# SI units
Th = 373.0
Tc = 273.0
efficiency = 1 - (Tc / Th)
print 'Th, Tc, efficiency : ', Th, Tc, "%5.2f" % efficiency
P1 = 101325.0  # 1 atm
n = 0.5  # 0.5 moles
R = 8.314462  # J/K.mol

# isothermal expansion
# PV=nRT : V1 = nRT1/P1
V1 = n * R * Th / P1
print ('P1, V1, TH = ' + "%7.1f" % P1 +
       '; ' "%5.2f" % (1000 * V1) + 'liters; ' + str(Th))
V2 = 2.0 * V1
f1 = n * R * Th / x  # aqui x= V
print 'f1(x) =', f1
integral1 = sy.integrate(f1, x)
print 'W1 = ', integral1, ' entre ', "%7.1f" % V1, ' y ', "%7.4f" % V2
area1 = sy.integrate(f1, (x, V1, V2))
print 'W1 = ', "%7.3f" % area1
P2 = P1 * V1 / V2
print ('P2, V2, Th = ' + "%7.1f" % P2 + '; ' "%5.2f" % (1000 * V2) + 'liters; ' + str(Th))
print '---------------------------------------------'

# adiabatic expansion
#a = V2
#b = V3
gamma = 1.66
print 'gamma = ', gamma
K = P2 * (V2 ** gamma)
print 'K = ', K
V3 = V2 * (Th / Tc) ** (1 / (gamma - 1))
```

```python
P3 = K / (V3 ** gamma)
f2 = K * (x ** (-1 * gamma))
print 'f2(x) =', f2
integral2 = sy.integrate(f2, x)
print 'W2 = ', integral2, ' entre ', "%7.4f" % V2, ' y ', "%7.4f" % V3
area2 = sy.integrate(f2, (x, V2, V3))
print 'W2 = ', "%7.3f" % area2
print ('P3, V3, Tc = ' + "%7.1f" % P3 +
       '; ' "%5.2f" % (1000 * V3) + 'liters; ' + str(Tc))
print '_____'

# isothermal compression
#P3 * V3 = P4 * V4 thus P4 = P3 * V3 / V4
#P4 V4 ** gamma = P1 V1 ** gamma = K1
K1 = P1 * V1 ** gamma
#P4 = K1 * V4 ** (-1 * gamma)
#V4 ** (1-gamma) = P3 * V3 / K1
V4 = (P3 * V3 / K1) ** (1 / (1 - gamma))
P4 = K1 * V4 ** (-1 * gamma)
f3 = n * R * Tc / x  # here x= V
print 'f3(x) =', f3
integral3 = sy.integrate(f3, x)
print 'W3 = ', integral3, ' entre ', "%7.4f" % V3, ' y ', "%7.4f" % V4
area3 = sy.integrate(f3, (x, V3, V4))
print 'W3 = ', "%7.3f" % area3
print ('P4, V4, Tc = ' + "%7.1f" % P4 + '; ' "%5.2f" % (1000 * V4) + 'liters; ' + str(Tc))
print '_____'

# adiabatic compression
#a = V4
#b = V1
#gamma = 1.66
print 'gamma = ', gamma
print 'K = ', K1
f4 = K1 * (x ** (-1 * gamma))
print 'f4(x) =', f4
integral4 = sy.integrate(f4, x)
print 'W4 = ', integral4, ' entre ', "%7.4f" % V4, ' y ', "%7.4f" % V1
area4 = sy.integrate(f4, (x, V4, V1))
print 'W4 = ', "%7.3f" % area4
print '_____'

print 'W = W1 + W2 + W3 + W4 = ', "%7.3f" % (area1 + area2 + area3 + area4)
print 'checkings: '
print ('W = efficiency * Q12 = ' + "%7.3f" % (efficiency * area1))
print ('Q12 absorbed by the hot reservoir during isothermal expansion = ' +
       "%7.3f" % area1 + ' = ' "%7.3f" % (n * R * Th * np.log(V2 / V1)))
print ('Q34 taken by the cold reservoir during isothermal expansion = ' +
       "%7.3f" % area3 + ' = ' "%7.3f" % (n * R * Tc * np.log(V4 / V3)))
print '|Q34| / Q12 = Tc / Th :'
print "%7.5f" % (np.abs(area3) / area1), ' = ', "%7.5f" % (Tc / Th)

#graph
pointsnum = 100
equis1 = np.linspace(V1, V2, pointsnum)
y1 = np.zeros(pointsnum, float)
```

9.6. APPLICATIONS OF THE INTEGRAL TO PHYSICS

```python
equis2 = np.linspace(V2, V3, pointsnum)
y2 = np.zeros(pointsnum, float)
equis3 = np.linspace(V3, V4, pointsnum)
y3 = np.zeros(pointsnum, float)
equis4 = np.linspace(V4, V1, pointsnum)
y4 = np.zeros(pointsnum, float)


def f1(z):  # isothermal expansion
    f1x = n * R * Th / z
    return f1x


def f2(z):  # adiabatic expansion
    f2x = K * (z ** (-1 * gamma))
    return f2x


def f3(z):  # isothermal compression
    f3x = n * R * Tc / z
    return f3x


def f4(z):  # adiabatic expansion
    f4x = K1 * (z ** (-1 * gamma))
    return f4x

for i in range(0, pointsnum):
    y1[i] = f1(equis1[i])
    y2[i] = f2(equis2[i])
    y3[i] = f3(equis3[i])
    y4[i] = f4(equis4[i])

plt.plot(equis1, y1, color='k', lw=2)
plt.fill_between(equis1, y1, 0, alpha=1.0, color='#A1BCC3')
plt.plot(equis2, y2, color='k', lw=2)
plt.fill_between(equis2, y2, 0, alpha=1.0, color='#A1BCC3')
plt.plot(equis3, y3, color='k', lw=2)
plt.fill_between(equis3, y3, 0, alpha=1.0, color='#FFFFFF')
plt.plot(equis4, y4, color='k', lw=2)
plt.fill_between(equis4, y4, 0, alpha=1.0, color='#FFFFFF')
plt.text(V1, 1.02 * P1, '1', horizontalalignment='center',
         fontsize=15, color='black', weight='bold')
plt.text(V2, 1.02 * P2, '2', horizontalalignment='left',
         fontsize=15, color='black', weight='bold')
plt.text(V3, 1.05 * P3, '3', horizontalalignment='left',
         fontsize=15, color='black', weight='bold')
plt.text(V4, 0.95 * P4, '4', horizontalalignment='center',
         verticalalignment='top',
         fontsize=15, color='black', weight='bold')
plt.ylabel('$P \quad (Nm^-2)$')
plt.xlabel('$V \quad (m^3)$')
plt.axhline(color='black', lw=1)
plt.axvline(color='black', lw=1)
plt.xlim(0, 1.1 * np.max(equis2))
plt.show()
```

9.7 The logarithm function

A logarithmic function is a function f, non-constant, differentiable and defined for all real number greater than zero, such that
$$f(xy) = f(x) + f(y)$$
The logarithmic function so defined has the following properties:

- $f(1) = 0$, because $f(1) = f(1 \cdot 1) = f(1) + f(1) = 2f(1)$
- $f(\frac{1}{x}) = -f(x)$, since $0 = f(1) = f(x \cdot \frac{1}{x}) = f(x) + f(\frac{1}{x})$
- $f(x/y) = f(x) - f(y)$
- $f'(x) = \frac{1}{x} \cdot f'(1)$ since:

$$\frac{f(x + \Delta x) - f(x)}{\Delta x} = \frac{f(\frac{x + \Delta x}{x})}{\Delta x} = \frac{f(1 + \frac{\Delta x}{x})}{\Delta x}$$

we multiply and divide the second member by x:

$$\frac{f(x + \Delta x) - f(x)}{\Delta x} = \frac{1}{x} \cdot \frac{f(1 + \frac{\Delta x}{x})}{\Delta x / x}$$

and since $f(1) = 0$ we can subtract it to the numerator:

$$\frac{f(x + \Delta x) - f(x)}{\Delta x} = \frac{1}{x} \cdot \frac{f(1 + \Delta x/x) - f(1)}{\Delta x / x}$$

and we calculate the limits when Δx approaches 0: the limit of the first member is the derivative $f'(x)$, and if we call $\alpha = \frac{\Delta x}{x}$, we find that α also approaches 0 and the limit of the second member is:

$$\lim_{\alpha \to 0} \frac{1}{x} \cdot \frac{f(1 + \alpha) - f(1)}{\alpha} = \frac{1}{x} \cdot f'(1)$$

On the other hand, $f'(1)$ cannot be zero, since if $f(1) = 0$ the function will be equal to zero at $x = 1$, and it would also have slope equal to zero, and the function f should be constant, which contradicts its definition. Choosing a value for $f'(1)$ is equivalent to choosing a base for the logarithm. The easiest choice is to take $f'(1) = 1$, and the derivative will be $f'(x) = \frac{1}{x}$ and in this case the logarithm is called natural logarithm.

Thus, a logarithmic function has to fulfil that $f(1) = 0$; and if we take $f'(1) = 1$, it must be $f'(x) = \frac{1}{x}$ for $x > 0$. This function is called natural logarithm and must the following form:

$$\ln x = \int_1^x \frac{dt}{t}$$

The graph represents $\ln e$ as the area under the curve $y = f(x)$ between $x = 1$ and $x = e$:

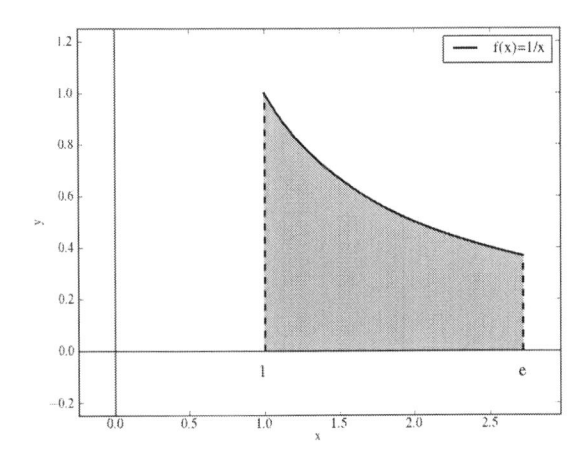

9.7. THE LOGARITHM FUNCTION

———————————— output of the program ————————————
```
f(x) =
1/z
F(x) =
log(z)
Total area calculated using Sympy:
A =    1.000
ln e = 1
```
——

```python
# -*- coding: utf-8 -*-
"""
Mathematics and Python Programming    www.pysamples.com
p9w.py
"""

import numpy as np
import matplotlib.pyplot as plt
import sympy as sy
from matplotlib import rc

rc('font', **{'family': 'serif', 'serif': ['Times']})
rc('text', usetex=True)
z = sy.symbols('z')
sy.init_printing(use_unicode=True)
a = 1.0
b = np.e
#sympy
funcion = 1 / z
print 'f(x) = '
print funcion
I = sy.integrate(funcion)
print 'F(x) = '
print I
Area = sy.integrate(funcion, (z, a, b))
print 'Total area calculated using Sympy:'
print 'A = ', "%6.3f" % Area
print 'ln e = 1'
#graph
pointsnum = 100
x = np.linspace(a, b, pointsnum)
f = np.zeros(pointsnum, float)
for i in range(0, pointsnum):
    f[i] = 1.0 / x[i]
plt.plot(x, f, 'k-', lw=2)
plt.fill_between(x, f, 0, alpha=1, color='#BDD0D7')
plt.plot([a, a], [0, 1], 'k-', lw=2)
plt.plot([b, b], [0, 1 / b], 'k-', lw=2)
plt.text(a, -0.1, '1', horizontalalignment='center',
         fontsize=15, color='black', weight='bold')
plt.text(b, -0.1, 'e', horizontalalignment='center',
         fontsize=15, color='black', weight='bold')
plt.legend(('f(x)=1/x',), loc='best')
plt.ylabel('y')
plt.xlabel('x')
plt.axhline(color='black', lw=1)
plt.axvline(color='black', lw=1)
plt.xlim(-0.25, b + 0.25)
plt.ylim(-0.25, 1.25)
plt.show()
```

The function thus defined has the properties of a logarithmic function:

- $\ln x$ is defined for all $x \in (0, \infty)$
- $\ln x$ is differentiable, and therefore continuous in $(0, \infty)$
- $\ln(xy) = \ln x + \ln y$
- $\ln 1 = 0$
- $\ln \frac{1}{x} = -\ln x$
- $\ln(x/y) = \ln x - \ln y$
- $(\ln x)' = \frac{1}{x}$
- $[\ln x^{p/q}]' = \frac{p}{q} \ln x$, with $p, q \in \mathbf{Q}$, since if we take $u = p/q$ and we apply the chain rule:

$$[\ln x^u]' = \frac{1}{x^u}(x^u)' = \frac{1}{x^u} u x^{u-1} = u\frac{1}{x}$$

$$[\ln x^{p/q}]' = \frac{p}{q} \cdot \frac{1}{x}$$

$$[\ln x^{p/q}]' = [\frac{p}{q} \ln x]'$$

And as both functions have the same derivative, they can only differ by a constant C:

$$\ln x^{p/q} = \frac{p}{q} \ln x + C$$

And since both functions are equal to zero for $x = 1$, we get $C = 0$.

- The function $\ln x$ is not bounded, its range is $(-\infty, \infty)$.
- The value of x for which $\ln x = 1$ is the number e, that is to say that the function $\ln x$ is the inverse of the exponential function

$$\ln e = 1$$

$$\ln e^y = y, \quad \forall y \in \mathbf{R}$$

- The Taylor series of $y = \ln(1 + x)$ is:

```
─────────────────────── output of p8d.py ───────────────────────
f = log(x + 1)
D1 = 1/(x + 1); D1(0) = 1.00
D2 = -1/(x + 1)**2; D2(0) = -1.00
D3 = 2/(x + 1)**3; D3(0) = 2.00
D4 = -6/(x + 1)**4; D4(0) = -6.00
D5 = 24/(x + 1)**5; D5(0) = 24.0
D6 = -120/(x + 1)**6; D6(0) = -120.
D7 = 720/(x + 1)**7; D7(0) = 720.
x - x**2/2 + x**3/3 - x**4/4 + x**5/5 - x**6/6 + x**7/7 + O(x**8)
```

9.7. THE LOGARITHM FUNCTION

The graphical representation of the logarithm function is:

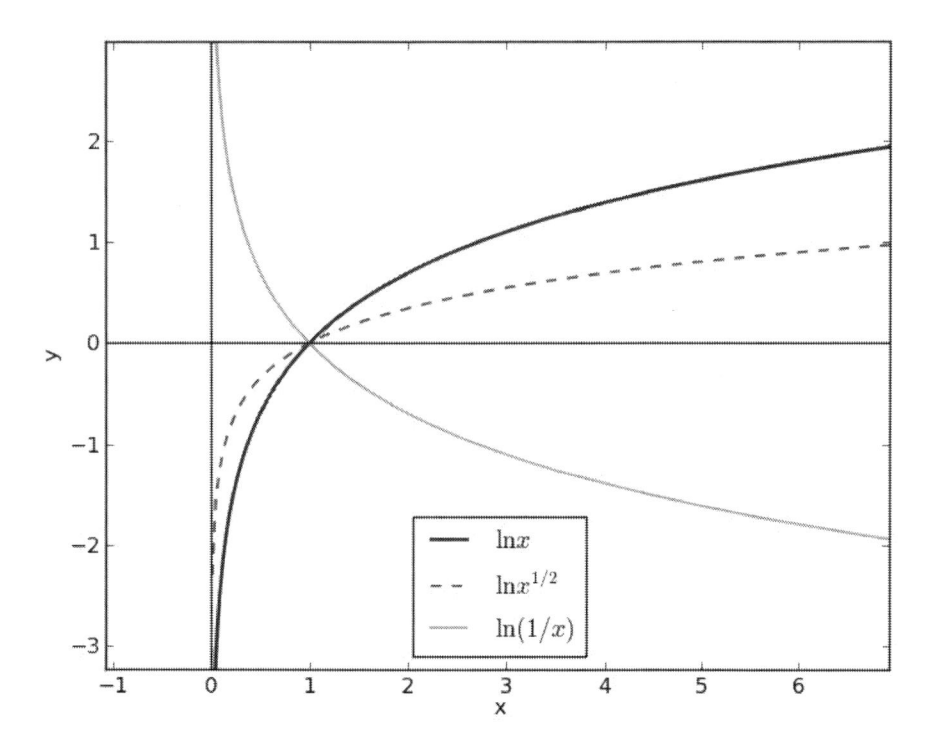

```
# -*- coding: utf-8 -*-
"""
Mathematics and Python Programming    www.pysamples.com
p9x.py
"""

import numpy as np
import matplotlib.pyplot as plt

pointsnum = 200
fig = plt.figure(facecolor='white')
x = np.linspace(0.01, 7.0, pointsnum)
y1 = np.zeros(pointsnum, float)
y2 = np.zeros(pointsnum, float)
y3 = np.zeros(pointsnum, float)
for i in range(0, pointsnum):
    y1[i] = np.log(x[i])
    y2[i] = np.log(x[i] ** (1.0 / 2.0))
    y3[i] = np.log(1 / x[i])
ax = fig.add_subplot(1, 1, 1, aspect='equal')
p1, = plt.plot(x, y1, 'b', lw=2, label='$\ln x$')
p2, = plt.plot(x, y2, 'r--', lw=1.5, label='$\ln x^{\\frac{1}{2}}$')
p3, = plt.plot(x, y3, 'y--', lw=1.5, label='$\ln 1/x}$')
plt.ylabel('y')
plt.xlabel('x')
plt.axhline(color='black', lw=1)
plt.axvline(color='black', lw=1)
plt.axis('equal')
plt.show()
```

9.8 Numerical integration

It is often the case that we need find the integral of a function for which there is no known method of integration, or there is no single function whose derivative is the given function. In the event that the function is unknown, we have only some of their values obtained experimentally.

Let's imagine an experiment to determine the area under a function, so that we have obtained the values of the function in 13 points at regular intervals of x, that we put in a matrix x of a program of Python:

```
# -*- coding: utf-8 -*-
"""
Mathematics and Python Programming    www.pysamples.com
p9r.py
numerical integration: left end of each interval
"""

import numpy as np
import matplotlib.pyplot as plt

x = [-2.0, -1.5, -1.0, -0.5, 0.0, 0.5, 1.0, 1.5, 2.0, 2.5, 3.0, 3.5, 4.0]
y = [0.0, 2.151, 2.646, 2.806, 2.828, 2.85, 3.0, 3.373, 4.0, 4.861, 5.916, 7.133, 8.485]
print 'x = ', x
print 'y = ', y
n = len(x)
print n, ' points; ', n - 1, ' intervals'
```

In principle we do not know which function these points follow, but we are interested in calculating the area under the curve that unites them.

The first method of integration: drawing rectangles between each pair of values of x, and we are going to take the value of y obtained for the lower end of that interval to be the height of that rectangle.

```
#left end
A1 = 0
for i in range(0, n - 1):
    delta = x[i + 1] - x[i]
    A1 += y[i] * delta
print 'A (left end) = ', "%8.3f" % A1
```

We're not going to limit ourselves to calculate only the numerical value of the area by each method, but we are also going to represent its implementation. We add to the previous code the following lines to represent the rectangles whose area the program is going to add, as well as a dashed line that approximates the probable shape of the unknown function, by joining the point of each rectangle that we take as a reference to calculate its area:

```
#graph
heights = np.zeros(n, float)


def bars(h):
    heights = h
    heights = np.append(heights, 0)
    plt.bar(x, heights, width=delta, alpha=0.4, color='#3E9ECB')

for i in range(0, n - 1):
    plt.plot([x[i], x[i + 1]], [y[i], y[i + 1]], 'k--', lw=0.7)
```

9.8. NUMERICAL INTEGRATION

```
heights = np.delete(y, -1)
bars(heights)
plt.plot(x, y, 'ko')
plt.ylabel('y')
plt.xlabel('x')
plt.axhline(color='grey', lw=1)
plt.axvline(color='grey', lw=1)
plt.show()
```

When we run the program, we get the numeric result of the area and its graphical representation. We can see that possibly the difference between the actual value of the area and the calculated by this method can be quite large since the rectangles are not too close to the line joining the points. The more data we have, the more precise this method will be.

```
──────────────────────── output of the program ────────────────────────
 x =   [-2.0, -1.5, -1.0, -0.5, 0.0, 0.5, 1.0, 1.5, 2.0, 2.5, 3.0, 3.5, 4.0]
 y =   [0.0, 2.151, 2.646, 2.806, 2.828, 2.85, 3.0, 3.373, 4.0, 4.861, 5.916, 7.133, 8.485]
13 points;  12 intervals
A (left end) =     20.782
────────────────────────────────────────────────────────────────────────
```

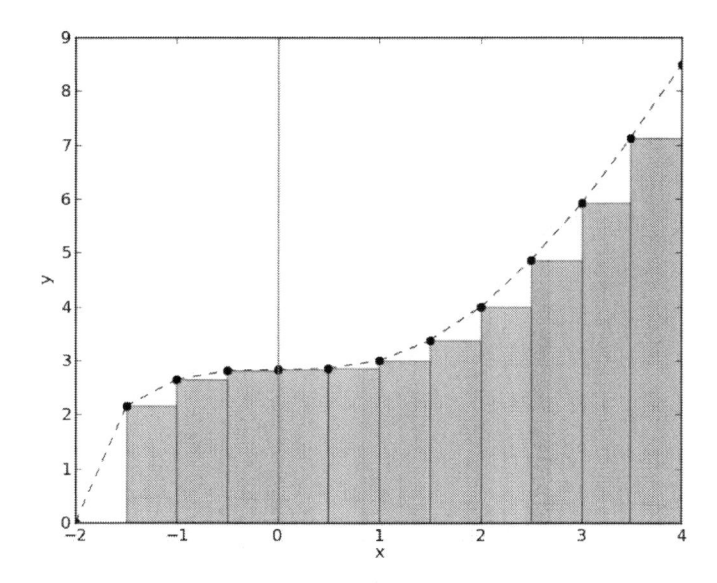

The second method consist of taking as the height of each rectangle, the value of y corresponding to the right end of each interval. The program is virtually the same as in the previous case, with a few modifications. We included here the whole program for greater clarity:

```
# -*- coding: utf-8 -*-
"""
Mathematics and Python Programming    www.pysamples.com
p9s.py
numerical integration: right end of each interval
"""

import numpy as np
import matplotlib.pyplot as plt

x = [-2.0, -1.5, -1.0, -0.5, 0.0, 0.5, 1.0, 1.5, 2.0, 2.5, 3.0, 3.5, 4.0]
y = [0.0, 2.151, 2.646, 2.806, 2.828, 2.85, 3.0, 3.373, 4.0, 4.861, 5.916, 7.133, 8.485]
print 'x = ', x
print 'y = ', y
n = len(x)
print n, ' points; ', n - 1, ' intervals'
```

```python
#right end
A2 = 0
for i in range(0, n - 1):
    delta = x[i + 1] - x[i]
    A2 += y[i + 1] * delta
print 'A (right end) = ', "%8.3f" % A2

#graph
heights = np.zeros(n, float)


def bars(h):
    heights = h
    heights = np.append(heights, 0)
    plt.bar(x, heights, width=delta, alpha=0.4, color='#3E9ECB')

for i in range(0, n - 1):
    plt.plot([x[i], x[i + 1]], [y[i], y[i + 1]], 'k--', lw=0.7)

heights = np.delete(y, 0)
bars(heights)
plt.plot(x, y, 'ko')
plt.ylabel('y')
plt.xlabel('x')
plt.axhline(color='grey', lw=1)
plt.axvline(color='grey', lw=1)
plt.show()
```

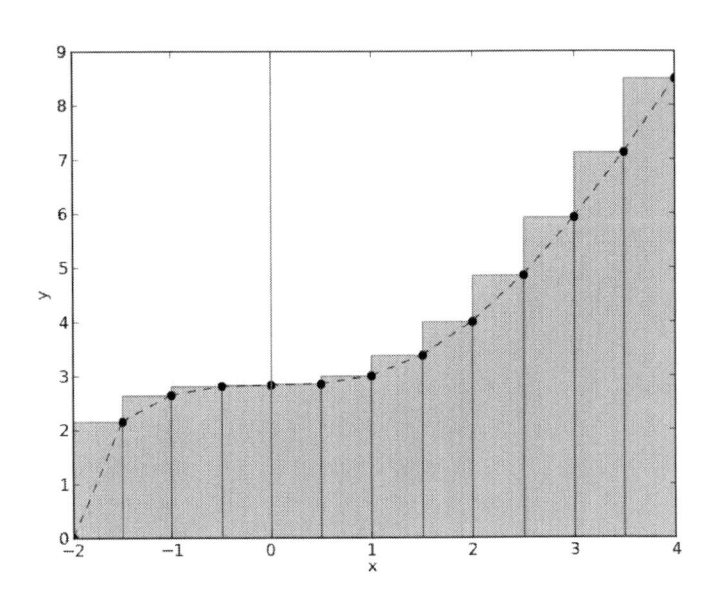

```
──────────────────────── output of the program ────────────────────────
x = [-2.0, -1.5, -1.0, -0.5, 0.0, 0.5, 1.0, 1.5, 2.0, 2.5, 3.0, 3.5, 4.0]
y = [0.0, 2.151, 2.646, 2.806, 2.828, 2.85, 3.0, 3.373, 4.0, 4.861, 5.916, 7.133, 8.485]
13 points; 12 intervals
A (right end) =    25.024
```

Since we have in this example an increasing function, its integration taking the left end of the intervals gives as a result an area possibly less than the real one, while the integral taking the right end is clearly greater than the actual area under the curve.

9.8. NUMERICAL INTEGRATION

The third method consist of grouping the three points in threes, and taking as the height of the rectangle the value of y corresponding to the mid-point of the three points. As we will see when we run the program, we get a value for the area which lies between the values obtained by the two previous methods. The program is the following:

```python
# -*- coding: utf-8 -*-
"""
Mathematics and Python Programming    www.pysamples.com
p9t.py
numerical integration: mid-point of each interval
"""

import numpy as np
import matplotlib.pyplot as plt


x = [-2.0, -1.5, -1.0, -0.5, 0.0, 0.5, 1.0, 1.5, 2.0, 2.5, 3.0, 3.5, 4.0]
y = [0.0, 2.151, 2.646, 2.806, 2.828, 2.85, 3.0, 3.373, 4.0, 4.861, 5.916, 7.133, 8.485]
print 'x = ', x
print 'y = ', y
n = len(x)
print n, ' points; ', (n - 1) / 2, ' intervals'

#mid-point
A3 = 0
x3 = []
y3 = []
for i in range(0, n - 2, 2):
    x3.append(x[i])
    y3.append(y[i + 1])
    delta = x[i + 2] - x[i]
    A3 += y[i + 1] * delta
print 'A (mid-point) = ', "%8.3f" % A3
#graph
heights = np.zeros(n, float)


def bars(h):
    heights = h
    plt.bar(x3, heights, width=delta, alpha=0.4, color='#3E9ECB')

for i in range(0, n - 1):
    plt.plot([x[i], x[i + 1]], [y[i], y[i + 1]], 'k-', lw=0.7)

heights = y3
bars(heights)
plt.plot(x, y, 'ko')
plt.ylabel('y')
plt.xlabel('x')
plt.axhline(color='grey', lw=1)
plt.axvline(color='grey', lw=1)
plt.show()
```

────────────────────────── output of the program ──────────────────────────
```
x =  [-2.0, -1.5, -1.0, -0.5, 0.0, 0.5, 1.0, 1.5, 2.0, 2.5, 3.0, 3.5, 4.0]
y =  [0.0, 2.151, 2.646, 2.806, 2.828, 2.85, 3.0, 3.373, 4.0, 4.861, 5.916, 7.133, 8.485]
13  points;  6  intervals
A (mid-point) =    23.174
```

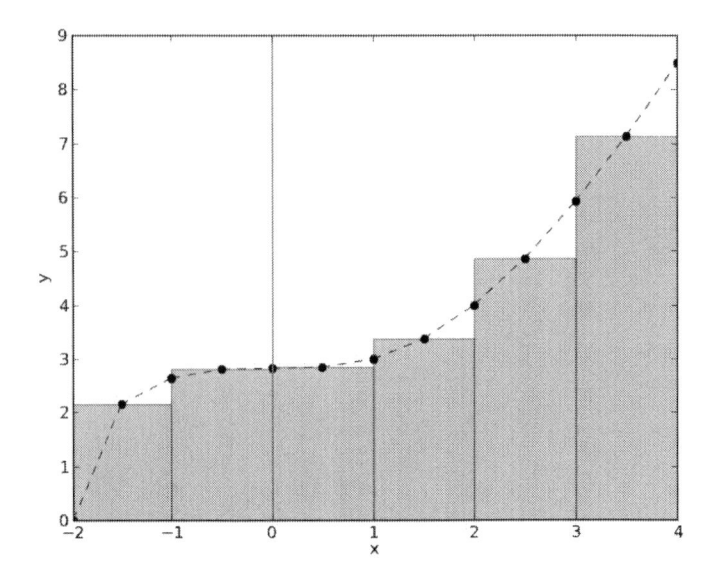

The fourth method of integration that we are going to study is to use trapezoids instead of rectangles. Each trapezoid has as its basis two consecutive dots, and its base is

$$\Delta = x_{i+1} - x_i$$

the area of each trapezoid is equal to:

$$\frac{(y_{i+1} + y_i) \cdot \Delta}{2}$$

This is the Python program:

```
# -*- coding: utf-8 -*-
"""
Mathematics and Python Programming    www.pysamples.com
p9u.py
numerical integration: trapezoids
"""

import matplotlib.pyplot as plt


x = [-2.0, -1.5, -1.0, -0.5, 0.0, 0.5, 1.0, 1.5, 2.0, 2.5, 3.0, 3.5, 4.0]
y = [0.0, 2.151, 2.646, 2.806, 2.828, 2.85, 3.0, 3.373, 4.0, 4.861, 5.916, 7.133, 8.485]
print 'x = ', x
print 'y = ', y
n = len(x)
print n, ' points; ', n - 1, ' intervals'

#trapezoid
A4 = 0
y4 = []
for i in range(0, n - 1):
    delta = x[i + 1] - x[i]
    y4.append((y[i] + y[i + 1]) / 2.0)
    A4 += y4[i] * delta
print 'A (trapezoid) = ', "%8.3f" % A4

#graph
plt.plot(x, y, 'k-', lw=0.7)
```

9.8. NUMERICAL INTEGRATION

```
plt.fill_between(x, y, 0, alpha=1, color='#3E9ECB')
for i in range(0, n - 1):
    plt.plot([x[i], x[i]], [0, y[i]], 'k-', lw=0.7)
plt.plot(x, y, 'ko')
plt.ylabel('y')
plt.xlabel('x')
plt.axhline(color='grey', lw=1)
plt.axvline(color='grey', lw=1)
plt.show()
```

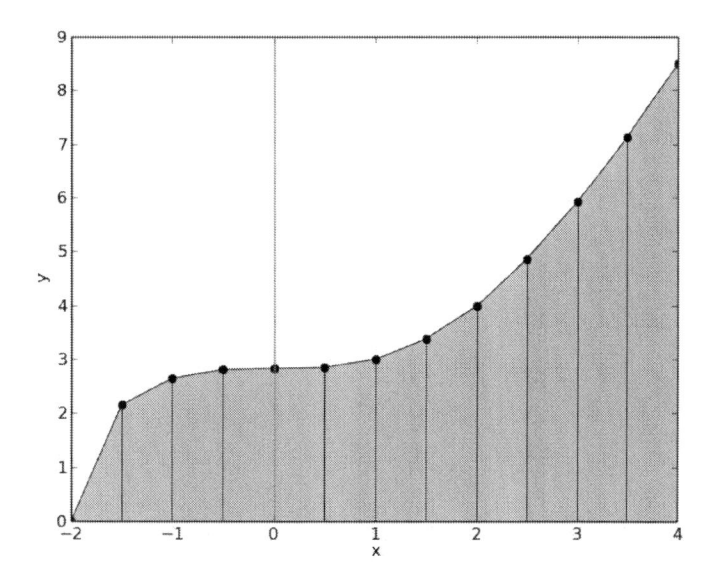

```
―――――――――――――――――――――――― output of the program ――――――――――――――――――――――――
x = [-2.0, -1.5, -1.0, -0.5, 0.0, 0.5, 1.0, 1.5, 2.0, 2.5, 3.0, 3.5, 4.0]
y = [0.0, 2.151, 2.646, 2.806, 2.828, 2.85, 3.0, 3.373, 4.0, 4.861, 5.916, 7.133, 8.485]
13 points;  12 intervals
A (trapezoid) =    22.903
```

Now the areas of the trapezoids are adjusted to the straight lines that joins the points. The approximation seems to be better than the previous methods, but the unknown function probably is not composed of straight line segments, but it will be a curve. The next method takes this fact into account.

The fifth method consists of making groups of three consecutive points, and then join them by a parabola and calculate the area under this parabola. The base of each interval is $\Delta = x_{i+2} - x_i$. The area of each of these intervals is

$$\frac{1}{6}[y_i + 4y_{i+1} + y_{i+2}]\Delta$$

This is the Python program:

```
# -*- coding: utf-8 -*-
"""
Mathematics and Python Programming     www.pysamples.com
p9v.py
numerical integration: parabolas
"""

import numpy as np
import matplotlib.pyplot as plt
```

```python
x = [-2.0, -1.5, -1.0, -0.5, 0.0, 0.5, 1.0, 1.5, 2.0, 2.5, 3.0, 3.5, 4.0]
y = [0.0, 2.151, 2.646, 2.806, 2.828, 2.85, 3.0, 3.373, 4.0, 4.861, 5.916, 7.133, 8.485]
print 'x = ', x
print 'y = ', y
n = len(x)
print n, ' points; ', (n - 1) / 2, ' intervals'

#parabola
A5 = 0
for i in range(0, n - 2, 2):
    delta = x[i + 2] - x[i]
    A5 += (y[i] + 4 * y[i + 1] + y[i + 2]) * delta / 6
print 'A (parabola) = ', "%8.3f" % A5


def drawparabola(j):
    A = np.array([[x[j] ** 2, x[j], 1.0],
                  [x[j + 1] ** 2, x[j + 1], 1.0],
                  [x[j + 2] ** 2, x[j + 2], 1.0]])
    #print A
    detA = np.linalg.det(A)
    if detA == 0:
        #the three points are in a straight line: draw trapezoid
        xrecta = [x[j], x[j + 2]]
        yrecta = [y[j], y[j + 2]]
        plt.plot(xrecta, yrecta, 'k-', lw=1.0)
        plt.fill_between(xrecta, yrecta, 0, alpha=1, color='#BDD0D7')
    else:
        B = np.array([y[j], y[j + 1], y[j + 2]])
        abc = np.linalg.solve(A, B)
        #calculates 10 points of the parabola
        xinterval = np.linspace(x[j], x[j + 2], 10)
        yinterval = np.zeros(10, float)
        for k in range(0, 10):
            yinterval[k] = (abc[0] * xinterval[k] ** 2 +
                            abc[1] * xinterval[k] + abc[2])
        plt.plot(xinterval, yinterval, 'k-', lw=1.0)
        plt.fill_between(xinterval, yinterval, 0, alpha=1, color='#BDD0D7')
        plt.plot([x[j], x[j]], [0, y[j]], 'k-', lw=0.7)

for interval in range(0, n - 2, 2):
    drawparabola(interval)

for i in range(0, n - 1):
    plt.plot([x[i], x[i + 1]], [y[i], y[i + 1]], 'k-', lw=0.7)
plt.ylabel('y')
plt.xlabel('x')
plt.axhline(color='grey', lw=1)
plt.axvline(color='grey', lw=1)
plt.show()
```

---------- output of the program ----------
```
x =  [-2.0, -1.5, -1.0, -0.5, 0.0, 0.5, 1.0, 1.5, 2.0, 2.5, 3.0, 3.5, 4.0]
y =  [0.0, 2.151, 2.646, 2.806, 2.828, 2.85, 3.0, 3.373, 4.0, 4.861, 5.916, 7.133, 8.485]
13  points;  6  intervals
A (parabola) =    22.994
```

9.8. NUMERICAL INTEGRATION

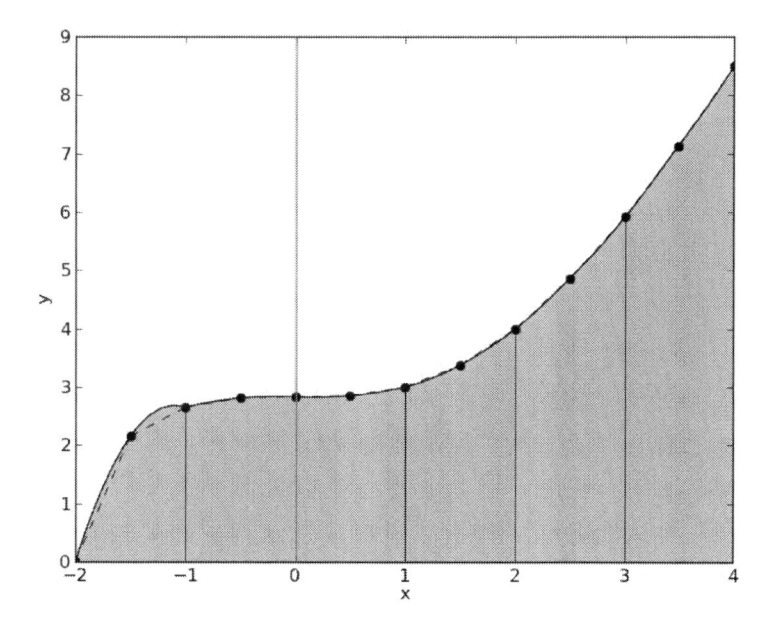

If we compare the values that we have obtained by the different methods:

1. Left end: 20.782

2. Right end: 25.024

3. Middle point: 23.174

4. Trapezoids: 22.903

5. Parabolas: 22.994

In fact, the data for this example have not been obtained experimentally, but I have taken the function $y = \sqrt{8 - x^3}$. If we run the program to calculate the sum of Riemann for this function in the interval $[-2, 4]$ taking intervals of amplitude less than 0.05, we can form an impression of how good each type of integration is for this particular function:

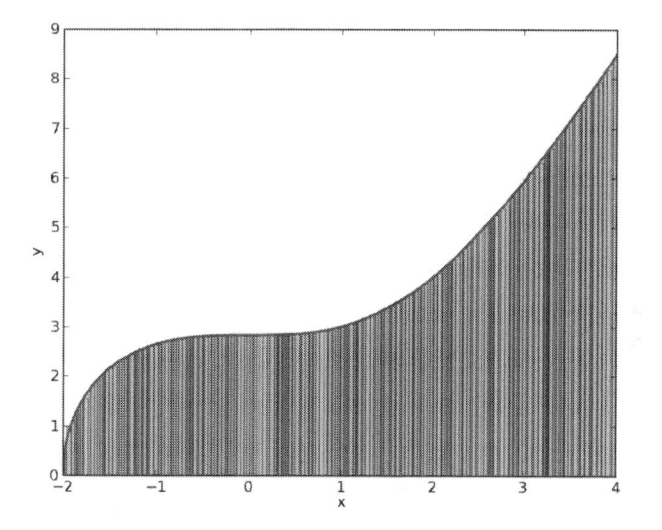

```
─────────────────────────── output of the program ───────────────────────────
[a, b] = [ -2.0 , 4.0 ]
delta =  0.05
number of intervals:  249
maximum length of an interval =  0.049805
Riemann Sum: 23.090
```

Bibliography for this chapter: [3], [12], [24], [31], [37], [38], [41], [45], [47], [46], [50], [53], [55], [56], [58]

10 | Vectors

10.1 Vector space

Let E be a set and P a field. If for all the elements of E, that we call vectors, the operations of addition and multiplication by the numbers of the body P are defined and the following axioms are met:

1. Any pair of vectors $\mathbf{x}, \mathbf{y}$ corresponds to a vector sum $\mathbf{x}+\mathbf{y}$ such that the set E with the operation of addition of vectors has the structure of an abelian group:

   - $\mathbf{x} + \mathbf{y} = \mathbf{y} + \mathbf{x}$
   - $\mathbf{x} + (\mathbf{y} + \mathbf{z}) = (\mathbf{x} + \mathbf{y}) + \mathbf{z}$
   - There is a single null vector $\mathbf{0}$ such that $\mathbf{x} + \mathbf{0} = \mathbf{0} + \mathbf{x} = \mathbf{x}, \quad \forall \mathbf{x} \in E$.
   - For all vector $\mathbf{x}$ there exists only one additive inverse $-\mathbf{x}$, such that $\mathbf{x} + (-\mathbf{x}) = \mathbf{0}$

2. For every element $\alpha \in P$, and every vector $\mathbf{x} \in E$ corresponds a product vector $\alpha \mathbf{x}$ such that

   - $\alpha(\beta \mathbf{x}) = (\alpha \beta) \mathbf{x}$
   - $1 \mathbf{x} = \mathbf{x}$

3. The operations of addition and multiplication comply with the following distributive properties:

   - $\alpha(\mathbf{x} + \mathbf{y}) = \alpha \mathbf{x} + \alpha \mathbf{y}$
   - $(\alpha + \beta) \mathbf{x} = \alpha \mathbf{x} + \beta \mathbf{x}$

Then it is said that the set E is a linear space (or vector space) over the field P. We call the elements of any vector space, vectors. The various types of vectors can look very different from the classic directed segments directed, for example, the set of polynomials of degree not greater than n with real coefficients, or the set of the continuous functions in a given interval $[a, b]$ are vector spaces. We will use bold letters to name vectors, and we will use numbers or Greek letters $\alpha, \beta, \ldots$ to name the scalars. In the event of signs matching, we will name the null vector $\mathbf{0}$ in bold, and the number zero 0 unbolded.

The following properties can be deduced from the definition of vector space:

- The null vector $\mathbf{0}$ is unique.

- The additive inverse vector $-\mathbf{x}$ is unique and equal to $-1\mathbf{x}$. The following expression is called difference of two vectors $\mathbf{x}$ and $\mathbf{y}$: $\mathbf{x} - \mathbf{y} = \mathbf{x} + (-\mathbf{y})$.

- For every vector $\mathbf{x}$: $0\mathbf{x} = \mathbf{0}$.

- For every real number α: $\alpha \mathbf{0} = \mathbf{0}$.

- If $\alpha \mathbf{x} = \mathbf{0}$, then at least one of the two factors must be null: $\alpha = 0$ or $\mathbf{x} = \mathbf{0}$.

10.1. VECTOR SPACE

A set consisting of an arbitrary number of vectors $e_1, e_2, ..., e_n$, of the space E, is called a system of vectors. If we perform multiplication and addition with the vectors of the system, we obtain a vector that is a linear combination of vectors of the system:

$$x = \alpha_1 e_1 + \alpha_2 e_2 + ... + \alpha_n e_n$$

The set $L(e_1, e_2, ..., e_n)$ consisting of all the possible linear combinations of the system vectors, which we obtain by assigning different values to the coefficients α_i, is called linear span or linear hull of the system vectors, and this linear span itself has the structure of a vector space, since the vectors x resulting of the linear combination of the system vectors comply the axioms, and both the null vector and the additive inverse vector belong to the linear span:

$$0 = 0e_1 + 0e_2 + ... + 0e_n$$

$$\text{-x} = (-\alpha_1)e_1 + (-\alpha_2)e_2 + ... + (-\alpha_n)e_n$$

Then, the linear span is the «minimum» vector space that contains the vectors of the system.

If for a given system of vectors, one of its vector, for example e_n can be expressed as a linear combination of the other vectors of the system, then the linear span of the system of n vectors can also be generated by combining only the other $n-1$ vectors. If we repeat this process of elimination of the vectors that can be expressed as linear combinations of the other vectors of the system, we finally obtain a system of vector in which none of its vectors can be expressed as a linear combination of the other vectors, or we get a system consisting only of one not null vector. Such a system is called a linearly independent system. In other words, the necessary and sufficient condition for a system of vectors to be linearly independent is:

$$\mathbf{x} = \alpha_1 e_1 + \alpha_2 e_2 + ... + \alpha_n e_n = 0$$

Then all the coefficients α_i are null. For a system to be linearly dependent is sufficient that one of the vectors is zero, or any of the vectors of the system can be expressed as a linear combination of the rest.

If two system of vectors are such that any vector of each system can be expressed as a linear combination of vectors from the other system, both systems are equivalent, and their linear spans match. The concept of equivalence of two systems of vectors is a relation of equivalence, and therefore meets the properties reflexive, symmetric and transitive. The linearly independent equivalent systems have the same number of vectors, which is called range of the system of vectors. By definition, the range of a null system of vectors is considered to be equal to zero..

Suppose that a vector system consists of n vector, then if we take any $n+1$ vectors of its linear span, these $n+1$ vectors will be linearly dependent system of vectors.

If a system of linearly independent vectors in a space E can have any number of vectors, the linear space is said to be of infinite dimension. We'll take care of the otherwise, that in which a system of linearly independent vectors, it can only have at the most n vectors. Then the space E is said to be of dimension n, and therefore any $n+1$ vectors will be linearly dependent. The space E of finite dimension n can be considered a linear span of a linearly independent system consisting of n of its vectors. and this system is called a basis of the space E.

Since the space E is a linear span of the basis $e_1, e_2, ..., e_n$, any vector $\mathbf{x} \in K$ can be expressed as a linear combination of vectors of the basis:

$$\mathbf{x} = \alpha_1 e_1 + \alpha_2 e_2 + ... + \alpha_n e_n$$

and the numbers α_i are called coordinates of the vector x on this basis (or also, vector components), and the vector x can be expressed as a row or column matrix:

$$\mathbf{x} = \begin{pmatrix} \alpha_1 & \alpha_2 & ... & \alpha_n \end{pmatrix}$$

Since for any vector x of the space E, its decomposition according to a given basis is unique. So that if we fix a basis, all of the vectors of the space are uniquely defined by their coordinates with respect to this basis, and the operations of addition and multiplication of vectors are reduced to operations of sum and product of matrices. That is to say, if:

$$\mathbf{x} = \begin{pmatrix} \alpha_1 & \alpha_2 & ... & \alpha_n \end{pmatrix}$$

$$\mathbf{y} = \begin{pmatrix} \beta_1 & \beta_2 & ... & \beta_n \end{pmatrix}$$

then

$$\mathbf{x} + \mathbf{y} = \begin{pmatrix} \alpha_1 + \beta_1 & \alpha_2 + \beta_2 & ... & \alpha_2 + \beta_n \end{pmatrix}$$

And the product of a vector by a scalar:

$$\lambda \mathbf{x} = \begin{pmatrix} \lambda \alpha_1 & \lambda \alpha_2 & ... & \lambda \alpha_n \end{pmatrix}$$

We define a function $E \times E \to \mathbf{R}$ which associates to each pair of vectors $\mathbf{x}, \mathbf{y}$ a real number $\mathbf{x} \cdot \mathbf{y}$. This function, which we call dot product (or scalar product), has the following properties $\forall \mathbf{x}, \mathbf{y}, \mathbf{z} \in E$:

- $\mathbf{x} \cdot \mathbf{x} \geq 0$

- Si $\mathbf{x} \cdot \mathbf{x} = 0$, then necessarily $\mathbf{x} = \mathbf{0}$

- $\mathbf{x} \cdot \mathbf{y} = \mathbf{y} \cdot \mathbf{x}$

- $\mathbf{x} \cdot (\mathbf{y} + \mathbf{z}) = \mathbf{x} \cdot \mathbf{y} + \mathbf{x} \cdot \mathbf{z}$

- For every real number α, it meets:

$$\alpha(\mathbf{x} \cdot \mathbf{y}) = \alpha \mathbf{x} \cdot \mathbf{y} = \mathbf{x} \cdot \alpha \mathbf{y}$$

- $\mathbf{0} \cdot \mathbf{0} = 0$

- Inequality of Cauchy-Buniakovski: $(\mathbf{x} \cdot \mathbf{y})^2 \leq (\mathbf{x} \cdot \mathbf{x})(\mathbf{y} \cdot \mathbf{y})$; equality is only found in the event that the two vectors $\mathbf{x}$ and $\mathbf{y}$ are collinear.

The vector space E in which a dot product has been defined, is called a metric space. Let's look at a metric space of dimension 3. In this space a basis will be determined by a system of three linearly independent vectors $\mathbf{e}_1, \mathbf{e}_2, \mathbf{e}_3$. Usually we choose these vectors so that in addition to be linearly independent, they fulfil that the dot product of any two of them is equal to 0 if the vectors are different ($\mathbf{e}_i \cdot \mathbf{e}_j = 0$); and equal to 1 if it is the product of a vector by itself $\mathbf{e}_i \cdot \mathbf{e}_i = 1$. The basis that satisfies this condition is called orthonormal. In a system of rectangular coordinates, these vectors are often called $\mathbf{i}, \mathbf{j}, \mathbf{k}$. Let's look at what would be the development of the dot product of two vectors whose rectangular coordinates were

$$\mathbf{a} = \begin{pmatrix} a_1 & a_2 & a_3 \end{pmatrix} = a_1 \mathbf{i} + a_2 \mathbf{j} + a_3 \mathbf{k}$$
$$\mathbf{b} = \begin{pmatrix} b_1 & b_2 & b_3 \end{pmatrix} = b_1 \mathbf{i} + b_2 \mathbf{j} + b_3 \mathbf{k}$$

Their dot product will be:

$$\mathbf{a} \cdot \mathbf{b} = (a_1 \mathbf{i} + a_2 \mathbf{j} + a_3 \mathbf{k}) \cdot (b_1 \mathbf{i} + b_2 \mathbf{j} + b_3 \mathbf{k})$$

and after some operations we get:

$$\mathbf{a} \cdot \mathbf{b} = a_1 \mathbf{i} \cdot b_1 \mathbf{i} + a_1 \mathbf{i} \cdot b_2 \mathbf{j} + a_1 \mathbf{i} \cdot b_3 \mathbf{k} + ... + a_3 \mathbf{k} \cdot b_3 \mathbf{k}$$

But in this expression all the dot products of any two different vectors of the basis are null: $\mathbf{i} \cdot \mathbf{j} = 0$, and if the vectors are equal, their dot product is $\mathbf{i} \cdot \mathbf{i} = \mathbf{j} \cdot \mathbf{j} = \mathbf{k} \cdot \mathbf{k} = 1$. And finally:

$$\mathbf{a} \cdot \mathbf{b} = a_1 b_1 + a_2 b_2 + a_3 b_3$$

Which coincides with the product of the coordinates of the two vectors, expressed as a product of an row matrix by a column matrix. Let us look at it with a simple Python program:

10.1. VECTOR SPACE

```
# -*- coding: utf-8 -*-
"""
Mathematics and Python Programming    www.pysamples.com
p10a.py
"""

import sympy as sy

a1, a2, a3, b1, b2, b3 = sy.symbols('a1, a2, a3, b1, b2, b3')
sy.init_printing(use_unicode=True)

rowvector = sy.Matrix([[1, 2, 3]])
columnvector = sy.Matrix([-1, 5, 10])
print 'a = ', rowvector
print 'b = '
print columnvector
print 'a b = ', rowvector * columnvector
print
a = sy.Matrix([[a1, a2, a3]])
b = sy.Matrix([b1, b2, b3])
print 'a = ', a
print 'b = '
print b
print 'a b = ', a * b
```

───────────────────────────── ejecución ─────────────────────────────
```
a =  Matrix([[1, 2, 3]])
b =
Matrix([[-1], [5], [10]])
a·b =  Matrix([[39]])

a =  Matrix([[a1, a2, a3]])
b =
Matrix([[b1], [b2], [b3]])
a·b =  Matrix([[a1*b1 + a2*b2 + a3*b3]])
```

Let's take a vector whose rectangular coordinates are:

$$\mathbf{r} = r_1\mathbf{i} + r_2\mathbf{j} + r_3\mathbf{k} \equiv \begin{pmatrix} r_1 & r_2 & r_2 \end{pmatrix}$$

The length or magnitude or norm of this vector is the positive real number $|\mathbf{r}| = \sqrt{r_1^2 + r_2^2 + r_3^2}$.

For example, in the next section we will represent the vector $\mathbf{r}(t) = t\mathbf{i} + \frac{t^2}{2}\mathbf{j} + \frac{t^3}{100}\mathbf{k}$, where the scalar t represents time. The length of the vector $\mathbf{r}(t)$ is the distance from the end of the vector to the origin of coordinates for each value of t. The following Python program calculates the length of $\mathbf{r}(t)$ for the values of $t = 0, 10, 20, ...90$:

```
# -*- coding: utf-8 -*-
"""
Mathematics and Python Programming    www.pysamples.com
p10b.py
"""

import numpy as np

for t in range(0, 101, 25):
    x = t
    y = t ** 2 / 2
    z = t ** 3 / 100
    longitud = np.sqrt(x ** 2 + y ** 2 + z ** 2)
    print '|r(', t, ')| = ', "%7.2f" % longitud
```

```
_____ output of the program _____
|r(   0 )| =       0.00
|r(  25 )| =     349.72
|r(  50 )| =    1768.47
|r(  75 )| =    5069.96
|r( 100 )| =   11180.79
```

For any vectors **a** and **b**, and any real number α, the length has the following properties:

- $|\mathbf{a}| = 0$ if and only if $\mathbf{a} = \mathbf{0}$
- $|\alpha \mathbf{a}| = |\alpha||\mathbf{a}|$
- $|\mathbf{a} \cdot \mathbf{b}| \leq |\mathbf{a}||\mathbf{b}|$
- $|\mathbf{a} + \mathbf{b}| \leq |\mathbf{a}| + |\mathbf{b}|$

The vectors whose length is equal to 1 are said to be normalized, and are called unit vectors. If we take a vector $\mathbf{a} = \begin{pmatrix} a_1 & a_2 & a_3 \end{pmatrix}$ if we need to normalize a we must find the unit vector **u** such that:

$$\mathbf{u} = \begin{pmatrix} \frac{a_1}{|\mathbf{a}|} & \frac{a_2}{|\mathbf{a}|} & \frac{a_3}{|\mathbf{a}|} \end{pmatrix} = \frac{1}{|\mathbf{a}|} \begin{pmatrix} a_1 & a_2 & a_3 \end{pmatrix}$$

The following Python program calculates the length and normalizes a vector:

```python
# -*- coding: utf-8 -*-
"""
Mathematics and Python Programming      www.pysamples.com
p10c.py
"""

import numpy as np


def mod(name, a):
    length = np.linalg.norm(a)
    print '|', name, '| = ', "%5.2f" % length
    return length


def normalizes(a):   # a is a vector
    length = mod('a', a)
    u = []
    u = a / length
    return u

a = np.array([3, -1, 5])
print 'a = ', a
u = normalizes(a)
print 'u = ', u
mod('u', u)
```

```
_____ output of the program _____
a =  [ 3 -1  5]
| a | =    5.92
u =  [ 0.50709255 -0.16903085  0.84515425]
| u | =    1.00
```

The angle $\{\mathbf{x}, \mathbf{y}\}$ formed by two non-zero vectors **x** and **y** of the space E is the angle defined by:

$$cos\{\mathbf{x}, \mathbf{y}\} = \frac{\mathbf{x} \cdot \mathbf{y}}{|\mathbf{x}||\mathbf{y}|}$$

The inequality of Cauchy-Buniakovski allows us to affirm that the expression call cosine of the angle between vectors is not greater in absolute value than the unit. The following Python program calculates the angle between two vectors:

10.1. VECTOR SPACE

```
# -*- coding: utf-8 -*-
"""
Mathematics and Python Programming    www.pysamples.com
p10d.py
"""

import numpy as np


a = np.array([0, 1])
b = np.array([1, 1])
#b = 2 * a.T # next example
mod_a = np.linalg.norm(a)
mod_b = np.linalg.norm(b)

print 'a = ', a
print 'b = '
print b

radians = np.arccos(np.dot(a,b) / (mod_a * mod_b))
dd = np.degrees(radians)
print 'angle = ' + str(radians) + ' radians = ' + str(dd) + ' decimal degrees'


def decdeg_dms(dd):
    dd = abs(dd)
    minutes,seconds = divmod(dd*3600,60)
    degrees,minutes = divmod(minutes,60)
    strgms = str(degrees) + '   ' + str(minutes) + "' " + "%5.2f" % seconds + "'''"
    return strgms

print 'angle{a,b} = ' + str(radians/ np.pi) + ' pi radians = ' + str(decdeg_dms(dd))
```

---------- several outputs of the program ----------
```
a =  [0 1]
b =
[1 1]
angle = 0.785398163397 radians = 45.0 decimal degrees
angle{a,b} = 0.25 pi radians = 45.0° 0.0'  0.00''

a =  [0 1]
b =
[-1 -3]
angle = 2.81984209919 radians = 161.565051177 decimal degrees
angle{a,b} = 0.89758361765 pi radians = 161.0° 33.0' 54.18''
```

Two vector **a** and **b** are collinear if there is a scalar α such that $\mathbf{a} = \alpha\mathbf{b}$. If in the previous program we change the following two lines:

```
#b = np.array([1, 1])
b = 2 * a.T
```

we get:

---------- output of the program ----------
```
a =  [0 1]
b =
[0 2]
angle = 0.0 radians = 0.0 decimal degrees
angle{a,b} = 0.0 pi radians = 0.0° 0.0'  0.00''
```

Two vectors are said to be orthogonal if its dot product is zero: $\mathbf{a} \cdot \mathbf{b} = 0$. For example, in the previous program we take:

```
a = np.array([[1, 0, 0]])
b = np.array([0, 5, 0])
```

--- output of the program ---
```
a =  [[1 0 0]]
b = 
[0 5 0]
angle = [ 1.57079633] radians = [ 90.] decimal degrees
angle{a,b} = [ 0.5] pi radians = [ 90.]° [ 0.]'  0.00''
```

Now take two non-collinear vectors a and b. We define the cross product (or vector product) of these two vectors as the application that assigns to this pair of vectors a, b, another vector c orthogonal to both vectors a and b, and such that the length of c is equal to the area of the parallelogram defined by a and b, and whose direction follows the right-hand rule. The cross product is designated as follows: $\mathbf{c} = \mathbf{a} \times \mathbf{b}$ or $\mathbf{c} = \mathbf{a} \wedge \mathbf{b}$.

From the condition that the length of c has to comply, it follows that $|\mathbf{c}| = |\mathbf{a}||\mathbf{b}|sin\alpha$, being α the angle formed by a and b reduced to a common origin. Since $\sin \alpha = -\sin(-\alpha)$, the vector product meets

$$\mathbf{a} \times \mathbf{b} = -\mathbf{b} \times \mathbf{a}$$

$$\mathbf{a} \times \mathbf{a} = 0$$

If we take the basis i, j, k, we find that $\mathbf{i} \times \mathbf{j} = \mathbf{k}$, $\mathbf{j} \times \mathbf{k} = \mathbf{i}$, and $\mathbf{k} \times \mathbf{i} = \mathbf{j}$. If we change the order of the factors, the cross product of both changes sign, and every cross product of a vector by the itself is null. If we express a and b in this basis we get

$$\mathbf{a} = a_1 \mathbf{i} + a_2 \mathbf{j} + a_3 \mathbf{k}$$

$$\mathbf{b} = b_1 \mathbf{i} + b_2 \mathbf{j} + b_3 \mathbf{k}$$

$\mathbf{a} \times \mathbf{b} = [a_1 b_1 (\mathbf{i} \times \mathbf{i}) + a_1 b_2 (\mathbf{i} \times \mathbf{j}) + a_1 b_3 (\mathbf{i} \times \mathbf{k})] + [a_2 b_1 (\mathbf{j} \times \mathbf{i}) + a_2 b_2 (\mathbf{j} \times \mathbf{j}) + a_2 b_3 (\mathbf{j} \times \mathbf{k})] + [a_3 b_1 (\mathbf{k} \times \mathbf{i}) + a_3 b_2 (\mathbf{k} \times \mathbf{j}) + a_3 b_3 (\mathbf{k} \times \mathbf{k})]$

Since all of the cross products of a vector by the itself are zero, then:

$$\mathbf{a} \times \mathbf{b} = a_1 b_2 \mathbf{k} - a_1 b_3 \mathbf{j} - a_2 b_1 \mathbf{k} + a_2 b_3 \mathbf{i} + a_3 b_1 \mathbf{j} - a_3 b_2 \mathbf{i}$$

$$\mathbf{a} \times \mathbf{b} = (a_2 b_3 - a_3 b_2)\mathbf{i} + (a_3 b_1 - a_1 b_3)\mathbf{j} + (a_1 b_2 - a_2 b_1)\mathbf{k}$$

That is equivalent to the development of the next determinant:

$$\mathbf{a} \times \mathbf{b} = \begin{vmatrix} \mathbf{i} & \mathbf{j} & \mathbf{k} \\ a_1 & a_2 & a_3 \\ b_1 & b_2 & b_3 \end{vmatrix}$$

Shown below are two examples solved by a Python program. They represent the three vectors a (blue), b (green), c (red) and the plane containing the vectors a and b:

--- output of the program ---
```
Represents Z coordinate scaled 1:10
a = [[ 1.5 -0.5  0. ]]
b = [[ 0.5  0.5  0. ]]
c = a X b = [[-0.  0.  1.]]
|c| = [ 1.]
Colors key: a(blue), b(green), c(red)
```

10.1. VECTOR SPACE

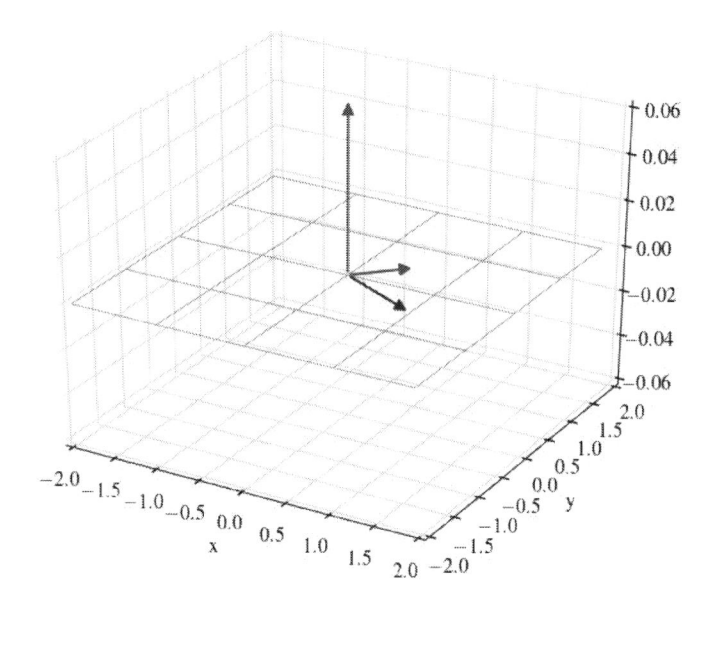

```
                            ──────── output of the program ────────
a =   [[-2 -1 -1]]
b =   [[1 3 1]]
c = a X b =   [[ 2  1 -5]]
|c| =  [ 5.47722558]
Colors key: a(blue), b(green), c(red)
```

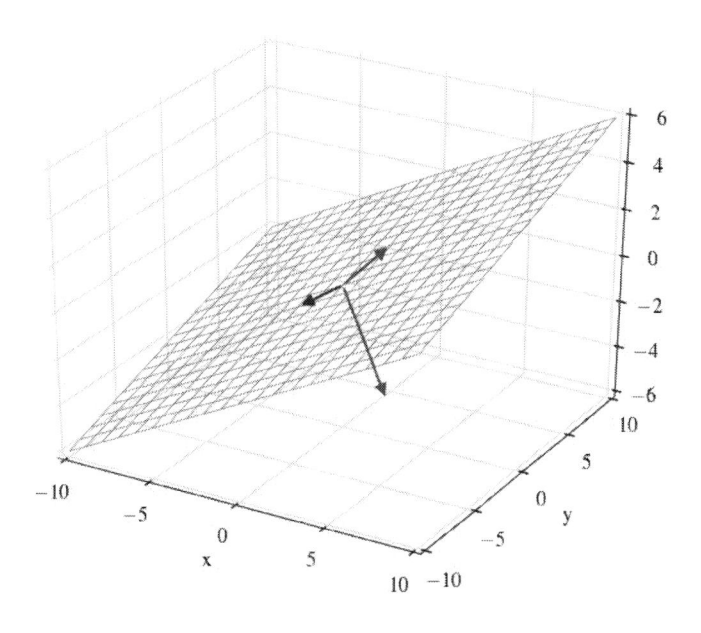

```python
# -*- coding: utf-8 -*-
"""
Mathematics and Python Programming     www.pysamples.com
p10e.py
"""

import matplotlib as mpl
from mpl_toolkits.mplot3d import Axes3D
import numpy as np
import matplotlib.pyplot as plt
```

```python
from matplotlib import rc
from matplotlib.patches import FancyArrowPatch
from mpl_toolkits.mplot3d import proj3d

rc('font', **{'family': 'serif', 'serif': ['Times']})
rc('text', usetex=True)

mpl.rcParams['legend.fontsize'] = 12

#==============================================================
# #example 1
# va = np.array([[1.5], [-0.5], [0]])
# vb = np.array([[0.5], [0.5], [0]])
# print 'Represents Z coordinate scaled 1:10'
#==============================================================

#example 2
va = np.array([[-2], [-1], [-1]])
vb = np.array([[1], [3], [1]])



vn = np.cross(va, vb, axis=0)
print 'a = ', np.transpose(va)
print 'b = ', np.transpose(vb)
print 'c = a X b = ', np.transpose(vn)
print '|c| = ', np.sqrt(vn[0] ** 2 + vn[1] ** 2 + vn[2] ** 2)
print 'Colors key: a(blue), b(green), c(red)'

fig = plt.figure()
ax = fig.gca(projection='3d')
ax.w_xaxis.set_pane_color((1.0, 1.0, 1.0, 1.0))
ax.w_yaxis.set_pane_color((1.0, 1.0, 1.0, 1.0))
ax.w_zaxis.set_pane_color((1.0, 1.0, 1.0, 1.0))

class Arrow3D(FancyArrowPatch):
    def __init__(self, xs, ys, zs, *args, **kwargs):
        FancyArrowPatch.__init__(self, (0, 0), (0, 0), *args, **kwargs)
        self._verts3d = xs, ys, zs

    def draw(self, renderer):
        xs3d, ys3d, zs3d = self._verts3d
        xs, ys, zs = proj3d.proj_transform(xs3d, ys3d, zs3d, renderer.M)
        self.set_positions((xs[0], ys[0]), (xs[1], ys[1]))
        FancyArrowPatch.draw(self, renderer)


#==============================================================
# #example 1
# x = np.arange(-2, 2, 0.1)
# y = np.arange(-2, 2, 0.1)
#==============================================================

#example 2
x = np.arange(-10, 10, 0.1)
y = np.arange(-10, 10, 0.1)
```

```
X, Y = np.meshgrid(x, y)

#plane containing point (0 0 0) and the end-points of the vectors a and b
zx = va[1] * vb[2] - va[2] * vb[1]
zy = va[2] * vb[0] - va[0] * vb[2]
zz = -1.0 * (va[0] * vb[1] - va[1] * vb[0])
Z = (zx * X + zy * Y) / zz


a = Arrow3D([0, va[0]], [0, va[1]], [0, va[2]], mutation_scale=20, lw=1.5,
            arrowstyle="-|>", color="b", linestyle='solid')
b = Arrow3D([0, vb[0]], [0, vb[1]], [0, vb[2]], mutation_scale=20, lw=1.5,
            arrowstyle="-|>", color="g", linestyle='solid')

#==============================================================================
# #example 1: represents Z coordinate scaled 1:10
# c = Arrow3D([0, vn[0]], [0, vn[1]], [0, 0.1 * vn[2]], mutation_scale=20, lw=1.5,
#             arrowstyle="-|>", color="r", linestyle='solid')  # ej. 1
#==============================================================================

#example 2
c = Arrow3D([0, vn[0]], [0, vn[1]], [0, vn[2]], mutation_scale=20, lw=1.5,
            arrowstyle="-|>", color="r", linestyle='solid')


ax.plot_wireframe(X, Y, Z, rstride=10, cstride=10, color='grey', lw='0.5')
ax.add_artist(c)
ax.add_artist(a)
ax.add_artist(b)
plt.xlabel('x')
plt.ylabel('y')
plt.show()
```

10.2 Equations of a straight line in the plane

Let E be a vector space on the field of the real numbers. A set A of elements is called affine space, and its elements are called points (and in this text we will write them in uppercase letters). Given any two points A, B, there is a single vector $\mathbf{a} = \text{AB} \in E$ from A to B (in this section we will dispense with the bold type to indicate the vector AB, with origin at the point A and end at the point B). The point A is called origin of the vector, and the point B, end of the vector. If A = B, then $\text{AA} = \mathbf{0} \in E$.

Let $\mathbf{v}_1, \mathbf{v}_2, ..., \mathbf{v}_n$ be a basis of the vector space E, and O a point in the set A. The set $\{\text{O}, \mathbf{v}_1, \mathbf{v}_2, ..., \mathbf{v}_n\}$ is called reference system of A. Any point of A is determined in a unique way in this system of reference, since $\text{OP}_1 = \mathbf{v}_1$, etc, or the vector $\mathbf{v}$ which unites the origin O with any given point P, will be a linear combination of vectors of the basis.

Consider a vector space of dimension 2. Their vectors have two coordinates $\mathbf{v} = \begin{pmatrix} v_1 & v_2 \end{pmatrix}$, and in a system of rectangular coordinates $\{\text{O}, \mathbf{i}, \mathbf{j}\}$ the points can be expressed through two coordinates: $\text{P} = (x, y)$ that correspond to the end point of the vector with origin in the origin of coordinates and end at the point P. The set Π of points determined by this system of coordinates is called plane.

Let's call now L a straight line in the plane, and $\text{P}_0(x_0, y_0)$ an arbitrary point of L. Any other point $\text{P}(x, y)$ of the straight line can be united with P_0 by a vector $\mathbf{r} = \text{P}_0\text{P} = \begin{pmatrix} x - x_0 & y - y_0 \end{pmatrix}$ which has the direction of the straight line and is called direction vector. Any non-zero vector

$\mathbf{v}_n = \begin{pmatrix} A & B \end{pmatrix}$ which is perpendicular to the straight line will fulfil $\mathbf{r} \cdot \mathbf{v}_n = 0$, then:

$$\begin{pmatrix} x - x_0 & y - y_0 \end{pmatrix} \cdot \begin{pmatrix} A & B \end{pmatrix} = 0$$
$$A(x - x_0) + B(y - y_0) = 0$$

and if we call $C = -Ax_0 - By_0$ we get the expression:

$$Ax + By + C = 0$$

which is the general form of the equation of the straight line.

Now take another specific point $P_1(x_1, y_1)$ of the straight line. The vector P_0P_1 is a multiple of the direction vector $\mathbf{r}$:

$$P_0P_1 = \lambda \mathbf{r}$$

being λ a real number. Therefore, the straight line is the affine subspace $\{P_0, \mathbf{r}\}$ of dimension one. Let P be any point on the line, and OP the vector from the origin to the point P. Then for any point on the line:

$$OP = OP_0 + P_0P$$

and since P_0P is on the straight line, it will be a multiple of its direction vector: $P_0P = \lambda \mathbf{r}$, being λ a parameter that can take any real value. Thus we come to an expression that is called vector equation of the straight line:

$$OP = OP_0 + \lambda \mathbf{r}$$

If $OP = \begin{pmatrix} x & y \end{pmatrix}$, $OP_0 = \begin{pmatrix} x_0 & y_0 \end{pmatrix}$, and $\mathbf{r} = \begin{pmatrix} r_x & r_y \end{pmatrix}$, we obtain the parametric equations of the straight line:

$$\begin{cases} x = x_0 + \lambda r_x \\ y = y_0 + \lambda r_y \end{cases}$$

The tangent of the angle formed by the straight line with the X axis is called slope of the line and is equal to:

$$m = \frac{y_1 - y_0}{x_1 - x_0}$$

The slope of the line is the same in all their points, then, for any point $P(x, y)$ of the straight line, the following equations will be also fulfilled:

$$\frac{y - y_0}{x - x_0} = \frac{y_1 - y_0}{x_1 - x_0}$$

$$\frac{y - y_0}{y_1 - y_0} = \frac{x - x_0}{x_1 - x_0}$$

$$\frac{x - x_0}{x_1 - x_0} = \frac{y - y_0}{y_1 - y_0}$$

and this equation is called continuous equation of the straight line.

If now we use the general equation of the straight line and we replace the coordinates of the two points P_0 and P_1 we obtain a system of two equations with two unknowns:

$$\begin{cases} Ax_1 + By_1 + C = 0 \\ Ax_0 + By_0 + C = 0 \end{cases}$$

By subtracting both:

$$A(x_1 - x_0) + B(y_1 - y_0) = 0$$

and divide by $(x_1 - x_0)$:

$$A + B\frac{y_1 - y_0}{x_1 - x_0} = 0$$

but this is just the slope m of the straight line:

$$A + Bm = 0$$

10.2. EQUATIONS OF A STRAIGHT LINE IN THE PLANE

$$m = \frac{-A}{B}$$

Therefore, if $A = 0$, this is a horizontal line; and if $B = 0$ this is a vertical line as the equation is simplified to $Ax = C$, i.e. $x =$ constant.

Suppose that $B \neq 0$. If we divide the general form of the equation of the straight line by $-B$, we get:

$$\frac{-A}{B}x - y - \frac{C}{B} = 0$$

$$y = \frac{-A}{B}x - \frac{C}{B}$$

$$y = mx - \frac{C}{B}$$

and if we call $n = -\frac{C}{B}$:

$$y = mx + n$$

this equation is called slope-intercept equation of the straight line.

Now we can see clearly that to find the equation of a straight line we need a point P_0 belonging to it, and in addition one of the three following data: the normal vector to the straight line; the direction vector; another point on the line. The following are examples of the three cases, solved with the Python program that is displayed below. When we run the program, we are asked the type of data that we are going to introduce:

```
─────────────────────────── output of the program ───────────────────────────
Problem type according to known data:
1 - P0 and the normal vector
2 - P0 and the direction vector
3 - P0 and P1

Input the type of problem:
```

The program calculates the equations of straight line and plot it in blue color. It also represents the point P_0 as a red dot. If you specify the point P_1, this one is represented as a black dot. It also represents the unit direction vector (in red) and the normal vector (in green). For convenience of printing, the parameter is shown with the letter t instead of the Greek letter λ. If it is a horizontal or vertical line, the program provides its equation but does not plot it.

```
─────────────────────────── output of the program ───────────────────────────
Input the type of problem: 1
Write point P0 coordinates, separated by one blank space:
3 5
P0:  [3.0, 5.0]

Write the coordinates of the normal vector, separated by one blank space:
1 0

vertical line x = 3.0
```

```
─────────────────────────── output of the program ───────────────────────────
Input the type of problem: 2
Write point P0 coordinates, separated by one blank space:
3 5
P0:  [3.0, 5.0]

Write the coordinates of the direction vector, separated by one blank space:
1 0

horizontal line y = 5.0
```

If the line is not horizontal nor vertical, the program provides its equations, and plots it:

```
─────────────────────────── output of the program ───────────────────────────
Input the type of problem: 1
Write point P0 coordinates, separated by one blank space:
1.5 2
P0:  [1.5, 2.0]

Write the coordinates of the normal vector, separated by one blank space:
1 3

Direction vector:  [-3.0, 1.0]
Normal vector:  [1.0, 3.0]
General form equation: 1.000x + 3.0 y + -7.500 = 0
Slope-intercept equation:  y = -0.333x + 2.500
Vector equation: OP = [1.5, 2.0] + t[-3.0, 1.0]
Parametric equations:
x =  1.5   +  -3.0 t
y =  2.0   +   1.0 t

Cuts axis X at point [7.500, 0]
Cuts axis Y at point [0, 2.500]
```

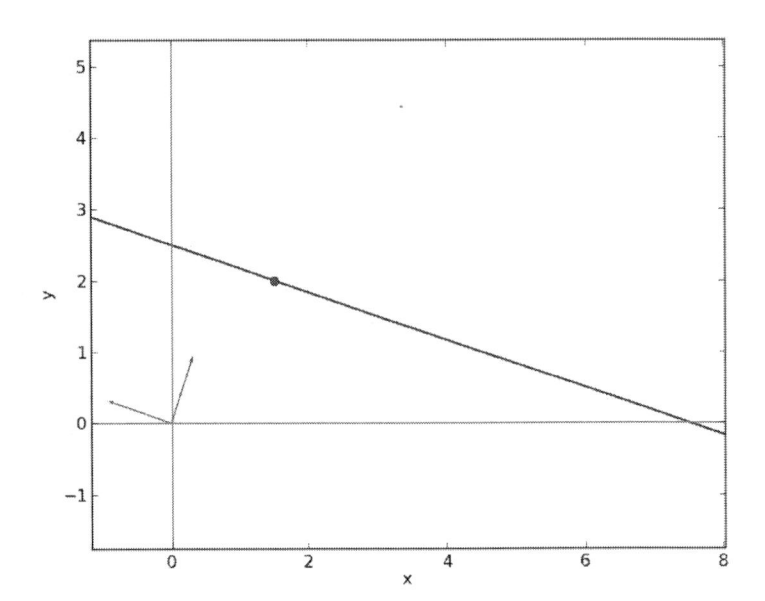

```
─────────────────────────── output of the program ───────────────────────────
Input the type of problem: 2
Write point P0 coordinates, separated by one blank space:
-3 -2.5
P0:  [-3.0, -2.5]

Write the coordinates of the direction vector, separated by one blank space:
2 1

Direction vector:  [2.0, 1.0]
Normal vector:  [-0.5, 1.0]
General form equation: -0.500x + 1.0 y + 1.000 = 0
Slope-intercept equation:  y = 0.500x + -1.000
Vector equation: OP = [-3.0, -2.5] + t[2.0, 1.0]
Parametric equations:
x =  -3.0   +   2.0 t
y =  -2.5   +   1.0 t

Cuts axis X at point [2.000, 0]
Cuts axis Y at point [0, -1.000]
```

10.2. EQUATIONS OF A STRAIGHT LINE IN THE PLANE

──────────────── output of the program ────────────────
```
Tipos de datos de partida:
Input the type of problem: 3
Write point P0 coordinates, separated by one blank space:
2 2
P0:  [2.0, 2.0]

Write the coordinates of point P1, separated by one blank space:
-1 -2.5
P1:  [-1.0, -2.5]

Direction vector:  [-3.0, -4.5]
Normal vector:  [-1.5, 1.0]
General form equation: -1.500x + 1.0 y + 1.000 = 0
Slope-intercept equation:  y = 1.500x + -1.000
Vector equation: OP = [2.0, 2.0] + t[-3.0, -4.5]
Parametric equations:
x =   2.0  +  -3.0 t
y =   2.0  +  -4.5 t

Cuts axis X at point [0.667, 0]
Cuts axis Y at point [0, -1.000]
```

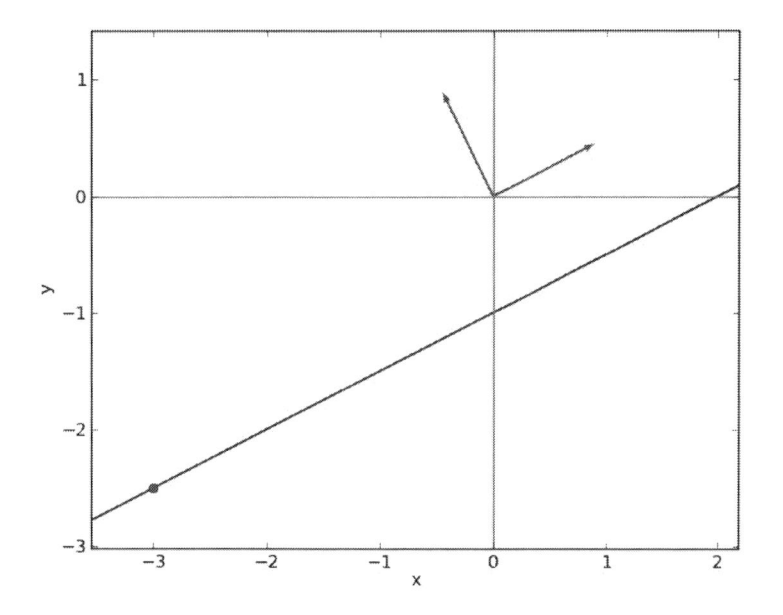

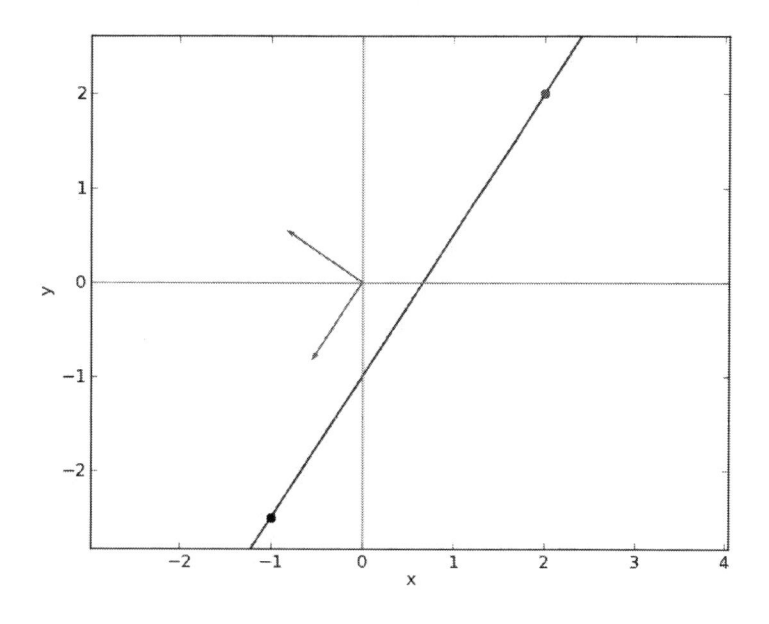

The code of the Python program is as follows:

```python
# -*- coding: utf-8 -*-
"""
Mathematics and Python Programming    www.pysamples.com
p10f.py
"""

import numpy as np
import matplotlib.pyplot as plt

print 'Problem type according to known data:'
print '1 - P0 and the normal vector'
print '2 - P0 and the direction vector'
print '3 - P0 and P1'
print
problem_type = int(raw_input('Input the type of problem: '))
print ('Write point P0 coordinates, separated by one blank space:')
sp0 = raw_input()
p0 = map(float, sp0.split())
print 'P0: ', p0
print
maxpoint = p0[0] + 1
minpoint = p0[0] - 1
oblique = False


def valid(director):   # continues if the line is oblique
    if r[0] == 0:
        oblique = False
        message = 'vertical line x = ' + str(p0[0])
    elif r[1] == 0:
        oblique = False
        message = 'horizontal line y = ' + str(p0[1])
    else:
        oblique = True
        message = 'oblique line'
    isoblique = {'oblique': oblique, 'message': message}
    return isoblique

if problem_type == 1:
    print ('Write the coordinates of the normal vector, separated by one blank space:')
    svn = raw_input()
    vn = map(float, svn.split())
    print
    if vn[0] == 0:
        oblique = False
        print 'horizontal line y = ' + str(p0[1])
    elif vn[1] == 0:
        oblique = False
        print 'vertical line x = ' + str(p0[0])
    else:
        oblique = True
        # a normal vector to r, choosing ry=1
        r = [-1.0 * vn[1] / vn[0], 1.0]
if problem_type == 2:
    print ('Write the coordinates of the direction vector, separated by one blank space:')
```

10.2. EQUATIONS OF A STRAIGHT LINE IN THE PLANE

```
        sr = raw_input()
        r = map(float, sr.split())
        print
        sigue = valid(r)
        oblique = sigue.get('oblique')
        if oblique:
            # un vector normal a r, eligiendo B=1
            vn = [-1.0 * r[1] / r[0], 1.0]
        else:
            print sigue.get('message')
if problem_type == 3:
    print ('Write the coordinates of point P1, separated by one blank space:')
    sp1 = raw_input()
    p1 = map(float, sp1.split())
    print 'P1: ', p1
    print
    if maxpoint < p1[0]:
        maxpoint = p1[0]
    if minpoint > p1[0]:
        minpoint = p1[0]
    r = [0, 0]
    r[0] = p1[0] - p0[0]
    r[1] = p1[1] - p0[1]
    sigue = valid(r)
    oblique = sigue.get('oblique')
    if oblique:
        vn = [-1.0 * r[1] / r[0], 1.0]   # a normal vector to r, choosing B=1
        maxpoint = max([p0[0], p1[0]]) + 1
        minpoint = min([p0[0], p1[0]]) - 1
    else:
        print sigue.get('message')


def geneq(vn, point):   # data: normal vector and one point in the line
    # vn and point are lineal arrays whose dimension is 2
    A = vn[0]
    B = vn[1]
    C = -A * point[0] - B * point[1]
    isoblique = {'A': A, 'B': B, 'C': C}
    return isoblique


def slopeq(ABC):   # data: la ecuacion general de la recta
    if ABC[1] != 0:
        m = -1.0 * ABC[0] / ABC[1]
        n = -1.0 * ABC[2] / ABC[1]
    else:
        m = 0
        n = 0
    isoblique = {'m': m, 'n': n}
    return isoblique


if oblique:
    parameter = unichr(0x3bb).encode('utf-8')
    print 'Direction vector: ', r
```

```
#vn = [-1.0 * r[1] / r[0], 1.0]   # un vector normal a r, eligiendo B=1
geneq = geneq(vn, p0)
ABC = [geneq.get('A'), geneq.get('B'), geneq.get('C')]
print 'Normal vector: ', vn
print ('General form equation: ' +
       "%.3f" % ABC[0] + 'x + ' + str(ABC[1]) +
       ' y + ' + "%.3f" % ABC[2] + ' = 0')
slopeq = slopeq(ABC)
print ('Slope-intercept equation: y = ' +
       "%.3f" % slopeq.get('m') +
       'x + ' + "%.3f" % slopeq.get('n'))
print ('Vector equation: OP = ' +
       str(p0) + ' + ' + parameter + str(r))
corteX = -1.0 * slopeq.get('n') / slopeq.get('m')
corteY = slopeq.get('n')
print 'Parametric equations:'
print 'x = ', p0[0], ' + ', r[0], parameter
print 'y = ', p0[1], ' + ', r[1], parameter
print
print ('Cuts axis X at point [' + "%.3f" % corteX + ', 0]')
print ('Cuts axis Y at point [0, ' + "%.3f" % corteY + ']')
#graph

def f(x):
    return slopeq.get('m') * x + slopeq.get('n')
ur = [0, 0]
modr = np.sqrt(r[0] ** 2 + r[1] ** 2)
ur[0] = r[0] / modr
ur[1] = r[1] / modr
plt.arrow(0, 0, ur[0], ur[1], width=0.01, fc='r',
          ec='none', length_includes_head=True, ls='solid')
uvn = [0, 0]
modvn = np.sqrt(vn[0] ** 2 + vn[1] ** 2)
uvn[0] = vn[0] / modvn
uvn[1] = vn[1] / modvn
plt.arrow(0, 0, uvn[0], uvn[1], width=0.01, fc='g',
          ec='none', length_includes_head=True, ls='solid')
minimos = [-1.1 * corteX, 1.1 * corteX, ur[0] - 1,
           uvn[0] - 1, minpoint - 1]
xmin = min(minimos)
maximos = [-1.1 * corteX, 1.1 * corteX, ur[0] + 1,
           uvn[0] + 1, maxpoint + 1]
xmax = max(maximos)

plt.plot([xmin, xmax], [f(xmin), f(xmax)], 'b-', lw=1.5)
plt.ylabel('y')
plt.xlabel('x')
plt.axhline(color='grey', lw=1)
plt.axvline(color='grey', lw=1)
plt.plot(p0[0], p0[1], 'ro')
if problem_type == 3:
    plt.plot(p1[0], p1[1], 'ko')
plt.xlim(xmin, xmax)
plt.axis('equal')
plt.show()
```

10.3 Vector Functions

If each value of a scalar argument λ is mapped to a vector $\mathbf{r}(\lambda)$, the vector $\mathbf{r}$ is called vector function of scalar argument.

If a vector function $\mathbf{r}(\xi)$ is defined for any point ξ of a region of the space, then it is said that in this region is defined a vector field. If the point is given by its coordinates

$$\xi = \xi_1 i + \xi_2 j + \xi_3 k$$

where ξ_1, ξ_2, ξ_3 can be constants or functions of the position $\xi_i = f(x, y, z)$, then to define a vector function $\mathbf{r}(\xi)$ is equivalent to defining the three scalar functions that provide their coordinates.

In Physics it is very frequent that the vector function is a time dependent function, which we write t:

$$\mathbf{r}(t) = x(t)i + y(t)j + z(t)k$$

. When the scalar value t changes, while maintaining the origin of the vector at a fixed point of the space that is taken as the origin of coordinates, the end of the vector $\mathbf{r}(t)$ describes a series of points, whose locus is called the curve of that vector function.

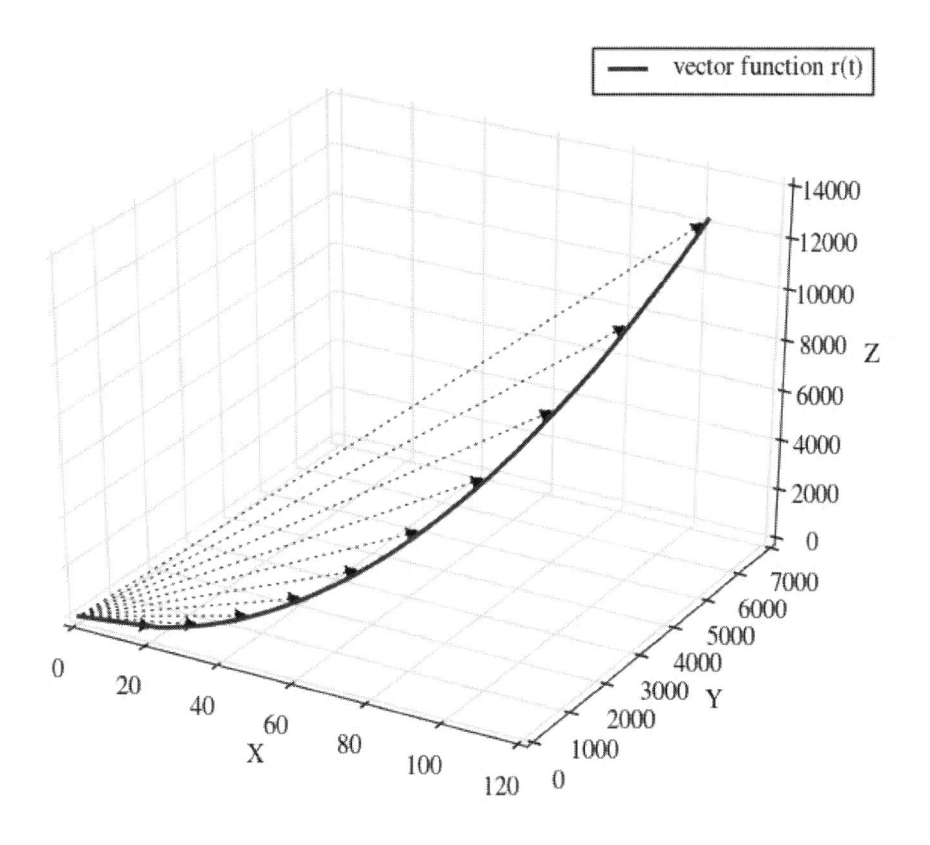

The graph represents the vector function

$$\mathbf{r}(t) = ti + \frac{t^2}{2}j + \frac{t^3}{100}k$$

in blue, and some vectors for certain values of t, with oriented segments whose origin is the origin of coordinates, and whose end is the value of $r(t)$:

```python
# -*- coding: utf-8 -*-
"""
Mathematics and Python Programming    www.pysamples.com
p10g.py
"""

import matplotlib as mpl
from mpl_toolkits.mplot3d import Axes3D
import numpy as np
import matplotlib.pyplot as plt
from matplotlib import rc
from matplotlib.patches import FancyArrowPatch
from mpl_toolkits.mplot3d import proj3d

rc('font', **{'family': 'serif', 'serif': ['Times']})
rc('text', usetex=True)

mpl.rcParams['legend.fontsize'] = 12


fig = plt.figure()
ax = fig.gca(projection='3d')
ax.w_xaxis.set_pane_color((1.0, 1.0, 1.0, 1.0))
ax.w_yaxis.set_pane_color((1.0, 1.0, 1.0, 1.0))
ax.w_zaxis.set_pane_color((1.0, 1.0, 1.0, 1.0))


class Arrow3D(FancyArrowPatch):
    def __init__(self, xs, ys, zs, *args, **kwargs):
        FancyArrowPatch.__init__(self, (0, 0), (0, 0), *args, **kwargs)
        self._verts3d = xs, ys, zs

    def draw(self, renderer):
        xs3d, ys3d, zs3d = self._verts3d
        xs, ys, zs = proj3d.proj_transform(xs3d, ys3d, zs3d, renderer.M)
        self.set_positions((xs[0], ys[0]), (xs[1], ys[1]))
        FancyArrowPatch.draw(self, renderer)


pointsnum = 111
x = np.zeros(pointsnum, float)
y = np.zeros(pointsnum, float)
z = np.zeros(pointsnum, float)
for t in range(20, pointsnum):
    x[t] = t
    y[t] = (t ** 2) / 2
    z[t] = (t ** 3) / 100
    if t % 10 == 0:
        a = Arrow3D([0, x[t]], [0, y[t]], [0, z[t]], mutation_scale=20, lw=0.75,
                    arrowstyle="-|>", color="k", linestyle='dotted')
        ax.add_artist(a)

ax.plot(x, y, z, label='vector function r(t)', lw=2)
ax.legend()
ax.set_xlabel('X')
ax.set_ylabel('Y')
```

```
ax.set_zlabel('Z')
plt.show()
```

10.4 Limit, continuity and derivative of a vector function

Let's take the vector function $\mathbf{r}(t)$. The constant vector $\mathbf{l}$ is the limit of the vector function when $t \to t_0$, and we write it as follows:
$$\lim_{t \to t_0} \mathbf{r}(t) = \mathbf{l}$$
if $\forall \varepsilon > 0$ there is any $\delta > 0$ such that when $|t - t_0| < \delta$ it fulfils $|\mathbf{r}(t) - \mathbf{l}| < \varepsilon$.

If $\mathbf{l} = \begin{pmatrix} l_1 & l_2 & l_3 \end{pmatrix}$ and $\mathbf{r} = \begin{pmatrix} r_1(t) & r_2(t) & r_3(t) \end{pmatrix}$, to say that the limit
$$\lim_{t \to t_0} \mathbf{r}(t) = \mathbf{l}$$
is tantamount to saying that the limit for each one of the components $r_i(t)$ is equal to the respective component l_i of the vector $\mathbf{l}$:
$$\lim_{t \to t_0} r_i(t) = l_i$$

Let's take a vector function $\mathbf{r}(t)$, which is defined in a neighbourhood of t_0, we say that the vector function is continuous at $t = t_0$ if
$$\lim_{t \to t_0} \mathbf{r}(t) = \mathbf{r}(t_0)$$

We say that a vector function $\mathbf{r}(t)$ is differentiable if the following limit exists:
$$\lim_{\Delta t \to 0} \frac{\Delta \mathbf{r}}{\Delta t} = \lim_{\Delta t \to 0} \frac{\mathbf{r}(t + \Delta t) - \mathbf{r}(t)}{\Delta t} = \frac{d\mathbf{r}(t)}{dt}$$

If $\mathbf{r} = \begin{pmatrix} r_1(t) & r_2(t) & r_3(t) \end{pmatrix}$, then
$$\frac{d\mathbf{r}(t)}{dt} = \begin{pmatrix} \frac{dr_1(t)}{dt} & \frac{dr_2(t)}{dt} & \frac{dr_3(t)}{dt} \end{pmatrix}$$

Let's look at an example in two dimensions.
$$\mathbf{r}(t) = \begin{pmatrix} r_1(t) & r_2(t) \end{pmatrix} = \begin{pmatrix} t - sent & 1 - cost \end{pmatrix}$$

The derivative is
$$\frac{d\mathbf{r}(t)}{dt} = \begin{pmatrix} 1 - cost & sent \end{pmatrix}$$

If we represent $\mathbf{r}(t)$ for $t \in [0, 2\pi]$ we obtain a cycloid. We already represented this curve in the previous chapter. The following Python program represents with dashed the cycloid, and in addition $\mathbf{r}(t)$ (in black) and its derivative (in green) for some values of t.

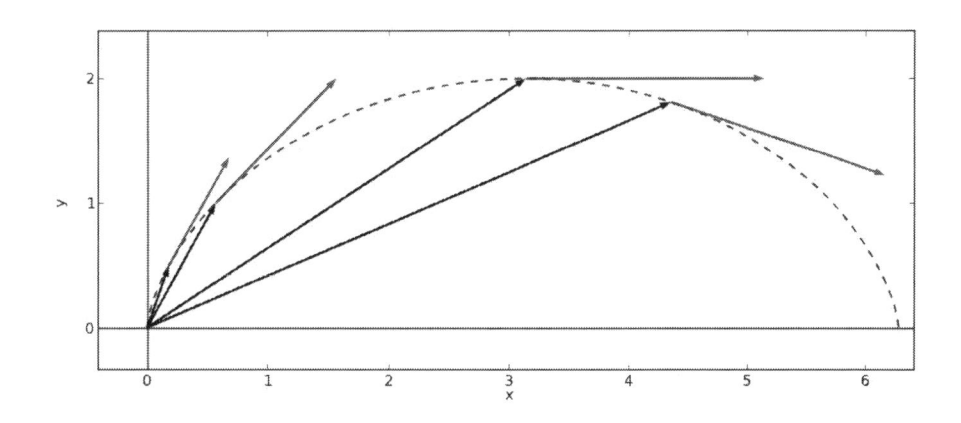

```python
# -*- coding: utf-8 -*-
"""
Mathematics and Python Programming    www.pysamples.com
p10h.py
"""

import numpy as np
import matplotlib.pyplot as plt

t1 = 0.0
t2 = 2 * np.pi
print 't1 = ' + "%5.3f" % t1
print 't2 = ' + "%5.3f" % t2 + ' = 2 pi'


def r1(t):
    fv = t - np.sin(t)
    return fv


def r2(t):
    gv = 1 - np.cos(t)
    return gv

print 'a = ' + "%5.3f" % r1(t1)
print 'b = ' + "%5.3f" % r2(t2)
pointsnum = 360
starting_t = 0
final_t = pointsnum
x = np.zeros(pointsnum, float)
y = np.zeros(pointsnum, float)
t = starting_t
while t < final_t:
    radians = np.deg2rad(t)
    x[t] = r1(radians)
    y[t] = r2(radians)
    t += 1
plt.plot(x, y, 'b-', lw=1.5)


def dr1(t):
    d1 = 1 - np.cos(t)
    return d1


def dr2(t):
    d2 = np.sin(t)
    return d2

#values of t to represent the vector
points = [np.pi / 3, np.pi / 2, np.pi, 6 * np.pi / 5]
for i in range(0, 4):
    t = points[i]
    plt.arrow(0, 0, r1(t), r2(t), width=0.01, fc='k', ec='none',
              length_includes_head=True, lw=1.0)
    plt.arrow(r1(t), r2(t), dr1(t), dr2(t), width=0.01, fc='g', ec='none',
```

10.4. LIMIT, CONTINUITY AND DERIVATIVE OF A VECTOR FUNCTION

```
                length_includes_head=True, lw=0.5)
plt.axhline(color='black', lw=1)
plt.axvline(color='black', lw=1)
plt.axis('equal')
plt.xlim(-0.5, 6.5)
plt.ylabel('y')
plt.xlabel('x')
plt.show()
```

If the parameter t is the time, the geometric and physical meaning of the derivative of the vector function $\mathbf{r}(t)$ is $\frac{d\mathbf{r}}{dt}$, that is to say, the speed of the end of the vector $\mathbf{r}$ describing the curve, and it is a vector tangent to the curve at each point.

$$\mathbf{v} = \frac{d\mathbf{r}}{dt}$$

The second derivative corresponds to the acceleration of the end of $\mathbf{r}$ describing the curve:

$$\mathbf{a} = \frac{d\mathbf{v}}{dt} = \frac{d^2\mathbf{r}}{dt^2}$$

The following Python program represents the vector function $\mathbf{r}(t)$ that describes the position of an object in a parabolic shot, and represents the vectors $\mathbf{r}(t)$ (in black), $\mathbf{v}(t)$ (in green) and $a(t)$ (in red) for some values of t:

──────────────────── output of the program ────────────────────

```
angle    ymax     throwrange    time
50.0     36.64    122.98        5.47)
```

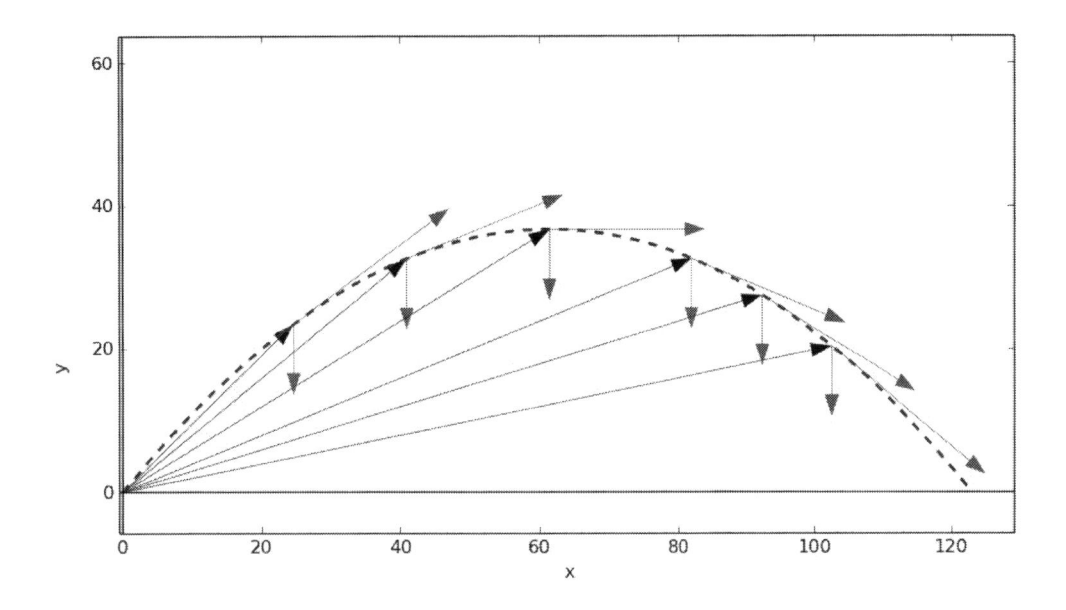

```
# -*- coding: utf-8 -*-
"""
Mathematics and Python Programming    www.pysamples.com
p10i.py
"""

import numpy as np
import matplotlib.pyplot as plt
```

```python
#physics:
y0 = 0.0
v0 = 35.0
g = -9.81
angle = 50.0
#vy = v0y - g * t
#vx = v0x
#x = v0x * t
#y = y0 + v0y*t + 0.5*g*t**2
#y = y0 + v0y*x/v0x + (0.5*g/v0x**2)*x**2
#ymax = y0 + v0y*tfloor/2 + 0.5*g*(tfloor/2)**2
maxrange = 0
maxheight = 0
print 'angle    ymax    throwrange    time'
alfa = np.deg2rad(angle)
v0y = v0 * np.sin(alfa)
tfloor = - 2 * v0y / g
v0x = v0 * np.cos(alfa)
throwrange = - (v0 ** 2) * np.sin(2 * alfa) / g
ymax = y0 + (v0y * tfloor / 2) + (0.5 * g * (tfloor / 2) ** 2)
if throwrange > maxrange:
    maxrange = throwrange
if ymax > maxheight:
    maxheight = ymax
print (str(angle) + "   %7.2f" % ymax + "      %7.2f" % throwrange + "    %7.2f" % tfloor)


def f(x, beta):
    v0x = v0 * np.cos(np.deg2rad(beta))
    #coefficients a0, a1,... an
    a = [y0, (np.tan(np.deg2rad(beta))), ((0.5 * g) / (v0x ** 2))]
    y = a[0]
    i = 1
    while i < len(a):
        y = y + a[i] * x ** i
        i += 1
    return y


numpoints = 300
x = np.linspace(0, np.ceil(maxrange), numpoints)
y = np.zeros(numpoints, float)
for i in range(0, numpoints):
    y[i] = f(x[i], angle)

plt.plot(x, y, 'b--', lw=2, label='$38$')
#plot r(t)
# values of t to represent the vector
points = [tfloor / 5, tfloor / 3, tfloor / 2,
          2 * tfloor / 3, 3 * tfloor / 4, 5 * tfloor / 6]


def xvector(time):
    return v0x * time
```

10.4. LIMIT, CONTINUITY AND DERIVATIVE OF A VECTOR FUNCTION

```
vx = v0x    # derivative of xvector is v0x
ax = 0.0    # derivative of vx es 0


def yvector(time):
    return y0 + v0y * time + 0.5 * g * time ** 2


def vy(time):
    return v0y + g * time

ay = g    # derivative of de vy is g

for i in range(0, 6):
    t = points[i]
    plt.arrow(0, 0, xvector(t), yvector(t), width=0.1, fc='k',
              ec='none', length_includes_head=True, lw=0.5)
    plt.arrow(xvector(t), yvector(t), vx, vy(t), width=0.1, fc='g',
              ec='none', length_includes_head=True, lw=0.5)
    plt.arrow(xvector(t), yvector(t), ax, ay, width=0.1, fc='r',
              ec='none', length_includes_head=True, lw=0.5)


plt.axhline(color='black', lw=1)
plt.axvline(color='black', lw=1)
plt.axis('equal')
plt.xlim(-0.5, 1.05 * throwrange)
plt.ylabel('y')
plt.xlabel('x')
plt.show()
```

The derivative of vector functions has the following properties (t is a scalar parameter):

- If **c** is a constant vector, then $\frac{d\mathbf{c}}{dt} = 0$

- $\mathbf{a} = \mathbf{a}(t)$ and $\mathbf{b} = \mathbf{b}(t)$ are vector functions, then

$$\frac{d(\mathbf{a}+\mathbf{b})}{dt} = \frac{d\mathbf{a}}{dt} + \frac{d\mathbf{b}}{dt}$$

$$\frac{d(\mathbf{a}\cdot\mathbf{b})}{dt} = \frac{d\mathbf{a}}{dt}\cdot\mathbf{b} + \mathbf{a}\cdot\frac{d\mathbf{b}}{dt}$$

- Let $\mathbf{a} = \mathbf{a}(t)$ be a vector function, and $\lambda(t)$ a scalar function of t, then:

$$\frac{d(\lambda\mathbf{a})}{dt} = \lambda\frac{d\mathbf{a}}{dt} + \frac{d\lambda}{dt}\mathbf{a}$$

In physics we have a concrete example: if according to the second law of Newton, the mass m depends on time according to a scalar function $m = m(t)$ and we express the acceleration by the vector function $\mathbf{a}(t)$:

$$\frac{d(m\mathbf{a})}{dt} = m\frac{d\mathbf{a}}{dt} + \frac{dm}{dt}\mathbf{a}$$

10.5 Scalar Fields

If at each point in a region of the space we can assign a scalar value through a scalar function $\phi(x, y, z)$, it is said that in this region is defined a scalar field. For example, if in each point of a flat metal plate we can know the temperature of that point, there is defined a scalar field of temperatures on that metal plate. In this case it is a plane scalar field scalar field and we call the set of points where the scalar field takes the same value forms a contour line.

Let's look at how to represent a scalar field using Python: suppose that in a stretch of coast the altitude of any point, as well as the depth of the sea, are given by $z = 0.03(x^2 - y^2)$, for a region around a point that is taken as the origin of coordinates. The contour lines for negative values of Z are dashed in the figure:

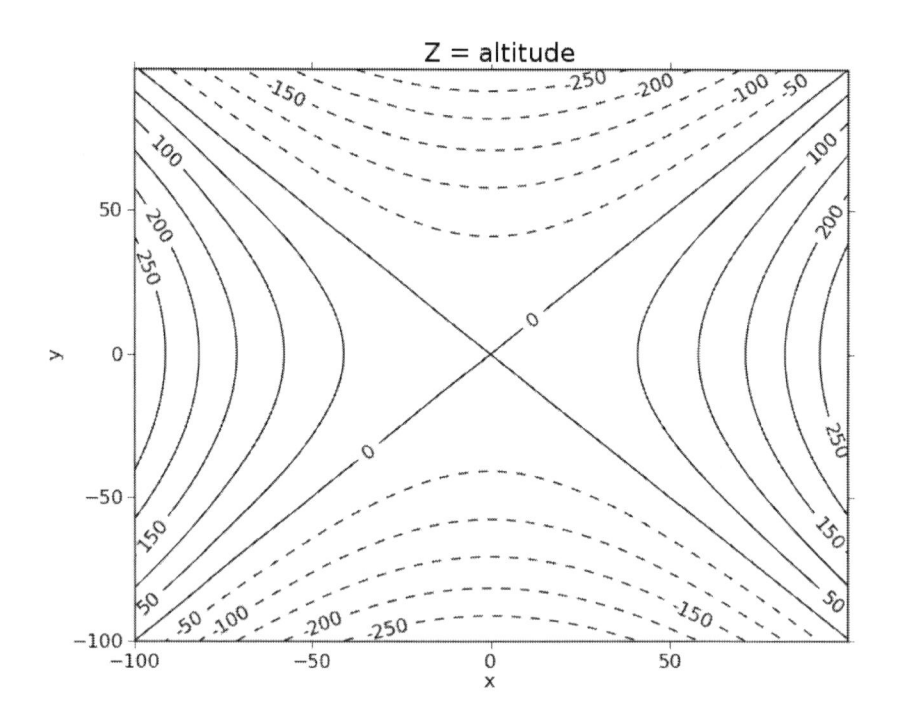

```
# -*- coding: utf-8 -*-
"""
Mathematics and Python Programming     www.pysamples.com
p10j.py
"""

import matplotlib
import numpy as np
import matplotlib.pyplot as plt

matplotlib.rcParams['xtick.direction'] = 'out'
matplotlib.rcParams['ytick.direction'] = 'out'

x = np.arange(-100, 100, 1.0)
y = np.arange(-100, 100, 1.0)
X, Y = np.meshgrid(x, y)
Z = 0.03 * (X ** 2 - Y ** 2)   # z = altitude
plt.figure()
CS = plt.contour(X, Y, Z, 12, colors='k', linewidth=1.0)
```

10.5. SCALAR FIELDS

```
plt.clabel(CS, inline=1, fontsize=12, fmt='%1.0f')
plt.title('Z = altitud')
plt.xlabel('x')
plt.ylabel('y')
plt.show()
```

We can get a 3D image of this scalar field also using Python:

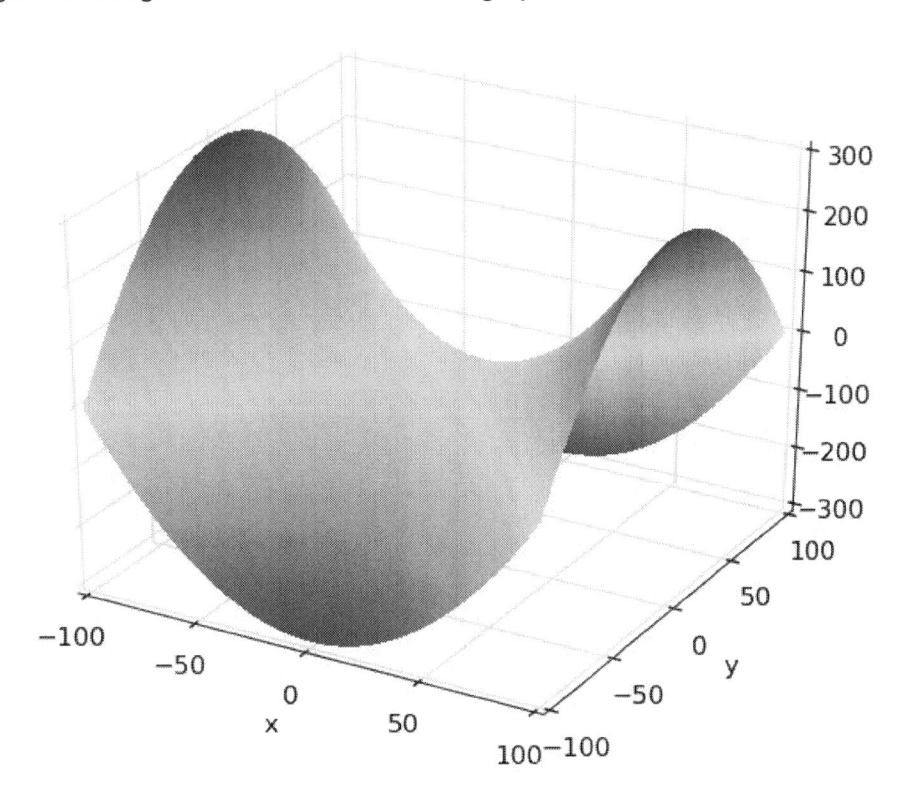

```
# -*- coding: utf-8 -*-
"""
Mathematics and Python Programming    www.pysamples.com
p10k.py
"""

from mpl_toolkits.mplot3d import Axes3D
import matplotlib.pyplot as plt
from matplotlib import cm
import numpy as np

fig = plt.figure()
ax = fig.add_subplot(111, projection='3d')
x = np.arange(-100, 100, 1.0)
y = np.arange(-100, 100, 1.0)
X, Y = np.meshgrid(x, y)
Z = 0.03 * (X ** 2 - Y ** 2)
ax.plot_surface(X, Y, Z, rstride=1, cstride=1, cmap=cm.coolwarm,
                linewidth=0, antialiased=False)
plt.xlabel('x')
plt.ylabel('y')
plt.show()
```

Another example is the field of electric potential $V = k\frac{q}{r}$ created by a point electric charge. The figure shows the field created by an electric charge $q = 2 \cdot 10^{-9}C$:

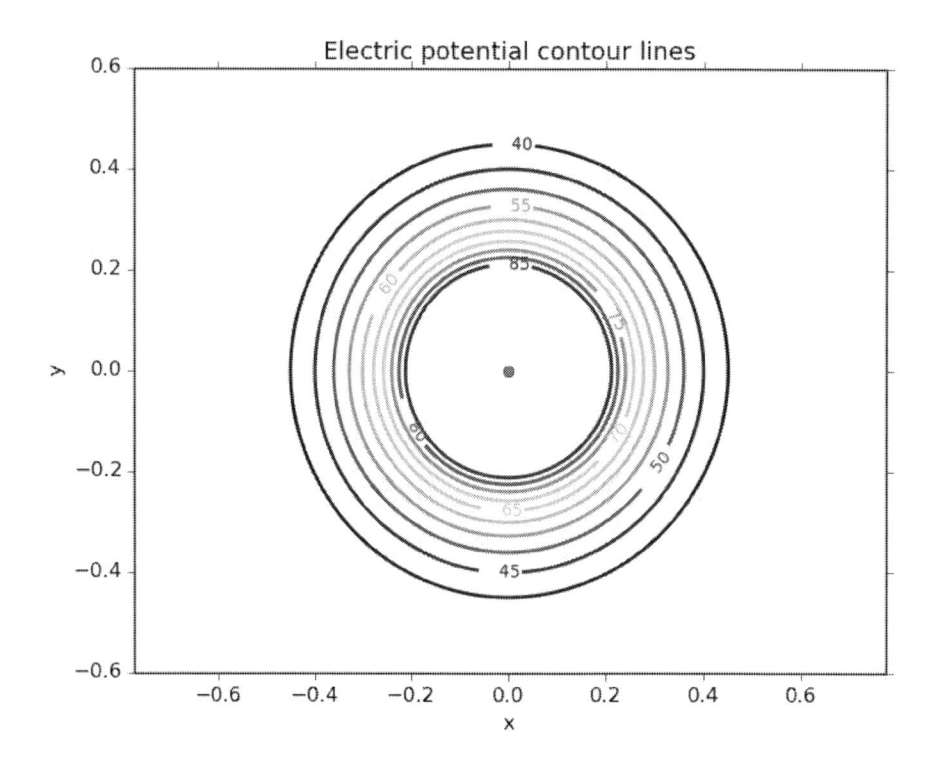

```
# -*- coding: utf-8 -*-
"""
Mathematics and Python Programming    www.pysamples.com
p101.py
"""

import matplotlib.pyplot as plt
import numpy as np

r = np.linspace(0.2, 0.5, 200)
t = np.linspace(0, 2 * np.pi, 200)
r, t = np.meshgrid(r, t)

k = 9.0 * 1e9   # constant
q = 2.0 * 1e-9  # electric charge (Coulombs)
z = k * q / r   # electric potential

CS = plt.contour(r * np.cos(t), r * np.sin(t), z, 10, linewidths=2)
plt.plot(0, 0, 'o', color='grey')
plt.clabel(CS, inline=1, fmt='%5.0f', fontsize=10)
plt.title('Electric potential contour lines')
plt.axis('equal')
plt.xlabel('x')
plt.ylabel('y')
plt.show()
```

If the scalar field is a space of dimension 3, the points that have the same scalar value, form a level surface: $f(x, y, z) = cte$.

10.6 Gradient

Let $\phi = f(x, y, z)$ be a scalar differentiable function. The gradient of the scalar field ϕ is a vector that is defined as follows:

$$\mathrm{grad}\phi = \frac{d\phi}{dx}\mathbf{i} + \frac{d\phi}{dy}\mathbf{j} + \frac{d\phi}{dz}\mathbf{k} = \begin{pmatrix} \frac{d\phi}{dx} & \frac{d\phi}{dy} & \frac{d\phi}{dz} \end{pmatrix}$$

The direction of the gradient is the growth of the function ϕ, and it is normal to the contour lines of the scalar field created by ϕ. The gradient indicates the direction and magnitude in which the scalar field presents its maximum variation at a certain point.

The following Python program calculates the gradient of a scalar function $z = 0.03 \cdot (x^2 - y^2)$, the scalar function that we represented in the previous section.

```
# -*- coding: utf-8 -*-
"""
Mathematics and Python Programming     www.pysamples.comm
p10m.py
"""

import sympy as sy

x, y, z, k = sy.symbols('x, y, z, k')
sy.init_printing(use_unicode=True)

k = 0.03
z = k * (x ** 2 - y ** 2)
print 'z = ', sy.simplify(z)


def gradient(a, dimension):   # a es una funcion escalar
    vectorgrad = []
    r = 0.0
    vectorgrad.append(sy.diff(a, x))
    if dimension == 2:
        vectorgrad.append(sy.diff(a, y))
    if dimension == 3:
        vectorgrad.append(sy.diff(a, z))
    for i in range(0, dimension):
        r += vectorgrad[i] ** 2
    length = sy.sqrt(r)
    result = {'vectorgrad': vectorgrad, 'length': length}
    return result

gradz = gradient(z, 2)
print 'grad z = ', gradz.get('vectorgrad')
modz = gradz.get('length')
print '|z| = ', modz
print '|z| = ', sy.simplify(modz)
```

output of the program
```
z =  0.03*x**2 - 0.03*y**2
grad z =  [0.06*x, -0.06*y]
|z| =  sqrt(0.0036*x**2 + 0.0036*y**2)
|z| =  0.06*sqrt(x**2 + y**2)
```

We can plot the contour lines and the gradient vector at several points of the lines $Z = 100$ and $Z = -100$, by slightly modifying the program we used to represent the contour lines:

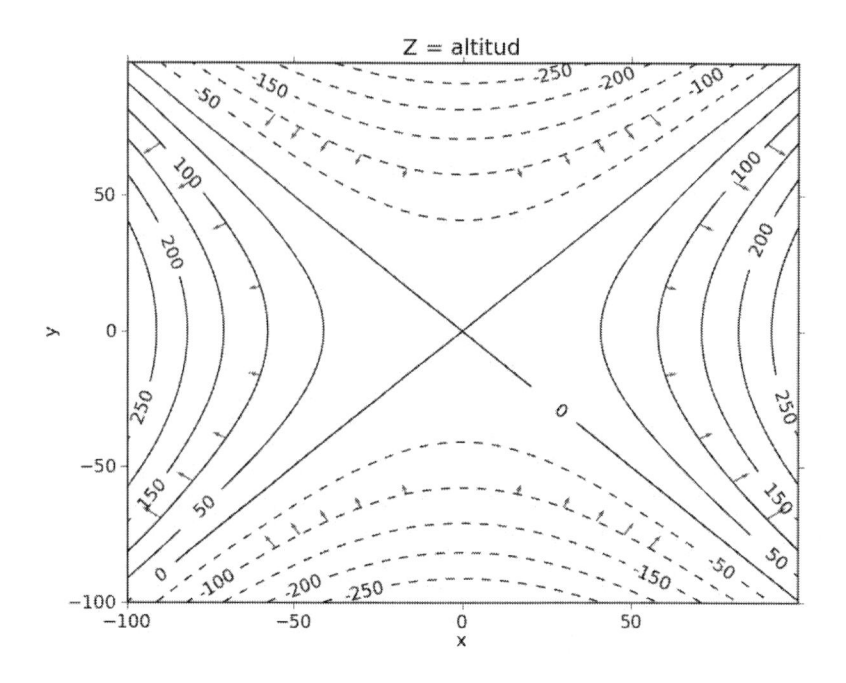

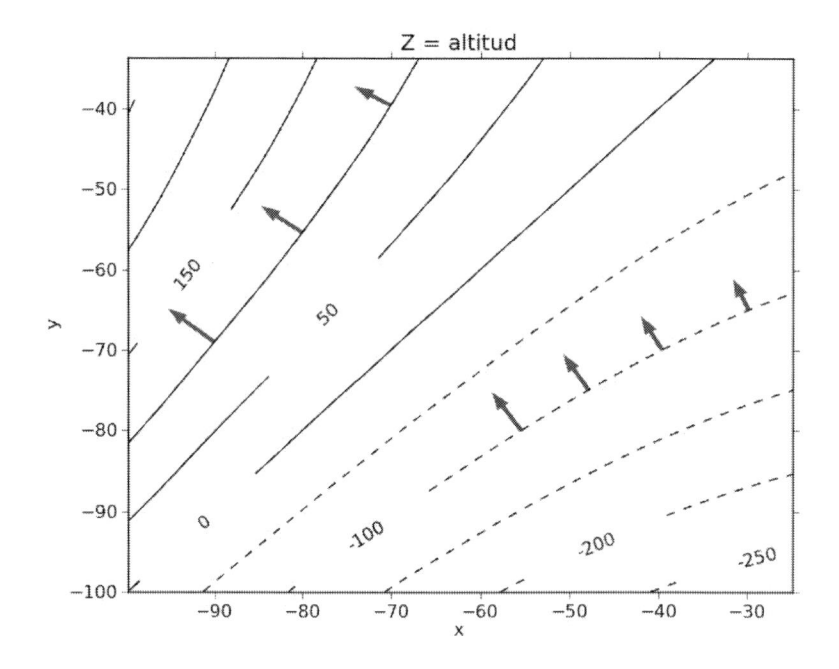

```
# -*- coding: utf-8 -*-
"""
Mathematics and Python Programming    www.pysamples.com
p10n.py
"""

import matplotlib
import numpy as np
import matplotlib.pyplot as plt

matplotlib.rcParams['xtick.direction'] = 'out'
matplotlib.rcParams['ytick.direction'] = 'out'
```

```python
x = np.arange(-100, 100, 1.0)
y = np.arange(-100, 100, 1.0)
X, Y = np.meshgrid(x, y)
Z = 0.03 * (X ** 2 - Y ** 2)   # z = altitude
plt.figure()
CS = plt.contour(X, Y, Z, 12, colors='k', linewidth=1.0)

plt.clabel(CS, inline=1, fontsize=12, fmt='%4.0f')
px1 = [-90, -80, -70, -60]
py1 = []
for i in range(0, 4):
    py1.append(np.sqrt(px1[i] ** 2 - (100 / 0.03)))
    plt.arrow(px1[i], py1[i], 0.06 * px1[i], -0.06 * py1[i],
              width=0.1, fc='r', ec='none',
              length_includes_head=True, lw=0.5)
    plt.arrow(px1[i], -py1[i], 0.06 * px1[i], 0.06 * py1[i],
              width=0.1, fc='r', ec='none',
              length_includes_head=True, lw=0.5)
    plt.arrow(-px1[i], py1[i], -0.06 * px1[i], -0.06 * py1[i],
              width=0.1, fc='r', ec='none',
              length_includes_head=True, lw=0.5)
    plt.arrow(-px1[i], -py1[i], -0.06 * px1[i], 0.06 * py1[i],
              width=0.1, fc='r', ec='none',
              length_includes_head=True, lw=0.5)

py2 = [-60, -65, -70, -75, -80]
px2 = []
for i in range(0, 5):
    px2.append(np.sqrt(py2[i] ** 2 + (-100 / 0.03)))
    plt.arrow(px2[i], py2[i], 0.06 * px2[i], -0.06 * py2[i],
              width=0.1, fc='r', ec='none',
              length_includes_head=True, lw=0.5)
    plt.arrow(px2[i], -py2[i], 0.06 * px2[i], 0.06 * py2[i],
              width=0.1, fc='r', ec='none',
              length_includes_head=True, lw=0.5)
    plt.arrow(-px2[i], py2[i], -0.06 * px2[i], -0.06 * py2[i],
              width=0.1, fc='r', ec='none',
              length_includes_head=True, lw=0.5)
    plt.arrow(-px2[i], -py2[i], -0.06 * px2[i], 0.06 * py2[i],
              width=0.1, fc='r', ec='none',
              length_includes_head=True, lw=0.5)

plt.title('Z = altitude')
plt.xlabel('x')
plt.ylabel('y')
plt.show()
```

10.7 Integration of a vector function

Let $\mathbf{r}(t)$ be a vector function of scalar argument t. The vector function $\mathbf{R}(t)$ whose derivative is equal to $\mathbf{r}(t)$ is called antiderivative function of $\mathbf{r}(t)$, that is to say:

$$\frac{d\mathbf{R}}{dt} = \mathbf{r}(t)$$

or what is the same:

$$\int \mathbf{r}(t)dt = \mathbf{R}(t) + cte$$

If $\mathbf{r}(t) = \begin{pmatrix} r_1(t) & r_2(t) & r_3(t) \end{pmatrix}$ then

$$\mathbf{R}(t) = \int \mathbf{r}(t)dt = \begin{pmatrix} \int r_1(t)dt & \int r_2(t)dt & \int r_3(t)dt \end{pmatrix}$$

Each component of $\mathbf{R}(t)$ is equal to the integral of the respective component $\mathbf{r}(t)$. If this is a definite integral between the values of $t = t_i$ and $t = t_f$ we have:

$$\int_{t_i}^{t_f} \mathbf{r}(t)dt = \mathbf{R}(t_f) - \mathbf{R}(t_i)$$

10.7.1 Line integral of a vector field

Let's take the vector function:

$$\mathbf{r} = r_1\mathbf{i} + r_2\mathbf{j} + r_3\mathbf{k} = \begin{pmatrix} r_1 & r_2 & r_3 \end{pmatrix}$$
$$d\mathbf{r} = dr_1\mathbf{i} + dr_2\mathbf{j} + dr_3\mathbf{k} = \begin{pmatrix} dr_1 & dr_2 & dr_3 \end{pmatrix}$$

which defines a curve C.

Let $\mathbf{F}$ be a continuous vector field:

$$\mathbf{F} = F_1\mathbf{i} + F_2\mathbf{j} + F_3\mathbf{k} = \begin{pmatrix} F_1 & F_2 & F_3 \end{pmatrix}$$

The integral

$$\int_C \mathbf{F} \cdot d\mathbf{r}$$

is called line integral (or also curve integral) of the field $\mathbf{F}$ along the curve C. The physical meaning of the curvilinear integral is the following: if the vector field is a force field

$$\mathbf{F} = F_x\mathbf{i} + F_y\mathbf{j} + F_z\mathbf{k} = \begin{pmatrix} F_x & F_y & F_z \end{pmatrix}$$

Which associates to each point in space a force vector that acts on an object, the curve integral

$$\int_C \mathbf{F} \cdot d\mathbf{r}$$

equals the work done to move the object along the curve C defined by the vector $\mathbf{r}$. Since

$$\mathbf{F} \cdot d\mathbf{r} = \begin{pmatrix} F_x & F_y & F_z \end{pmatrix} \cdot \begin{pmatrix} dr_1 \\ dr_2 \\ dr_3 \end{pmatrix}$$

$$\mathbf{F} \cdot d\mathbf{r} = F_x dr_1 + F_y dr_2 + F_z dr_3$$

and we get the following curve integral:

$$\int_C \mathbf{F} \cdot d\mathbf{r} = \int_C F_x dr_1 + F_y dr_2 + F_z dr_3$$

$$\int_C \mathbf{F} \cdot d\mathbf{r} = \int_C F_x dr_1 + \int_C F_y dr_2 + \int_C F_z dr_3$$

If we chose a positive sense of displacement along the curve C to the travelling between two points A and B, the integral changes sign if we reverse the direction of movement:

$$\int_{AB} \mathbf{F} \cdot d\mathbf{r} = -\int_{BA} \mathbf{F} \cdot d\mathbf{r}$$

If the curve C is closed, the curve integral of the vector field $\mathbf{F}$ is written:

$$\oint_C \mathbf{F} \cdot d\mathbf{r}$$

and the integral is called circulation of the vector $\mathbf{F}$ around C.

10.7.2 Circulation of a conservative vector field

Let's see what happens if the vector function to integrate vector is the gradient of a scalar field $\phi(x, y, z)$ along a curve that is defined by the vector r:

$$\mathrm{grad}\phi = \frac{d\phi}{dx}\mathbf{i} + \frac{d\phi}{dy}\mathbf{j} + \frac{d\phi}{dz}\mathbf{k} = \begin{pmatrix} \frac{d\phi}{dx} & \frac{d\phi}{dy} & \frac{d\phi}{dz} \end{pmatrix}$$

$$d\mathbf{r} = dr_1\mathbf{i} + dr_2\mathbf{j} + dr_3\mathbf{k} = \begin{pmatrix} dr_1 & dr_2 & dr_3 \end{pmatrix}$$

$$\mathrm{grad}\phi \cdot d\mathbf{r} = \begin{pmatrix} \frac{d\phi}{dx}dr_1 & \frac{d\phi}{dy}dr_2 & \frac{d\phi}{dz}dr_3 \end{pmatrix}$$

This expression is called total differential of the scalar function ϕ:

$$d\phi = \mathrm{grad}\phi \cdot d\mathbf{r}$$

and the curve integral is:

$$\int_{AB} \mathrm{grad}\phi \cdot d\mathbf{r} = \int_{AB} d\phi = \phi(B) - \phi(A)$$

In a closed curve, the end point coincides with the initial point and therefore the gradient has the following important property:

$$\oint_C \mathrm{grad}\phi \cdot d\mathbf{r} = 0$$

Therefore, for any vector field **F** which can be expressed as the gradient of a scalar function: $\mathbf{F} = \mathrm{grad}\phi$, its curve integral between two points does not depend on the way C followed, but only the start and end points, and as a result, the circulation of **F** along a closed line will be zero. Then it is said that the vector field **F** is conservative.

If $\mathbf{F} = \mathrm{grad}\phi$ is a force field, the work done by **F** between two points does not depend on the path, but only the start and end points; and the work carried out along a closed line C, will be null. This type of fields are called conservative force fields. Not all vector fields are conservative, for example, the circuit law of Ampere for static electric fields establishes that

$$\oint_l \frac{\mathbf{B}}{\mu_0} \cdot d\mathbf{l} = i$$

being i the net electric static current that flows through the surface limited by the curve l, **B** is the magnetic field and μ_0 is a constant.

10.7.3 Curl of a vector field

Let's consider a system of rectangular coordinates, whose unit vectors are $\mathbf{i}, \mathbf{j}, \mathbf{k}$. If a field is not conservative, its circulation around a closed trajectory is not equal to zero. If we make the area enclosed by the curve approach zero, the limit

$$\lim_{\Delta s_1 \to 0} \frac{\oint_l \mathbf{F} \cdot d\mathbf{l}}{\Delta s_1}$$

is defined as the component of a vector called curl of **F**, which we write rot**F**, in the direction of unit vector **i**. In this way, the curl vector of **F** is equal to:

$$\mathrm{rot}\mathbf{F} = \mathbf{i} \lim_{\Delta s_1 \to 0} \frac{\oint_l \mathbf{F} \cdot d\mathbf{l}}{\Delta s_1} + \mathbf{j} \lim_{\Delta s_2 \to 0} \frac{\oint_l \mathbf{F} \cdot d\mathbf{l}}{\Delta s_2} + \mathbf{k} \lim_{\Delta s_3 \to 0} \frac{\oint_l \mathbf{F} \cdot d\mathbf{l}}{\Delta s_3}$$

It is clear that in the case that **F** were conservative, its curl would be zero, i.e. if $\mathbf{F} = \mathrm{grad}\phi$, we have

$$\mathrm{rot}\mathbf{F} = \mathrm{rot}(\mathrm{grad}\phi) = 0$$

The curl of the gradient is equal to zero.

Therefore, the curl expresses the rotation of a vector field, per unit area, and is a vector whose direction is that of the thumb of the right hand, if we close the fist with the other fingers in the direction of the movement along the curve. Each component of the vector is normal to unit area perpendicular to that axis.

On the basis of the definition of curl with curve integrals, it can be proved that if $\mathbf{F} = F_x\mathbf{i} + F_y\mathbf{j} + F_z\mathbf{k}$, the curl of the field $\mathbf{F}$ is:

$$\mathbf{rot}\mathbf{F} = \left(\frac{\partial F_z}{\partial y} - \frac{\partial F_y}{\partial z}\right)\mathbf{i} + \left(\frac{\partial F_x}{\partial z} - \frac{\partial F_z}{\partial x}\right)\mathbf{j} + \left(\frac{\partial F_y}{\partial x} - \frac{\partial F_x}{\partial y}\right)\mathbf{k}$$

The following program calculates the curl of a vector expressed in rectangular coordinates $\mathbf{v} = v_x\mathbf{i} + v_y\mathbf{j} + v_z\mathbf{k}$. The output for three different vectors is shown.

```python
# -*- coding: utf-8 -*-
"""
Mathematics and Python Programming     www.pysamples.com
p10o.py
"""

import sympy as sy

x, y, z = sy.symbols('x, y, z')
sy.init_printing(use_unicode=True)

vector1 = sy.Matrix([[x + z, y + z, x ** 2 + z]])
vector2 = sy.Matrix([[x, y, z]])
vector3 = sy.Matrix([[z, x, y]])


def curl(v):
    rot = []
    rotx = sy.simplify(sy.diff(v[2], y) - sy.diff(v[1], z))
    roty = sy.simplify(sy.diff(v[0], z) - sy.diff(v[2], x))
    rotz = sy.simplify(sy.diff(v[1], x) - sy.diff(v[0], y))
    rot.append(rotx)
    rot.append(roty)
    rot.append(rotz)
    length = sy.sqrt(rotx ** 2 + roty ** 2 + rotz ** 2)
    result = {'curl': rot, 'length': length}
    return result

print 'v = ', vector1
rotv = curl(vector1)
print 'rot v = ', rotv.get('curl')
print '|rot v| = ', rotv.get('length')
print
print 'v = ', vector2
rotv = curl(vector2)
print 'rot v = ', rotv.get('curl')
print '|rot v| = ', rotv.get('length')
print
print 'v = ', vector3
rotv = curl(vector3)
print 'rot v = ', rotv.get('curl')
print '|rot v| = ', rotv.get('length')
```

10.7. INTEGRATION OF A VECTOR FUNCTION

```
―――――――――――――――――――――――― output of the program ――――――――――――――――――――――――
v    =   [x + z, y + z, x**2 + z]
rot v =  [-1, -2*x + 1, 0]
|rot v| =   sqrt((-2*x + 1)**2 + 1)

v    =   [x, y, z]
rot v =  [0, 0, 0]
|rot v| =  0

v    =   [z, x, y]
rot v =  [1, 1, 1]
|rot v| =  sqrt(3)
```
―――

To understand the physical meaning of the curl, suppose we have a channel of water in which there is a speed vector field such that the velocity of the water is greater, the closer we come to the surface. The vector field would be of this type, by placing the axis X at the bottom of the channel, the coordinate Y expresses the distance from the bottom:

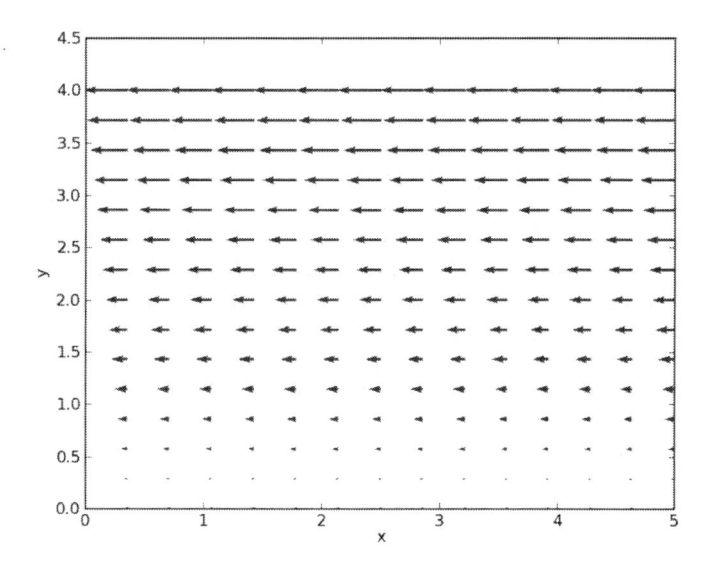

$$\mathbf{v} = -3y\mathbf{i} + 0\mathbf{j} + 0\mathbf{k} = \begin{pmatrix} -3y & 0 & 0 \end{pmatrix}$$

We can visualize this plane vector field using Python:

```
# -*- coding: utf-8 -*-
"""
Mathematics and Python Programming     www.pysamples.com
p10q.py
"""

import matplotlib.pyplot as plt
import numpy as np

numpuntos = 15
x = np.linspace(0, 5, numpuntos)
y = np.linspace(0, 4, numpuntos)
X, Y = np.meshgrid(x, y, sparse=False)
U, V = -3 * Y, 0 * X

plt.quiver(X, Y, U, V, color='b')
plt.xlim(0, 5)
plt.ylim(0, 4.5)
plt.xlabel('x')
plt.ylabel('y')
```

```
plt.show()
```

If you immerse a paddle wheel in the water stream, it will begin to rotate due to the velocity of the water in its upper part is greater than that of the water under the wheel. In our case, the water moves to the left with a speed $\mathbf{v} = -3y\mathbf{i}$, and the top of the wheel is moved to the left, so that the wheel is rotated anticlockwise. If we put the right fist with the fingers closed toward the left, the thumb is pointing up: this is the direction of the curl vector, which we can calculate with the program we used in the previous examples:

```
──────────────────────────── output of the program ────────────────────────────
v   =  [-3*y, 0, 0]
rot v  =  [0, 0, 3]
|rot v|  =  3
```

10.7.4 Multiple integrals

Consider a function $f(x, y)$, and a bounded and closed domain D with a border of zero area. We divide the domain in r partial subdomains through arbitrary curves with null area. Each partial domain D_i has an area A_i and we chose an arbitrary point $p_i = (x_i, y_i)$ in each domain. We define the integral sum σ of that function in the domain D with that partition:

$$\sigma = \sum_{i=i}^{r} f(x_i, y_i) A_i$$

For each division in partial domains, there will be a partial domain having a greater surface area than the others. We will write this maximum partial surface of that partition as A_{max}.

The function $f(x, y)$ is said to be Riemann integrable in the domain D if there is the limit of the integral sums σ when the maximum partial surface A_{max} approaches zero. This limit is called double integral of the function $f(x, y)$ extended to the domain D, and is written

$$\iint_D f(x, y) dA$$

If the domain D is bounded, closed and such that any straight line parallel to the axis OY cuts the border of the domain at the most in two points, then the double integral can be reduced to simple integrals:

$$\iint_D f(x, y) dx dy = \int_{x_1}^{x_2} dx \int_{y_1}^{y_2} f(x, y) dy$$

and if we change the order of integration:

$$\iint_D f(x, y) dx dy = \int_{y_1}^{y_2} dy \int_{x_1}^{x_2} f(x, y) dx$$

being x_1 and x_2 the minimum and maximum x-coordinates of the domain, and y_1, y_2 its minimum and maximum y-coordinates.

If $F(x, y)$ is an antiderivative function for the variable x, then

$$\iint_D f(x, y) dx dy = \int_{y_1}^{y_2} [F(x_2, y) - F(x_1, y)] dy$$

These results can be generalized for a 3D domain, and in that case we have a volume integral:

$$\iiint_V f(x, y, z) dV = \iiint_V f(x, y, z) dx dy dz$$

$$\iiint_V f(x, y, z) dV = \int_{x_1}^{x_2} [\int_{y_1}^{y_2} [\int_{z_1}^{z_2} f(x, y, z) dz] dy] dx$$

10.7.5 Flux of a vector field. Divergence

Consider a vector field $\mathbf{A} = A_x\mathbf{i} + A_y\mathbf{j} + A_z\mathbf{k}$ whose magnitude and direction represent the volume of fluid that passes in the unit of time through a element of unit area perpendicular to $\mathbf{A}$. Let S be a closed surface that we draw within the vector field, $d\mathbf{s}$ is a normal vector directed toward the outside of an element of surface ds. The amount of fluid that comes out of the surface through an element ds in the time unit will be the differential flow of fluid $\mathbf{A} \cdot d\mathbf{s}$. The sign of the flow toward the inside of the surface will be negative. If we add the differential flows through all of the elements of surface we get the net flow of fluid toward the outside of the surface in the unit of time:

$$\psi = \oint_S \mathbf{A} \cdot d\mathbf{s}$$

If in the interior of the surface there are no sources or sinks of the vector field, the net flow will be zero, as it will as much flow outward as it comes in. Otherwise, this integral provides a measure of the strength of the sources (or sink) contained within the volume enclosed by the surface.

For example, for the magnetic field $\mathbf{B}$, the integral law of Maxwell states:

$$\oint_S \mathbf{B} \cdot d\mathbf{s} = 0$$

Which implies that there are no sources of magnetic field (physically there exist no free magnetic charges), and the flow lines of the magnetic field lines are always closed lines. In contrast, the integral law of Maxwell for the electric field $\mathbf{E}$ states:

$$\oint_S \epsilon_0 \mathbf{E} \cdot d\mathbf{s} = q$$

being ϵ_0 a constant and q the electric charge contained inside the volume enclosed by the surface.

It is important the relationship between flow and the unit of enclosed volume:

$$\frac{1}{V} \oint_S \mathbf{A} \cdot d\mathbf{s}$$

If this ratio has finite limit when the volume V collapses toward a point p, then this limit is called divergence of the vector field $\mathbf{A}$ at the point p

$$div\mathbf{A} = \lim_{\Delta V \to 0} \frac{\oint_S \mathbf{A} \cdot d\mathbf{s}}{\Delta V}$$

The points p for which $div\mathbf{A} > 0$ are called sources of the vector field; the points for which the divergence is negative are called sinks of the vector field: $div\mathbf{A} < 0$. If the divergence of the vector field is zero at all points in a certain region: $div\mathbf{A} = 0$, it is said that the field is solenoidal, (or also tubular) in that region.

For example, for the electric and magnetic fields, the laws of Maxwell establish:

$$div\mathbf{B} = 0 \qquad div(\epsilon_0 \mathbf{E}) = \rho$$

being ρ the density of electric charge.

In fluid dynamics, the continuity equation establishes that:

$$div(\rho\mathbf{v}) = -\frac{\partial \rho}{\partial t}$$

being $\mathbf{v}$ the velocity of the fluid and ρ its density. If the fluid can be considered as incompressible, the former equation becomes $div(\rho\mathbf{v}) = 0$

In rectangular coordinates, it can proved that the divergence is:

$$div\mathbf{A} = \frac{\partial A_x}{\partial x} + \frac{\partial A_y}{\partial y} + \frac{\partial A_z}{\partial z}$$

Let's look at some examples calculated with Python. The graphics in 3D show the net flow through a closed surface.

```
# -*- coding: utf-8 -*-
"""
Mathematics and Python Programming    www.pysamples.com
p10r.py
"""

import sympy as sy

x, y, z = sy.symbols('x, y, z')
sy.init_printing(use_unicode=True)

vector1 = sy.Matrix([[x ** 2 + y, y ** 2 + z, z ** 2 + x]])


def divergence(v):
    return sy.diff(v[0], x) + sy.diff(v[1], y) + sy.diff(v[2], z)

print 'v = ', vector1
print 'div v = ', divergence(vector1)
```

The following program has been used for the 3D representations. We open Mayavi2 and run the following Python program. Mayavi IDE allows to change colors of the 3D image, rotate it, etc.

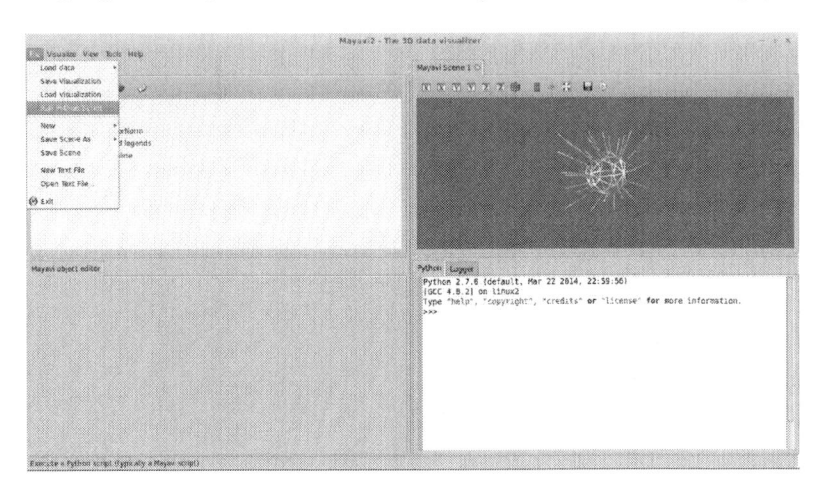

```
# -*- coding: utf-8 -*-
"""
Mathematics and Python Programming    www.pysamples.com
p10s.py
"""

import numpy as np
from mayavi import mlab

x, y, z = np.mgrid[-5:5, -5:5, -5:5]
u = x
```

10.7. INTEGRATION OF A VECTOR FUNCTION

```
v = y
w = z
obj = mlab.flow(u, v, w)
mlab.show()
```

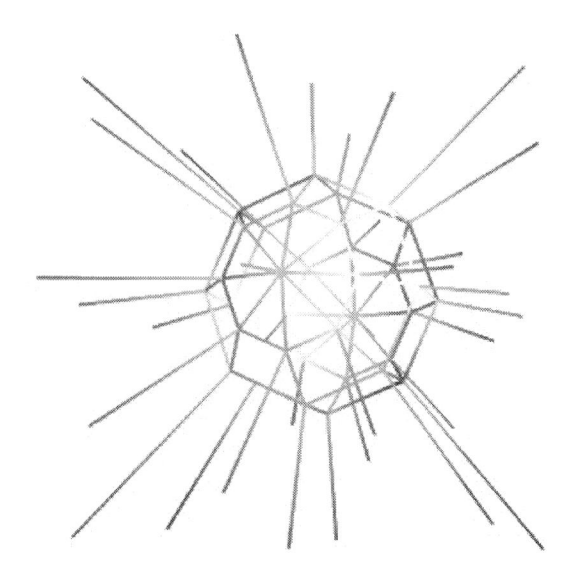

────────────── output of the program ──────────────
```
v = [x, y, z]
div v = 3
```

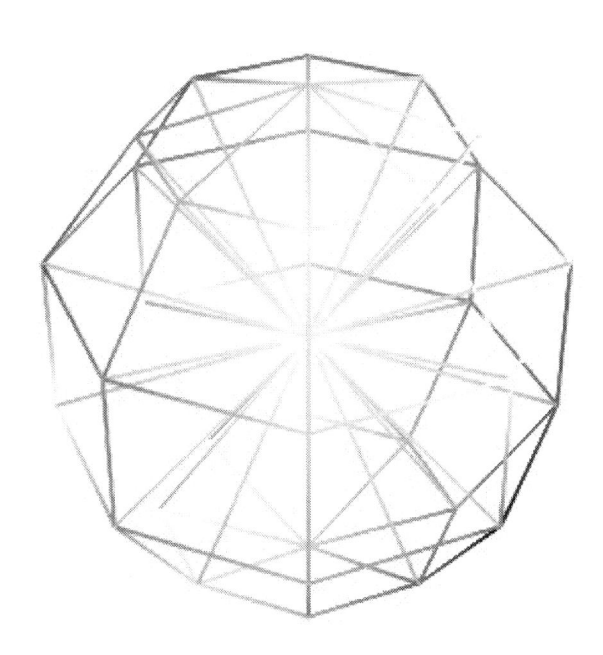

────────────── output of the program ──────────────
```
v = [-x, -y, -z]
div v = -3
```

10.8 Field lines. Examples of vector fields

The field lines of a vector field are those lines that have the property that the vector field is tangent to the line at each point. The calculation of the equations of these lines requires solving differential equations, which are beyond the objectives of this book, but Python allows us to easily represent a vector field both by drawing a number of vectors at different points, or by means of flow lines. Now we are going to see some examples of a vector field

$$\mathbf{A} = A_x\mathbf{i} + A_y\mathbf{j} + A_z\mathbf{k}$$

in which we will use the Python programs that we have developed along this chapter, as well as a new program to show the lines. The representations are shown by both methods:

```
# -*- coding: utf-8 -*-
"""
Mathematics and Python Programming    www.pysamples.com
p10t.py
"""

import matplotlib.pyplot as plt
import numpy as np

Y, X = np.mgrid[-5:5:200j, -5:5:200j]
U = -Y
V = X
#U = X
#V = Y
speed = np.sqrt(U * U + V * V)
lw = 0.2 + 2 * speed / speed.max()
plt.streamplot(X, Y, U, V, density=0.6, color='b', linewidth=lw, arrowsize=2, arrowstyle='-|>')
plt.xlabel('x')
plt.ylabel('y')
plt.show()
```

1. The field $\mathbf{A} = -y\mathbf{i} + x\mathbf{j} + 0\mathbf{k}$. We are going to represent the field, and to calculate its curl and divergence with the programs that we have developed this chapter. The length of the vector will be $|\mathbf{A}| = \sqrt{y^2 + x^2} = r$. The length of the vector increases as we move further away from the source. This field has $div\mathbf{A} = 0$, and the vector $rot\mathbf{A}$ is perpendicular to the plane, and oriented upwards:

─────────────── output of the program ───────────────

```
v =  [-y, x, 0]
rot v =  [0, 0, 2]
|rot v| =  2

v =  [-y, x, 0]
div v =  0
```

10.8. FIELD LINES. EXAMPLES OF VECTOR FIELDS

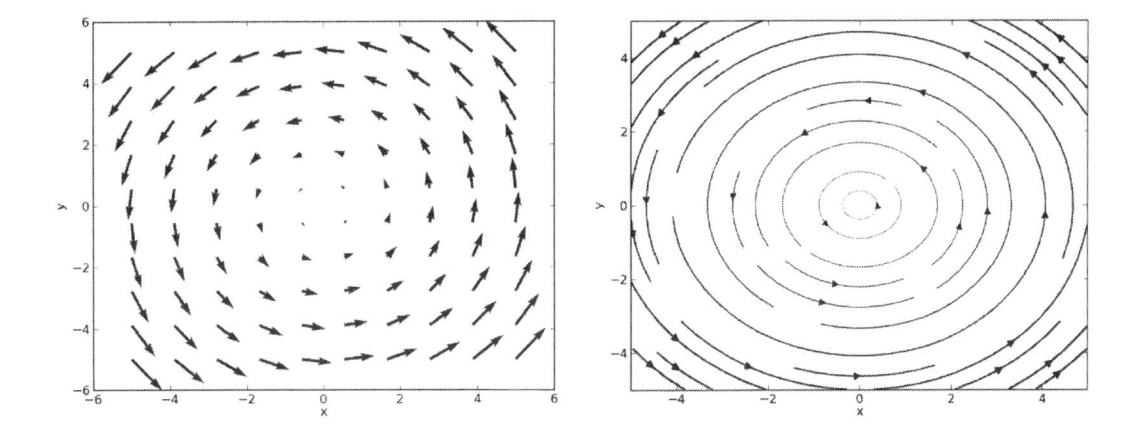

2. The field $\mathbf{A} = x\mathbf{i} + y\mathbf{j} + 0\mathbf{k}$. The length of the vector will be $|\mathbf{A}| = \sqrt{x^2 + y^2} = r$. This field has $div\mathbf{A} > 0$, which indicates that there are sources of the field, and $rot\mathbf{A} = 0$:

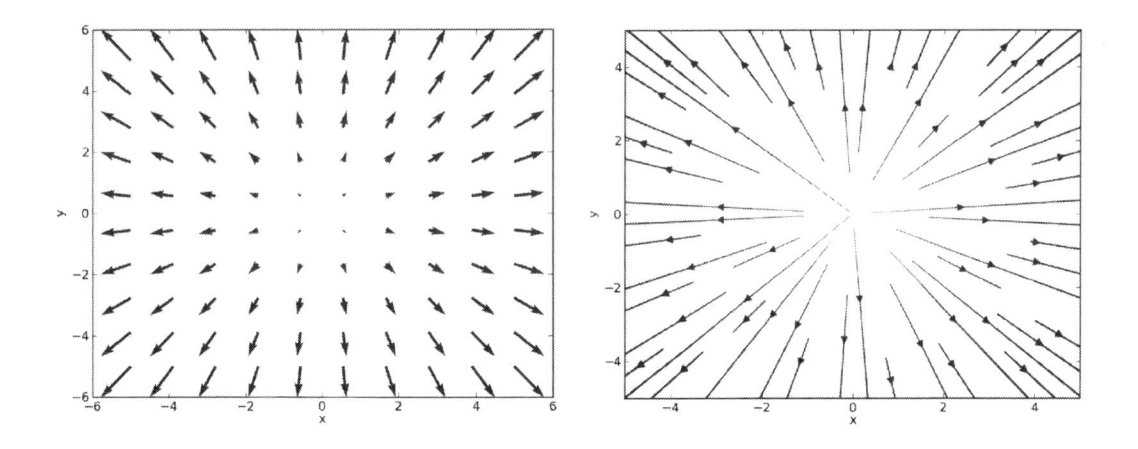

```
_____ output of the program _____
v    = [x, y, 0]
rot v = [0, 0, 0]
|rot v| = 0

v    = [x, y, 0]
div v = 2
```

Since $rot\mathbf{A}$ is zero, $\mathbf{A}$ is the gradient of a scalar field $\phi = \frac{1}{2}(x^2 + y^2)$, and the vector field $\mathbf{A}$ is normal at each point to the contour lines of the scalar field ϕ and it is oriented according to increasing values of ϕ:

```
_____ output of the program _____
z      =  0.5*x**2 + 0.5*y**2
grad z =  [1.0*x, 1.0*y]
|z|    =  sqrt(1.0*x**2 + 1.0*y**2)
|z|    =  sqrt(x**2 + y**2)
```

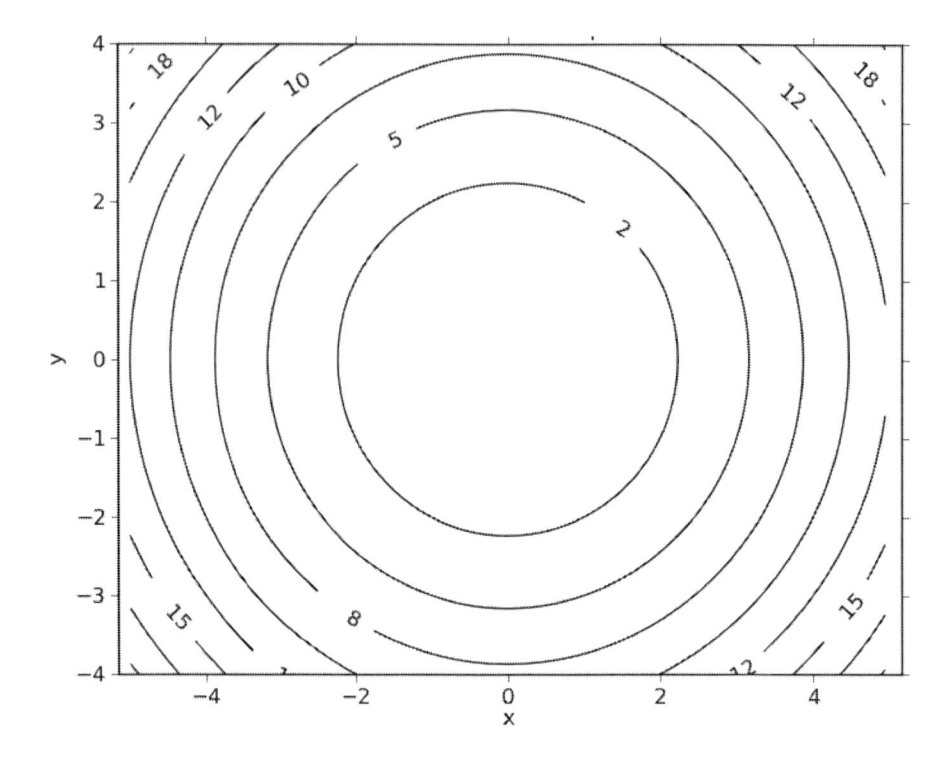

Bibliography for this chapter: [20], [21], [23], [24], [25], [26], [28], [34], [40], [44], [57], [61]

Bibliography

[1] R.A. Adams y C. Essex - Calculus, a Complete Course. Pearson, 2010.

[2] A.D. Alexandrov, A.N. Kolmogorov, M.A. Lavrentiev - Mathematics, its Content, Methods and Meaning. Dover, 1999.

[3] V.V. Amelkin - Ecuaciones Diferenciales en la Práctica. URSS, 2003.

[4] L.T. Ara y M.E. Ríos - Álgebra Lineal y Cálculo Integral. Santander, 1965.

[5] B. Baule - Tratado de Matemáticas Superiores para Ingenieros y Físicos. Labor, 1949.

[6] A.H. Beiler - Recreations in the Theory of Numbers. Dover, 1966.

[7] V.G. Boltianski - La Envolvente. Editorial Mir, 1977.

[8] R.S. Borden - A Course in Advanced Calculus. Dover, 1998.

[9] R.C. Buck - Advanced Calculus. Mc Graw Hill, 1978.

[10] F. Calvo - Estadística Aplicada. Ediciones Deusto, 1990.

[11] C. Chapman - Real Mathematical Analysis. Springer, 2002.

[12] F. Coquillat - Cálculo Integral. Tebar Flores, 1997.

[13] B. Demidovich - Problemas y Ejercicios de Análisis Matemático. Paraninfo, 1985.

[14] J. de Lorenzo - Introducción al Estilo Matemático. Tecnos, 1989.

[15] A. Efimov y B. Demidovich - Problemas de las matemáticas superiores. Editorial Mir, 1983.

[16] H.B. Enderton - Elements of Set Theory. Academic Press, 1977.

[17] J.B. Fraleigh - Álgebra Abstracta. Addison-Wesley Iberoamericana, 1985.

[18] J.A. Fernández Viña - Análisis matemático. Tecnos, 1994.

[19] M. García y R. Bronte - Problemas de Álgebra y Analítica. 1978.

[20] E. García Camarero - Álgebra lineal, 1967.

[21] L.I. Golovina - Álgebra lineal y algunas de sus aplicaciones, Editorial Mir, 1967.

[22] Horst R. Beyer - Calculus: A modern, rigorous approach. Louisiana State University, 2003.

[23] J.M. Iñíguez Almech - Matemáticas para Quimicos. Editorial Labor, 1941.

[24] V. Ilín, E. Pozniak - Fundamentos del Análisis Matemático. Editorial Mir, 1991.

[25] C.T.A. Johnk - Ingeniería Electromagnética, Campos y Ondas. Limusa, 1992.

[26] G.Joos, I.M. Freeman - Theoretical Physics, Dover, 1986.

[27] E. Kamke - Theory of Sets. Dover, 1950.

[28] D. Kletenik - Problemas de Geometría Analítica. Editorial Mir, 1986.

[29] K. Knopp - Theory and Application of Infinite Series. Blackie and Son, 1951.

[30] K. Knopp - Infinite Sequences and Series. Dover, 1956.

[31] K. Knopp - Theory of Functions. Dover, 1996.

[32] A.N. Kolmogorov y S.V.Fomin - Introductory Real Analysis. Dover, 1970.

[33] A. I. Kostrikin - Introducción al Álgebra. Editorial Mir, 1983.

[34] M.L. Krasnov, A.I. Kiseliov, G.I. Makarenko - Análisis Vectorial. Editorial URSS, 2005.

[35] R.E. Larson, R.P. Hostetler - Cálculo y Geometría Analítica. McGraw-Hill, 1989.

[36] D.R. LaTorre - Calculus Concepts. Cengage Learning, 2012.

[37] L. Leithold - The Calculus with Analytic Geometry. Harper and Row, 1976.

[38] A. López, A. de la Villa - Geometría diferencial. CLAGSA, 1997.

[39] A. Luzarraga - Problemas Resueltos de Álgebra Lineal. Barcelona, 1970.

[40] J.E. Marsden, A.J. Tromba - Vector Calculus. Freeman and Company, 2003.

[41] W.J. Moore - Química Física, Urmo, 1978.

[42] J. Martínez Salas - Elementos de matemáticas, 1979.

[43] T. Needham - Visual Complex Analysis. Oxford University Press, 1998.

[44] J.M. Pacheco, C. Durán - Curso de Matemáticas 1. Gómez Puig Ediciones, 1979.

[45] N. Piskunov - Cálculo diferencial e Integral. Editorial Mir, 1983.

[46] P. Puig Adam - Cálculo Integral. Gómez Puig Ediciones, 1979.

[47] L.S. Pontriaguin - Análisis Infinitesimal. URSS, 2011.

[48] L.S. Pontriaguin - Generalizaciones de los números. URSS, 2005.

[49] V. Quesada - Curso y Ejercicios de Estadística. Editorial Alhambra, 1989.

[50] M. Rosenlicht - Introduction to Analysis. Dover, 1968.

[51] W. Rudin - Real and Complex Analysis. McGraw-Hill, 1970.

[52] L. Scharf - A First Course in Electrical and Computer Engineereing. Rice University - CONNEXIONS, 2009.

[53] S.L. Salas y E. Hille - Calculus. Editorial Reverté, 1995.

[54] A. A. Sazánov - El Universo Tetradimensional de Minkowski. Editorial Mir, 1990.

[55] V.S. Shipachev - Fundamentos de las Matemáticas Superiores. Editorial Mir, 1991.

[56] V.I. Smirnov - A Course of Higher Mathematics. Addison Wesley, 1964.

[57] M.R. Spiegel - Cálculo superior. McGraw-Hill, 1987.

[58] P.A. Tipler - Física. Reverté, 1985.

[59] I. Vinográdov- Fundamentos de la Teoría de los Números. Editorial Mir, 1977.

[60] N.N. Vorobiov - Teoría de las Series. Rubiños, 1995.

[61] V.V. Voevodin - Álgebra Lineal. Editorial Mir, 1982.

[62] S.Y. Yan - Number Theory for Computing. Springer, 2002.

[63] A.V. Zhúkov - El Omnipresente Número π. Editorial URSS, 2005.

Printed in Great Britain
by Amazon